Frontiers in agri-food supply chains

Frameworks and case studies

It is widely recognised that agriculture is a significant contributor to global warming and climate change. Agriculture needs to reduce its environmental impact and adapt to current climate change whilst still feeding a growing population, i.e. become more 'climate-smart'. Burleigh Dodds Science Publishing is playing its part in achieving this by bringing together key research on making the production of the world's most important crops and livestock products more sustainable.

Based on extensive research, our publications specifically target the challenge of climate-smart agriculture. In this way we are using 'smart publishing' to help achieve climate-smart agriculture.

Burleigh Dodds Science Publishing is an independent and innovative publisher delivering high quality customer-focused agricultural science content in both print and online formats for the academic and research communities. Our aim is to build a foundation of knowledge on which researchers can build to meet the challenge of climate-smart agriculture.

For more information about Burleigh Dodds Science Publishing simply call us on +44 (0) 1223 839365, email info@bdspublishing.com or alternatively please visit our website at www.bdspublishing.com.

Related titles:

The sustainable intensification of smallholder farming systems
Print (ISBN 978-1-78676-430-0); Online (ISBN 978-1-78676-432-4, 978-1-78676-433-1)

Developing smart agri-food supply chains: Using technology to improve safety and quality
Print (ISBN 978-1-78676-749-3); Online (ISBN 978-1-78676-751-6, 978-1-78676-752-3)

Advances in agri-food robotics
Print (ISBN 978-1-80146-277-8); Online (ISBN 978-1-80146-278-5, 978-1-80146-279-2)

Consumers and food: Understanding and shaping consumer behaviour
Print (ISBN 978-1-80146-354-6); Online (ISBN 978-1-80146-355-3, 978-1-80146-356-0)

Chapters are available individually from our online bookshop:
https://shop.bdspublishing.com

climate•SMART•publishing
IN AGRICULTURAL SCIENCE

BURLEIGH DODDS SERIES IN AGRICULTURAL SCIENCE
NUMBER 137

Frontiers in agri-food supply chains

Frameworks and case studies

Edited by Professor Sander de Leeuw, Dr Renzo Akkerman and Dr Rodrigo Romero-Silva, Wageningen University, The Netherlands

burleigh dodds
SCIENCE PUBLISHING

Published by Burleigh Dodds Science Publishing Limited
82 High Street, Sawston, Cambridge CB22 3HJ, UK
www.bdspublishing.com

Burleigh Dodds Science Publishing, 1518 Walnut Street, Suite 900, Philadelphia, PA 19102-3406, USA

First published 2024 by Burleigh Dodds Science Publishing Limited

Library of Congress Control Number: 2023952207

British Library Cataloguing in Publication Data
A catalogue record for this book is available from the British Library

ISBN 978-1-80146-271-6 (print)
ISBN 978-1-80146-273-0 (PDF)
ISBN 978-1-80146-272-3 (ePub)
ISSN 2059-6936 (print)
ISSN 2059-6944 (online)

DOI: 10.19103/AS.2023.0122

Typeset by Deanta Global Publishing Services, Dublin, Ireland

Contents

Part 2 Assessing agri-food supply chain performance: Tools and techniques

Series list

Title	Series number
Achieving sustainable cultivation of maize - Vol 1 From improved varieties to local applications Edited by: Dr Dave Watson, CGIAR Maize Research Program Manager, CIMMYT, Mexico	001
Achieving sustainable cultivation of maize - Vol 2 Cultivation techniques, pest and disease control Edited by: Dr Dave Watson, CGIAR Maize Research Program Manager, CIMMYT, Mexico	002
Achieving sustainable cultivation of rice - Vol 1 Breeding for higher yield and quality Edited by: Prof. Takuji Sasaki, Tokyo University of Agriculture, Japan	003
Achieving sustainable cultivation of rice - Vol 2 Cultivation, pest and disease management Edited by: Prof. Takuji Sasaki, Tokyo University of Agriculture, Japan	004
Achieving sustainable cultivation of wheat - Vol 1 Breeding, quality traits, pests and diseases Edited by: Prof. Peter Langridge, The University of Adelaide, Australia	005
Achieving sustainable cultivation of wheat - Vol 2 Cultivation techniques Edited by: Prof. Peter Langridge, The University of Adelaide, Australia	006
Achieving sustainable cultivation of tomatoes Edited by: Dr Autar Mattoo, USDA-ARS, USA and Prof. Avtar Handa, Purdue University, USA	007
Achieving sustainable production of milk - Vol 1 Milk composition, genetics and breeding Edited by: Dr Nico van Belzen, International Dairy Federation (IDF), Belgium	008
Achieving sustainable production of milk - Vol 2 Safety, quality and sustainability Edited by: Dr Nico van Belzen, International Dairy Federation (IDF), Belgium	009
Achieving sustainable production of milk - Vol 3 Dairy herd management and welfare Edited by: Prof. John Webster, University of Bristol, UK	010
Ensuring safety and quality in the production of beef - Vol 1 Safety Edited by: Prof. Gary Acuff, Texas A&M University, USA and Prof. James Dickson, Iowa State University, USA	011
Ensuring safety and quality in the production of beef - Vol 2 Quality Edited by: Prof. Michael Dikeman, Kansas State University, USA	012
Achieving sustainable production of poultry meat - Vol 1 Safety, quality and sustainability Edited by: Prof. Steven C. Ricke, University of Arkansas, USA	013
Achieving sustainable production of poultry meat - Vol 2 Breeding and nutrition Edited by: Prof. Todd Applegate, University of Georgia, USA	014

Acknowledgement

We wish to acknowledge the following for their help in reviewing a particular chapter:

- Chapter 11: Dr Giannis Tsoulfas, Agricultural University of Athens, Greece and Professor Jill Hobbs, University of Saskatchewan, Canada

Chapter 1

Studying agri-food supply chains: an analytical framework

Sander de Leeuw, Renzo Akkerman and Rodrigo Romero-Silva, Wageningen University, The Netherlands

1 Introduction

Food supply chains are an essential part of our society. From the primary production of agricultural products to the retail and foodservice settings through which food products are distributed, these extensive supply chains provide a wide range of products to satisfy daily demands for human nutrition. In addition, these food supply chains also provide a significant contribution to the economy and to the employment across the world (Beckman and Countryman, 2021).

We probably live in one of the most interesting times to study agri-food supply chains (AFSCs). Demand for food is expected to rise significantly over the next few decades, both as a result of population growth and even more so due to the rise in average income levels. Increases in agricultural food production (mostly due to improved yields) have so far been keeping up with increasing demand,[1] even though food insecurity remains a major problem in many parts of the developing and developed world. Matching food supply and demand is thus one of the key global challenges for the future.

1 https://ourworldindata.org/yields-vs-land-use-how-has-the-world-produced-enough-food-for-a-growing-population.https://hbr.org/2016/04/global-demand-for-food-is-rising-can-we-meet-it.

http://dx.doi.org/10.19103/AS.2023.0122.01

In addition to this challenge of rising demand, AFSCs regularly have to deal with external events that can lead to significant supply chain disturbances. Since the start of the COVID-19 pandemic around the turn of the year 2019–2020, for example, the media have almost continually reported problems with supply chains and resulting disruptions in food supplies. Examples include the closure of meat processing facilities due to COVID-19 outbreaks among staff (e.g. *Guardian*, 2020) or the occurrence of agricultural surpluses (with resulting losses and waste) due to mismatches in supply and demand and interruptions in supply chain operations (e.g. *Los Angeles Times*, 2020; *Washington Post*, 2020). The war in Ukraine has made us realize that the development of globalized agri-food chains has made many countries heavily dependent on agriculture in particular regions. The war has both disrupted production in Ukraine itself and closed the Black Sea, resulting in shortages in grain and consequently considerable price increases for both grain and commodities using grains such as bread.[2] Ukraine is also a major exporter of sunflower oil, which became a scarce commodity in many places soon after the start of the war.[3] The resulting price increases in sunflower oil have since then had an effect on pricing and availability of products that use sunflower oil as an ingredient such as snacks.

Partly due to these events, the design and development of effective global strategies for managing the supply chain of agri-food products has become an even more complex and challenging issue. An added dimension is ever-increasing consumer requirement (which now encompass not just price and quality but other issues such as environmental and social stability or animal welfare), which require responding to almost continuous changes in lifestyle and dietary preferences.

AFSCs have unique characteristics and challenges for which we need an appropriate analytical framework. In addressing these issues, we first discuss AFSC characteristics, after which we provide a framework for designing AFSCs.

2 Agri-food supply chains: a brief review

An AFSC comprises various stakeholders representing the different stages related to agricultural production, post-harvest, storage, processing, distribution and retail operations, and linkages between these components (Behzadi et al., 2018). The management and design of AFSCs has received considerable attention from both researchers and practitioners over recent decades. Although AFSCs have similarities with other manufacturing supply chains, they have specific characteristics that make the management of these chains more challenging than other manufacturing supply chains (Onggo

2 https://www.nbcnews.com/news/world/russia-ukraine-war-grain-blockade-global-food-crisis-rcna25910.
3 https://time.com/6155095/sunflower-oil-russia-ukraine/.

et al., 2019). Some examples are fluctuations in raw material supply (e.g. due to weather conditions which are becoming more unpredictable and extreme with climate change), food price volatility, food waste, food and nutrition security, and specific power and governance issues due to regulation frameworks governing food production (Ali et al., 2019). In addition, each of the AFSC stakeholders faces specific and unique challenges compared to other types of supply chains when it comes to running their operations in the supply chain (Mccullough et al., 2010). Farmers face issues with resource availability and fertilizer and pesticide use, amongst others (Yadav et al., 2022). Wholesalers and retailers face challenges such as variability in quality and quantity of farm-based inputs, seasonality in supply, and requirements for managing perishable food products in transport and storage, while shelf-life constraints and related raw material/finished product quality deterioration over time are an issue in all parts of the AFSC (van der Vorst, 2000).

The foremost issue that emerges both in practice and in the academic literature relates to food losses and waste (FLW). It is often argued that FLW accounts for one-third of total food produced, with the UN arguing that food waste is a particular problem at consumer level in developed countries in particular (UNEP, 2021). FLW is not only a problem because of loss of valuable food product but also because many people still do not have access to sufficient food or a balanced diet. Food insecurity has been on the rise in recent years, even in developed countries (FAO, 2021). A number of initiatives aim to tackle this, including food banks in countries such as the Netherlands which redirect surplus food to low-income families, simultaneously contributing to ameliorating food insecurity and FLW. They typically rely on donated food (e.g. provided free by retail outlets and consumers) and only rarely use financial donations to buy food.

FLW also represents a loss of resources, including water, inputs, energy and land (Scherhaufer et al., 2018). Water use for food production and processing is the biggest factor affecting global water scarcity (Weinzettel and Pfister, 2019). The FAO has established that 70% of freshwater withdrawals are for food production. In addition to the energy required in food production (either directly in for example fuel for farm machinery or indirectly for example in the energy required to manufacture inputs such as fertilizers and pesticides), several types of food (particularly fresh produce) are also often transported over long distances and may require refrigeration to stay fresh, all of which requires fuel or electricity. These resources could have been used more productively had FLW been prevented.

Closely related to the FLW issue is sustainability. It is well known that agriculture is a major contributor to greenhouse gas (GHG) emissions (Poore and Nemeck, 2018). In AFSCs the dominant share of emissions (around 80%) comes from farming, representing up to 24% of global GHG emissions

(Kucukvar and Samadi, 2015). However, it has been estimated that transport accounts for about 20% of food systems-related emissions (Li et al., 2022). In fact, one could argue that that FLW should be viewed from the perspective of sustainability to get a correct view of its impact: wasting 1 kg of meat has a much bigger impact on the environment than wasting 1 kg of vegetables, simply because of the higher level of inputs needed to produce meat (Van Rooijen et al., 2024). Reviews of AFSC reviews such as Yadav et al. (2022) mainly focus on the environmental and economic aspects of sustainability. The social implications of sustainability (such as effects on labour conditions or farmer behavior) are far less studied.

An often-quoted issue in AFSCs relates to food safety and food quality. In 2012, for example, Europe witnessed a major horsemeat scandal with research showing that consumers were particularly concerned about the authenticity (as well as safety) of meat products (Agnoli et al., 2016). Traceability of food has become a topic of concern that is often raised in research (cf. Yadav et al., 2022). However, food traceability systems are still relatively undeveloped compared to other supply chains (Feng et al., 2020). Blockchain technology, which is a distributed and decentralized system of time-stamped and linked blocks of information, is considered a promising technology to help increase transparency (Feng et al., 2020).

All these factors mean that there are many important and distinctive characteristics to be taken into account when designing AFSCs which, in other types of supply chain, would not play such an important role. These particularly relate to quality decay and perishability of food that should be accounted for in areas of supply chain design such as network modelling, distribution handling strategies, or inventory management. Several models have been developed to address different aspects of these challenges (see Yadav et al. (2021) for examples). The remainder of this chapter focuses on a design approach to address key challenges in AFSCs.

3 The development of the concept of 'supply chains'

An early reference to the notion of supply chains is by Professor Jay Forrester in his seminal book *Industrial Dynamics* (1961), which he referred to as supply pipelines. Many people point to the work of military theorists such as Sun Tzu (*The Art of War*) or Von Clausewitz (*On War*) on the challenges in supplying armies as the origin of thinking about supply chains. It is less well known that commercial organizations picked up the idea of building and managing relatively complex AFSCs many years ago. Amongst others, the East Indies Company based in Amsterdam (in the Netherlands) created their own supply network in the seventeenth and eighteenth centuries, building and operating a global shipping fleet to transport spices and other valuable food products from sources of production in Asia to consumer markets in Europe.[4] Between 1700 and 1730, the Zaanstreek area north of Amsterdam was already a logistics hub before the term had been invented. It was the shipbuilding heart of Europe with, at its height, a production output of almost one seagoing ship built every 2 days. Many principles that we see in supply chains were already applied back then. As an example, saw mills were situated close to the wharfs to avoid carrying heavy wooden components over long distances. More distant factories produced canvas for the sails or biscuits for food during the voyage since both were relatively light and could be transported easily over greater distances. In these more rural areas labour was also cheaper to keep costs down. Circular supply chain principles were also already applied: waste timber, for example, was used to build wharfs which also helped manage costs. In addition, a range of wood suppliers were used to ensure critical supplies kept coming in (cf. Daly et al., 2021).

Typically, most food supply chains in this period were still local or regional, with networks of local farms supplying neighbouring towns and cities. The development of railroads, advances in shipping (e.g. the development of steam power) and early refrigeration technologies in the nineteenth and early twentieth centuries stimulated the movement of food over longer distances (e.g. import of grain from North America and meat and dairy products from South America into Europe). However, it took until well into the twentieth century for supply chains to really explode globally. In particular, the development of the container made shipping over long distances more economical, making way for the development of regional and global third-party logistic networks (Notteboom and Rodrigue, 2006). It is now hard to imagine global food transport without the use of (in particular refrigerated) containers. Continual innovations in food packaging have also further expanded options for transporting food globally (see Risch, 2009, for an historical overview of developments in food packaging).

4 Food systems versus food supply chains

The notion of managing supply chains has long been acknowledged in food and agriculture research, albeit with more policy-oriented approach often focused on the concept of a 'food system'. The International Food Policy Research Institute (IFPRI) defines a food system as '...the sum of actors and interactions along the food value chain–from input supply and production of crops, livestock, fish, and other agricultural commodities to transportation, processing, retailing, wholesaling, and preparation of foods to consumption

4 This is further described in the canon of the Netherlands (https://www.canonvannederland.nl/en/noord-holland/zaanstreek/scheepsbouw).

and disposal'[5]. Even though this definition seems to have food supply chains at its core, food systems research does not usually focus on the practical issues in management of flows of goods from farm to consumer, which is where AFSC research can make a contribution.

There are two main ways in which a food systems perspective and a food supply chain perspective differ: scope and scale. Scope refers to the actors involved. Early supply chain research mostly considered the decision-making processes of commercial firms. The growing focus on sustainability has required a broader approach. In their seminal paper, Kleindorfer et al. (2005) call this the extended supply chain. The inclusion of environmental and social impacts in addition to classic economic considerations requires including the domains of government policy/regulation and societal expectations, meaning that actors such as governments, consumers, employees and non-governmental organizations (NGOs) also play a role.

This brings us to the second way the food systems perspective and the food supply chain perspective differs: scale. Typically, the food system perspective takes a more macro view as opposed to a micro view of businesses. This means that studies of food systems typically look at developments at a sector or product level, often including the biological systems that food production interacts with (e.g. by looking at areas such as how agriculture affects, or is affected by, the physical, chemical and biological health of the soils in which crops are grown). As an example, an important recent discussion is on the role of circular economy principles in the sustainability of food systems, in which different land use scenarios can be analysed on a national or even global level (see Van Zanten et al., 2023, for an analysis on the EU level). How this then translates to individual businesses and supply chains operating in the food system is typically not the focus of such studies. Clearly, both the food systems perspective and the food supply chain perspective are important contributors to the body of knowledge on the food sector.

Figure 1 provides a picture of a generalized AFSC, which consists of farming activities, processing activities (which relate to preparing the farm products for consumption where needed), distribution from farm and/or processing facilities to food service and retail outlets where consumers obtain food for consumption. Different companies in AFSCs cooperate strategically in one or more areas as legally independent and autonomous organizations. As a result, companies may be active in different AFSCs and play a different role in each. It is also important to note that AFSCs are not just focused on ensuring a safe, efficient and cost-effective flow of food products to the end consumer. They must also consider problems such as FLW which highlight broader resource management issues that must be addressed in AFSCs (Akkerman and Cruijssen, 2024).

5 See https://www.ifpri.org/topic/food-systems for more on food systems.

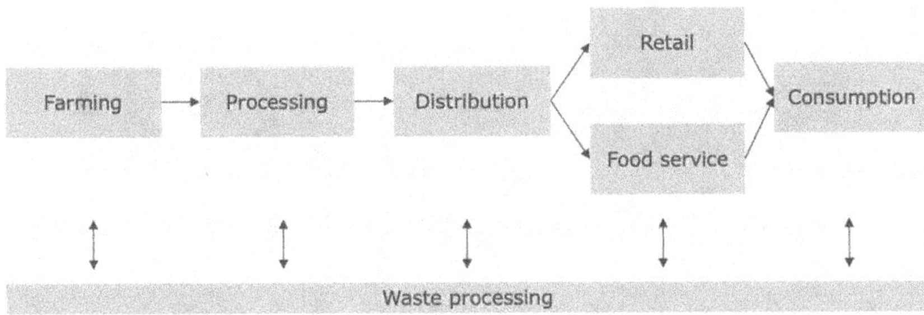

Figure 1 Systems approach to managing supply chains. Source: Adapted from: van der Vorst (2000) and de Leeuw (1988).

5 Designing agri-food supply chains: a framework

Frameworks for the design and management of supply chains have been a topic of research for a few decades. Bemelmans (1982) was one of the first to develop a framework for information systems management in supply chains. Fundamental to his work was the need to develop processes and control in supply chains first, before starting to design information systems to manage them. Van der Vorst (2000) based his supply chain analysis framework on this concept and combined it with the systems approach of de Leeuw (1988). De Leeuw (1988) uses a systems approach and differentiates between the managed system, the managing system and the information system. Transformations take place in the managed system, which is where services or products are made. The information system is used to take stock of the relevant data that are generated during transformation. This results in key performance information or other types of information that is necessary to measure whether the output of the system meets the targets set. The managing system focuses on realizing this output by means of control variables (decision variables such as production volume), taking into account environmental factors (e.g. disruptions or illness). Each aspect influences key performance outcomes.

6 The structure of the book

This book is split into three parts. In Part 1, we first outline the state of development of AFSCs across the globe. In Chapter 2, Schoenherr et al. discuss the development of AFSCs in the Americas, after which Bourlakis et al. provide an overview of developments in northern and western Europe in Chapter 3. Ruiz et al. discuss the state of development and challenges in Southern and Eastern Europe in Chapter 4. Teng et al. close this section in Chapter 5 with an overview of developments in Asia.

In Part 2, following the structure of Fig. 2, we outline the challenges and opportunities in each of the parts. In Chapter 6, Aramyan et al. discuss the

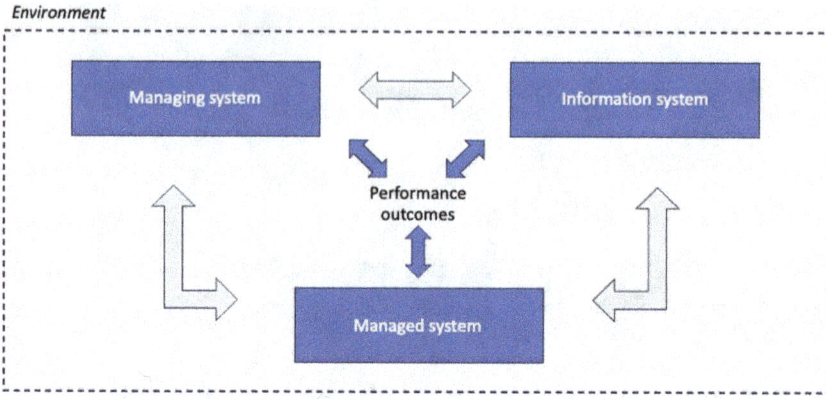

Figure 2 Agri-food supply chain model.

measurement and management performance in AFSCs, after which Soysal et al. review the literature on AFSC infrastructure in Chapter 7. In Chapter 8, Transchel et al. discuss developments in planning and control of AFSCs, after which information system developments are discussed in Chapter 9 by Schiefer.

Part 3 is devoted to case studies on how to improve AFSCs across the globe. In Chapter 10, Soleiro et al. discuss the Americas, after which Akkerman et al. discuss developments in Europe based on a case study of the COVID-19 pandemic in Chapter 11. Jordaan et al. discuss opportunities in Africa in Chapter 12, and Li et al. do this for Asia in Chapter 13. Concluding remarks and future research opportunities in AFSCs are provided in Chapter 14.

7 References

Agnoli, L., Capitello, R., De Salvo, M., Longo, A. and Boeri, M. (2016). Food fraud and consumers' choices in the wake of the horsemeat scandal. *British Food Journal*. 118(8), 1898–1913.

Akkerman, R. and Cruijssen, F. (2024). Food loss, food waste, and sustainability in food supply chains. In: Bouchery, Y., Corbett, C. J., Fransoo, J. C. and Tan, T. (eds). *Sustainable Supply Chains: A Research-Based Textbook on Operations and Strategy* (2nd edn.). Springer, in press.

Ali, S. M., Moktadir, M. A., Kabir, G., Chakma, J., Rumi, M. J. U. and Islam, M. T. (2019). Framework for evaluating risks in food supply chain: implications in food wastage reduction. *J. Clean. Prod.* 228, 786–800. https://doi.org/10.1016/j.jclepro.2019.04.322.

Beckman, J. and Countryman, A. M. (2021). The importance of agriculture in the economy: impacts from COVID-19. *American Journal of Agricultural Economics* 103(5), 1595–1611.

Behzadi, G., O'Sullivan, M. J., Olsen, T. L. and Zhang, A. (2018). Agribusiness supply chain risk management: a review of quantitative decision models. *Omega* 79, 21–42. https://doi.org/10.1016/j.omega.2017.07.005.

Bemelmans, T. M. A. (1982). *Bestuurlijke informatiesystemen en automatisering*. Kluwer bedrijfswetenschappen (In Dutch).

Daly, A., Domínguez-Delmás, M. and van Duivenvoorde, W. (2021). Batavia shipwreck timbers reveal a key to Dutch success in 17th-century world trade. *PLoS ONE* 16(10), e0259391.

FAO (2021). Transforming food systems for food security, improved nutrition and affordable healthy diets for all. *The State of Food Security and Nutrition in the World*. Available at: http://www.fao.org/3/cb4474en/online/cb4474en.html#chapter-51.

Feng, H., Wang, X., Duan, Y., Zhang, J. and Zhang, X. (2020). Applying blockchain technology to improve agri-food traceability: a review of development methods, benefits and challenges. *Journal of Cleaner Production* 260, 121031.

Forrester, J. W. (1961). *Industrial Dynamics*. Productivity Press, Cambridge, MA.

Guardian (2020). 'Chaotic and crazy': meat plants around the world struggle with virus outbreaks. *The Guardian*, May 11, 2020.

Kleindorfer, P. R., Singhal, K. and van Wassenhove, L. N. (2005). Sustainable operations management. *Production and Operations Management* 14(4), 482–492.

Kucukvar, M. and Samadi, H. (2015). Linking national food production to global supply chain impacts for the energy-climate challenge: the cases of the EU-27 and Turkey. *Journal of Cleaner Production* 108, 395–408.

Li, M., Jia, N., Lenzen, M., Malik, A., Wei, L., Jin, Y. and Raubenheimer, D. (2022). Global food-miles account for nearly 20% of total food-systems emissions. *Nature Food* 3(6), 445–453. https://doi.org/10.1038/s43016-022-00531-w.

Los Angeles Times (2020). Rotting food. Hungry masses. Chaotic supply chains. Coronavirus upends the U.S. food system. *Los Angeles Times*, May 5, 2020.

McCullough, E., Stamoulis, K. and Pingali, P. (Eds) (2010). *The Transformation of Agri-Food Systems: Globalization, Supply Chains and Smallholder Farmers*. Earthscan, London.

Notteboom, T. and Rodrigue, J.-P. (2006). Containerisation, box logistics and global supply chains: the integration of ports and liner shipping networks. *International Journal of Maritime Economics* 10(1–2), 152–174.

Onggo, B. S., Panadero, J., Corlu, C. G. and Juan, A. A. (2019). Agri-food supply chains with stochastic demands: a multi-period inventory routing problem with perishable products. *Simulation Modelling Practice and Theory* 97, 101970. https://doi.org/10.1016/j.simpat.2019.101970.

Poore, J. and Nemecek, T. (2018). Reducing food's environmental impacts through producers and consumers. *Science* 360(6392), 987–992.

Risch, S. J. (2009). Food packaging history and innovations. *Journal of Agricultural and Food Chemistry* 57(18), 8089–8092.

Scherhaufer, S., Moates, G., Hartikainen, H., Waldron, K. and Obersteiner, G. (2018). Environmental impacts of food waste in Europe. *Waste Management* 77, 98–113.

UNEP (2021). *Food Waste Index Report 2021*. UNEP. Available at: https://www.unep.org/resources/report/unep-food-waste-index-report-2021.

van der Vorst, J. G. A. J. (2000). Effective food supply chains: generating, modelling and evaluating supply chain scenarios. Doctoral dissertation.

Van Rooijen, M. A., Gerdessen, J. C., Claassen, G. D. H. and de Leeuw, S. L. J. M. (2024). Optimizing household food waste: the impact of meal planning, package sizes, and performance indicators. Working paper, Wageningen University, Wageningen, the Netherlands.

van Zanten, H. H. E., Simon, W., van Selm, B., Wacker, J., Maindl, T. I., Frehner, A., Hijbeek, R., van Ittersum, M. K. and Herrero, M. (2023). Circularity in Europe strengthens the sustainability of the global food system. *Nature Food* 4(4), 320–330.

Washington Post (2020). Extra food is rotting on farms while Americans go hungry. *The Washington Post*, June 10, 2020.

Weinzettel, J. and Pfister, S. (2019). International trade of global scarce water use in agriculture: modeling on watershed level with monthly resolution. *Ecological Economics* 159, 301–311.

Yadav, V. S., Singh, A. R., Gunasekaran, A., Raut, R. D. and Narkhede, B. E. (2022). A systematic literature review of the agro-food supply chain: challenges, network design, and performance measurement perspectives. *Sustainable Production and Consumption* 29, 685–704.

Part 1

Food supply chains in different regions

Chapter 2

Agri-food supply chains in the United States

Tobias Schoenherr, Grant Kanak, Amy Montalbano and Sneh Patel, Michigan State University, USA

1 Introduction

1.1 Agri-food supply chains

Agri-food supply chains involve a multitude of stakeholders and are challenged by a variety of dynamics, rendering these supply chains complex and multifaceted. Some of these challenges include the variability, seasonality, and perishability of agricultural products, necessitating an orchestrated management of their

http://dx.doi.org/10.19103/AS.2023.0122.04

flows. How these supply chains are designed is also influenced by dynamic institutional forces, such as consumer demands for sustainably-sourced and/or local produce, as well as policy and business interventions. This necessitates viewing agri-food supply chains as complex adaptive systems (e.g. Conti et al., 2023) characterized by consumers influencing the behaviour of agri-businesses (e.g. consumers demanding organic produce) and vice versa (e.g. agri-businesses introducing new types of offerings appreciated by consumers) (National Research Council, 2015). The interdependence of these and other stakeholders in the agri-food environment shapes how these supply chains should be designed and managed.

Agricultural supply chains generally consist of the following stakeholders (Fig. 1) (National Research Council, 2015; KPMG, 2013):

- Suppliers of raw agricultural commodities, such as seeds and fertilizers;
- Farmers (primary producers), who use their resources to grow crops and produce meat and dairy products;
- Primary processors (first-line handlers), who "aggregate, store, and provide initial processing of commodities before shipping them to wholesalers or the processing and manufacturing sectors" (National Research Council, 2015, p. 32) (Hemphill, 2023, p. 70);
- Wholesalers, distributors, and retailers, who market and sell the agricultural products;
- Consumers, who buy and consume the agricultural products;
- governmental agencies, non-governmental organizations (NGOs), and regulators that monitor and regulate the agricultural supply chain; and
- logistics companies and customs brokers in the case of international supply chains,

As an illustrative example serves the apple supply chain originating in the Canadian province of Quebec, as discussed by York (2023). Fruit production farms, including those for apples, are generally small, but require a large workforce, especially during the harvest season. While 70% of the crop is generally sold as fresh apples, the remaining 30% are processed into juice, cider, pies, apple sauce, and other products. The number of processors is small,

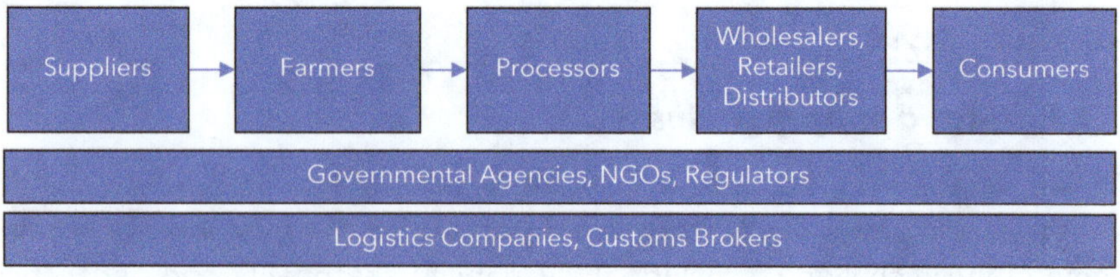

Figure 1 The agri-food supply chain.

reducing the bargaining power of farmers. To increase their leverage, farmers can organize, for instance inthe Apple Growers Federation of Quebec.

1.2 Focus on the U.S. agri-food supply chain

This chapter focuses on the agri-food supply chains in the USA. This focus is important not only due to the scale and scope of this industry in the USA, but also due to the dependency of other nations on U.S. agricultural output, as described in later parts of this chapter. Having a thorough understanding of the U.S. agri-food supply chain enables those who manage it or are dependent on it to proactively address any risks that may lead to disruptions. In the following we proceed with an overall description of the trends and developments that have shaped the current state of the U.S. agri-food supply chain, also covering some of the major stakeholders as outlined in Fig. 1. The stakeholders we focus on include farms of different sizes, farm suppliers, food processors, retailers, and food banks.The ensuing sections then review local versus national/international supply chains. This is followed by an overview of the primary international trade flows, emphasizing also the agri-food supply chains that exist between the two countries neighbouring the USA, i.e. Mexico and Canada. A further major part of this chapter then focuses on the current challenges in the U.S. agri-food supply chain which include the impact of the coronavirus disease 2019 (COVID-19) pandemic, food safety, labour shortages, droughts, and financial pressures. Some of the recent developments are then outlined, focusing on technological innovations, operational changes, sustainability, and temporary agricultural workers. A concluding section completes the chapter.

2 Developments in the U.S. agri-food supply chain

In a 2006 simulation study by the Potsdam Institute for Climate Impact Research, the U.S. heartland and central Europe were identified as two of the most fertile areas worldwide (Dutia, 2014). This can be attributed to the size and quality of the available farmland, water resources, and favourable climatic growing zones (National Research Council, 2015). This advantageous position is responsible for the scale of the U.S. agri-food sector (cf. Yih et al., 2015).

In 2017, the USA had 2 million farms farming 900 million acres (https://www.nass.usda.gov/AgCensus/). In 2021, U.S. farms contributed just under $165 billion to the American economy, representing around 0.7% of national gross domestic product. However, the *overall* agri-food sector (including farm-related industries,e.g. food and beverage manufacturing) contributed around $1264 billion to the economy, representing around 5.5% of U.S. GDP in 2022 (https://www.ers.usda.gov/data-products/ag-and-food-statistics-charting-the-essentials/ag-and-food-sectors-and-the-economy). Sectors within or linked to the agri-food value chain include food and beverage manufacturing, the food retail sector, the food service/catering sector, forestry, and textiles. In 2022, 2.6 million people were employed directly on farms, 1.7 million worked in food and beverage manufacturing, just under 13 million worked in the food services sector, with just over 22 million working in the agri-food sector as a whole, representing over 10% of the overall U.S. workforce. Around 23% of U.S. agricultural produce and processed food is exported, making the USA one the world's largest exporters of food. U.S. agricultural exports reached $196 million in 2022 (though, as discussed later, this figure reduced in 2023). Growth has particularly been in high value-added dairy and cereal products, processed meats, fruits, and vegetables. Leading export markets include East Asia (especially China), Canada, and Mexico. Overall, these statistics on the scale and scope of the U.S. agri-food supply chain illustrate its importance.

However, as with most industrialized nations, the agri-food supply chain in the USA has seen significant changes over recent decades. Specifically, agri-food supply chains have been experiencing a consolidation not only of farms but also of other stakeholders in the supply chain. This has been driven by the imperative to be more productive and profitable in the face of both internal competition and broader economic pressures in the U.S. economy as well as competition from producers in other countries (National Research Council, 2015; York, 2023). As a result, the number of farms has declined continuously since 1982 and, as noted earlier, is estimated to be around 2 million in 2022, with an average farm size of 446 acres (USDA, 2023a). Overall, while farm inputs have remained the same, total agricultural outputs have increased significantly, primarily due to increases in total factor productivity (USDA, 2023a).

The number of small farms has reduced most notably, dropping from being responsible for half of the U.S. food production in 1991 to a quarter in 2017, and farms with 200–999 acres decreasing by 44% from 1987 to 2012 (Semuels, 2019). Contributing to this development are lower commodity prices, climate change-induced weather fluctuations, and the need to invest in technology to be able to produce more efficiently, leading to increased farm debt and bankruptcies (Semuels, 2019). However, not only small farms in the agri-food supply chain were impacted by bankruptcies, but also major players like Dean's Foods, triggered by a change in consumer behaviour and private-label dairy products (Lucas, 2019). The bankruptcy of Dean's Foods also had implications for small farmers, since now, all of a sudden, one of their major customers was no longer buying their milk (Semuels, 2019).

Additional statistics on the current state of the U.S. agri-food supply chain are provided by the Economic Research Service of the U.S. Department of Agriculture (USDA), which classifies farms based on their annual gross cash farm income (GCFI) (USDA, 2023c). Applying this revenue measure, small family farms have an annual GCFI of less than $350 000. This category includes retirement farms (whose operators are retired but still farm on a small scale), off-farm occupations (whose operators report a primary occupation other than farming), and farming-occupation farms (whose operators report farming as their primary operation but who report only low to moderate sales). Based on the USDA classification, mid-sized family farms have an annual GCFI between \$350 000 and \$999 999, while large family farms have an annual GFCI over \$1 000 000. While small family farms represent 89.2% of the total number of farms according to a USDA report in 2022, they only produced 20.4% in production value. Large and very large farms, which represent 2.7% of the total number of farms, produced 46.1% in production value (USDA, 2023c).

Coinciding with this consolidation of farms was a decrease in farm employment. As such, while the U.S.' agricultural output almost tripled between 1984 and 2017, farm employment declined by 81% and labour hours worked fell by 83% (Wang et al., 2022). Today, only about 1% of U.S. employment is in agriculture (BLS, 2023). This is the result of the "great agricultural transition", which has also been described as "one of the most dramatic changes in the US economy and society in the past century" (Lobao and Meyer, 2001, p. 103).

The upstream supply chain, consisting of farm suppliers, has also experienced a significant consolidation, in particular the seed industry. After several mergers and acquisitions, four companies are now responsible for three quarters of corn and soybean seeds (BASF, Bayer, Corteva and ChemChina-owned Syngenta) (Douglas, 2023). Other agricultural segments, such as the poultry industry, experienced vertical integration, where integrators are now responsible for feeding, hatching, and processing - steps that had been conducted by independent stakeholders in the past (National Research Council, 2015).

We now proceed with a discussion of supply chain stakeholders that are downstream to the farmers, such as food processors, retailers, and food banks who market or supply food to consumers. Significant consolidation has also occurred here. An investigation by Guardian and Food and Water Watch (Lakhani et al., 2021) indicated that at least 50% of the market for 79% of the groceries is controlled by four or fewer firms, and that the top firms controlled at least 75% of the market share for close to a third of the shopping items. For example, PepsiCo controls 88% of the dip market (with products like Tostitos, Lay's and Fritos), 94% of sodas are owned by three companies (Coca-Cola, PepsiCo, and Keurig Dr Pepper), 84% of canned tuna is owned by three companies (Dongwon Industries, FSF Fisheries Company, and Thai Union), as is 73% of the cereal market (General Mills, Kellogg, and Post). Another illustrative example from the investigative report is Anheuser-Busch InBev, which owns more than 600 brands worldwide. Within the USA, the largest food manufacturing companies by revenue, together with their revenues and primary product categories in 2021/2022 included the following (Statista, 2023b)

- PepsiCo ($47.6 billion; chips, oatmeal, cereal, soda);
- Tyson Foods ($45.1 billion; chicken, beef, pork);
- JBS USA ($33.7 billion; beef, pork);
- Nestlé ($31.3 billion; baby food, chocolates, ready-to-eat meals, coffee, pizza, pet food);
- Kraft Heinz ($20.4 billion; sauces, deli meats, cheese, beverages); and
- Anheuser-Busch InBev ($16.3 billion; beverages).

While the primary objective of food processors is to provide food that is safe, other considerations also guide the design of food processing operations, including convenience (referring to the time economy of food products), health (referring to the safe consumption of food products), emotions (referring to sensorial quality of food products), function (referring to eating habits and occasions for consuming specific food products), and sustainability (referring to the sustainable production of food products) (Knorr and Watzke, 2019). The range of these principles highlights the increasing challenges for food processors meeting the increasing and more varied consumer expectations about the food they prefer (York, 2023), instead of a simple focus on process optimization. New processes include minimal processing, e.g. providing pre-cut and prepared food products for meal preparations, both for restaurants and as retail products for consumers lacking the time and (sometimes) skills to cook from scratch. Technologies have also been leveraged to improve palatability, shelf-life, and transportability of processed foods (Knorr and Watzke, 2019).

Further downstream are the retailers selling agri-food products to consumers. Similar consolidation has taken place at this level of the supply chain, with the U.S. retail market being dominated by just four (Walmart, Costco, Kroger, and Ahold Delhaize), which together make up 65% of the retail grocery market (Lakhani et al., 2021). This development was also spurred by the establishment of retail 'supercentres' in the 1990s, which introduced food products to the traditionally non-food big box stores. This change was enabled by greater retailer-supplier collaboration and retailer-owned distribution centres, the leveraging of point-of-sales data for effective replenishment and planning, lower food prices (from economies of scale), and restructuring to streamline operations (National Research Council, 2015).

All these stages of the agri-food supply chain generate surplus food which food banks aim to recover. Food banks are generally non-profit organizations that acquire donated food that would have been otherwise wasted. This food is then made available to those in need through outlets such as food pantries or soup kitchens (The Global Food Banking Network, 2023). The leading provider of charitable food assistance in the USA is Feeding America. The non-profit organization serves every state in the country, with 200 food banks and more than 60 000 agencies (Feeding America, 2022). In 2021, the organization distributed

5.5 billion meals to 53 million people (Feeding America, 2022). A grassroots approach to addressing food insecurity is the establishment of community-driven organizations that grow healthy foods themselves, coupled with educational programmes. While food insecurity is declining, it is still prevalent due to structural problems, including unemployment, pay inequity, and issues relating to the food environment (Cafer et al., 2019). Factors contributing to the food environment include proximity to stores, food prices, availability of food and nutrition assistance programmes, and community characteristics (USDA, 2023e). Rural communities, for example, have been found to be disadvantaged by higher food prices due to less competition amongst retailers (Cafer et al., 2019).

3 Types of agri-food supply chains: local food supply chains

While there is an increasing trend and preference towards locally produced agricultural products, this segment is still rather small. Local marketing and direct-to-consumer or intermediated sales have been pursued by about 8% of U.S. farms (American Farm Bureau Federation, 2023; cf. Kapala, 2022). A 2015 USDA survey on local food marketing practices identified 34.5% of local food being sold directly to consumers, while 39.1% was sold to intermediate markets and institutions and 26.4% was sold directly to retailers (Martinez and Park, 2021). Amongst the local food sold directly to consumers, the main channels were

- on-site farm stores or stands (43.7%);
- farmers' markets (23.5%);
- off-site stores or roadside stands (7.8%); and
- community supported agriculture (7.5%).

These local and regional agri-food supply chains are obviously shorter than national and international agri-food supply chains (especially in the case of on-site farm stores and stands), which can have a positive impact on carbon emissions needed for transportation and logistics. As such, consumers also frequently associate these locally produced agricultural products as being fresher and of higher quality (Carbone, 2017).

However, it is important to note that recent research comparing short food supply chains (SFSCs) and long food supply chains (LFSCs) has cast doubt on this assumption of SFSCs being associated with lower environmental impact. For instance, analyzing SFSCs using life cycle assessment (LCA), Majewski et al. (2020) found that LFSCs had fewer negative environmental impacts per unit of produce in terms of fossil fuel energy consumption, greenhouse gas emissions

(GHGEs), and pollution than SFSCs. They concluded that environmental impacts of food distribution are not only determined by geographical distance between producer and consumer but also depend on a range of other factors, including the quality and management of supply chain infrastructures. Reviewing the literature on SFSCs, both Chiffoleau and Dourian (2020) and Doernberg et al. (2022) found general agreement about the social benefits of SFSCs (e.g. for local communities of producers and consumers) but differences in views on their environmental benefits. One drawback of SFSCs is that participants in these chains, whether producers or consumers, can accumulate a large number of journeys moving relatively small quantities of food, producing higher total emissions per unit of produce than LFSCs which can be highly efficient in resource use and consolidated distribution of large quantities of produce, despite the greater distances involved (cf. Malak-Rawlilowska et al. 2019; Vittersø et al., 2019).

The popularity of farmers' markets, where mostly locally grown agricultural products can be purchased by consumers, has been growing in popularity over recent decades. The number of farmers' markets grew from 1755 in 1994 to 8771 in 2019, representing an annual growth of almost 7% (USDA, 2022, cf Cicatiello, 2020). Most of this increase happened before 2011 with only modest growth in recent years (USDA, 2022). This lower growth in recent years can be attributed to supermarkets increasingly featuring local produce grown in the state in response to greater consumer interest in how and where food is produced, with weekly circulars to consumers frequently having a dedicated section about such offerings. This increasing availability of local produce at retailers (enabled through intermediated or direct sales of farmers to these grocers) can be attributed to the only modest growth of farmers' markets in recent years (Martinez and Park, 2021).

Another direct-to-consumer model is community supported agriculture (CSA), where consumers purchase 'shares' of a farm's harvest in advance. The model originated in Japan in the 1960s, with the first CSA in the USA established in 1984 (Duckworth, 2023, cf. Chiffoleau and Dourian, 2020). The price of a share varies based on location, the length of the harvest season, and the variety and quantity of the produce. With the purchase of a share, the consumer invests in the farm and provides up-front financing to, for example, purchase seeds, fertilizer, and equipment. It also provides the farm with greater certainty about its income. In exchange, the consumer receives a certain quantity and variety of produce on a weekly basis, which is determined by what is currently grown. The produce can then be picked up at the farm directly, at a farmers' market (if the CSA farm is there), or is delivered by the farm to the consumer.

What has also been growing in popularity are 'farm-to-table (also known as 'farm-to-fork') offerings, where restaurants source their produce from local farms. With consumers potentially knowing the farm and/or its location, this offers a unique value proposition of consuming produce grown in one's 'backyard'. The benefits of this local supply chain include fresher produce (frequently picked on the same day that it is consumed), better traceability (e.g. knowing the particular farm on which produce has been grown), potentially lower carbon emissions due to shorter transportation requirements, and supporting the local economy.

Another form of the 'farm-to-table' consists of farms offering culinary experiences based on their own produce, essentially necessitating no transportation of the produce at all and guaranteeing greatest freshness. The origins of 'farm-to-table' can be traced back to the 1960s and 1970s when U.S. consumers were becoming increasingly dissatisfied with what they felt to be bland, processed foods from the major food processors (Taylor, 2023). The movement however did not gain traction until the 2000s. Challenges include the need for farm restaurants to frequently change their menus to accommodate produce availability depending on the season, higher meal prices to cover the higher costs (making this offering unaffordable for some), and adverse publicity from sometimes debatable claims about how far produce is locally sourced, and produce freshness (Taylor, 2023).

4 Types of agri-food supply chains: national supply chains

National and international agri-food supply chains continue to dominate over local and regional supply chains, given significant differences in scale and scope, a reality which is not expected to change in the near future. National and international agri-food supply chains are complex, and can take on various forms. Carbone (2017) identified six types of agri-food supply chains including

- supply chains that are driven by retailers (highlighting the central governance needed to manage the supply chain's complexity);
- supply chains driven by global processing companies (highlighting quality and innovation as a competitive advantage);
- supply chains driven by cooperatives (highlighting the benefits for farmers, including access to knowledge, services and inputs);
- supply chains focused on geographical indication (highlighting the brand equity of a product coming from a particular region);
- short supply chains (highlighting the attractiveness of local production); and
- supply chains driven by specialized, high-quality retailers (highlighting their ability to integrate various other supply chains, including geographical indication and short supply chains).

Due to primary production locations, population centres, and transportation infrastructure, as well as socio-political dynamics, dominant national flows of agricultural products in the USA can be identified. Lin et al. (2019) for instance identified

- large trade flows from the Great Lakes to the New York area;
- from the corn/soybean belt (primarily in the Midwest) to New Orleans; and
- from the grain belt to California.

The complexity of the national (U.S.) agri-food supply chain has also been illustrated by Konar (2019), who mapped food flows between all 3000 counties in the 50 states in the USA, yielding 9.5 million links.

Counties with the most food outflow and inflow were located in California, due to large urban areas and the productive Central Valley (Konar, 2019). The Central Valley plays a key role in the U.S. agri-food supply chain. While only comprising less than 1% of U.S. farmland, it is responsible for producing one quarter of the country's food (by value), which includes 40% of fruits, nuts, and other table foods sold in the USA (USGS, 2023). Altogether, more than 250 crops are grown in the Central Valley, amounting to a value of approximately $17 billion per year (USGS, 2023). When it comes to the production of vegetables, California was the leading state in 2020 with an estimated 435 922 000 cwt (Statista, 2023a). States with the largest areas planted with wheat in 2022 included (Statista, 2023)

- Kansas (7 300 000 acres);
- North Dakota (6 195 000 acres); and
- Montana (5 460 000 acres).

This reflects a wider development in U.S. agriculture with increasing specialization in producing particular products in specific States.

The movement of agri-food products is supported by a complex logistics network. For example, grain is transported via

- rail (61% in 2016);
- road (25%); and
- waterways (14%).

Specifically, during 2022, a total of 1.5 million carloads of grain were transported by Class 1 railroads (i.e. those with annual revenues exceeding $250 million) and, as of early 2023, the grain car fleet was comprised of 273 000 cars with 1.4 billion cubic feet capacity (AAR, 2023).

In addition, while road transportation via trucks has traditionally been a reliable option to move produce rapidly and precisely from a particular origin to a specific destination. Increasing fuel prices and labour shortages have created significant challenges for this method of transportation. As such, the producer price index for general freight trucking (long-distance truckload) has increased to an index value of 188 in March 2023 (based on the start of the index in 2004) which is, however, down from its maximum of 211 in March 2022 (PPI, 2023). Given issues such as labour shortages, reliability issues have also increased, resulting in delays and spoilage. This is why food producers are considering building or increasing their own fleets of trucks, primarily with the objective to improve customer service (Kapadia, 2022).

Inland waterways also play a significant role in the U.S. agri-food supply chain, in particular the Mississippi River System (which is comprised of the Mississippi, Arkansas, Illinois, Ohio, and Tennessee Rivers, and the Gulf Intracoastal Waterway) (Agribusiness Consulting, 2019). While transportation via water is significantly cheaper than other modes of transportation, the infrastructure is aging and in need of further investments. As such, delays due to broken infrastructure (e.g. dock closures or lock breakdowns) can increase the cost (and reduce the profit) for farmers and make waterways less competitive.

5 International trade flows in agricultural produce

With U.S. agricultural exports and imports having increased significantly over the last quarter century, the USA is one of the largest traders of agricultural products (after the European Union) (USDA, 2023b). U.S. agricultural trade is forecasted to be $184.5 billion in exports and $199.0 billion in imports in 2023 (Kenner et al., 2023). As such, international trade flows play a major role in the U.S. agri-food supply chain. U.S. farmers have been relying on exports to sustain prices and revenues, and exports of high-value products like diary products and meats, as well as fruits and vegetables, have been growing (USDA, 2023b). Exports have been enabled by the reduction of agricultural trade barriers. However, compared to the prior forecast made in November 2022, exports of all major agricultural commodities, such as corn, sorghum, and soybeans, were projected to be lower (USDA, 2023b). Major macroeconomic variables impacting the value of exports for agricultural products include fluctuations in the exchange rate for the U.S. dollar and in the real gross domestic product (GDP) growth rate (USDA, 2023b).

Amongst the recipients of exported goods, China is the largest market for the USA, receiving $34 billion in exports, followed by Mexico and Canada with the receipt of $28 and $27.5 billion in exports, respectively (Kenner et al., 2023). Looking at the average magnitude of exports between 2019 and 2021, China was the biggest recipient (with a share of 15.6%), followed by Canada (with a share of 14.9%) and Mexico (with a share of 13.5%). Other major recipients included Japan (with a share of 8.1%), the Eurozone (with a share of 6.9%), South Korea (with a share of 5.3%), and Indonesia (with a share of 1.8%) (Kenner et al., 2023).

Looking more closely on what is forecasted to be exported, amongst the $184.5 billion in exports forecasted in February 2023

- $43.8 billion were forecasted to be from grains and feeds;
- $43.4 billion from oilseeds and products;
- $40.5 billion from livestock, poultry, and dairy; and
- $39.0 billion from horticultural products.

Categories with lower values included sugar and tropical products (such as pineapples) ($6.0 billion), cotton ($5.9 billion), ethanol ($3.6 billion), and seeds ($1.6 billion) (Kenner et al., 2023).

Amongst the $199.0 billion in imports forecasted in February 2023, $99.9 billion were forecasted from the following (Kenner et al., 2023):

- horticultural products, which primarily consisted of fresh ($18.4 billion) and processed ($8.3 billion) fruits, and fresh ($11.8 billion) and processed ($7.3 billion) vegetables; as well as
- distilled spirits ($12.1 billion).

Other major categories included

- sugar and tropical products ($30.2 billion), including coffee and related products ($9.9 billion);
- sweeteners and related products (($7.0 billion); and
- livestock, dairy, and poultry ($26.8 billion), including livestock and meats ($20.8 billion), and beef and veal ($9.2 billion).

The remaining categories included

- grains and feed ($20.6 billion); and
- oilseeds and products ($17.6 billion).

Overall, imports are dominated by consumer-oriented products, with some of the fastest-growing agricultural products including cashews and peppers (from Vietnam); fruits (from Peru); pepper, sesame seed, and vegetable extracts (from India); soft drinks (from Switzerland); and tropical and essential oils (from Singapore) (USDA, 2023b).

Amongst the $199.0 billion in imports forecasted in February 2023, Mexico was the largest importing country ($46.4 billion), followed by Canada ($37.3

billion) and the European Union ($ 35.9 billion). Other major importers included Brazil ($6.6 billion), Indonesia ($5.7 billion), Colombia ($4.4 billion), Peru ($4.4 billion), and Australia ($3.7 billion) (Kenner et al., 2023).

6 Regional agricultural trade flows: the USA, Mexico and Canada

Due to Mexico and Canada bordering the USA, as well as their leading role as importing countries, a closer look at their agri-food supply chains is appropriate. The trade between these countries is enabled by the United States-Mexico-Canada Agreement (USMCA), which replaced the North American Free Trade Agreement (NAFTA) in 2020.

Based on USDA statistics (USDA, 2023f), in 2022, the USA purchased about 80% of Mexico's $46.6 billion exports of agricultural products, with the USA supplying about 70% of Mexico's $41.4 billion imports of agricultural produce. Mexico accounted for 14.5% of the U.S. agricultural exports and for 22.8% of its agricultural imports. Primary agricultural commodities moving from the USA to Mexico include grains and oilseeds, while the USA primarily imported vegetables, fruit, beverages, and distilled spirits from Mexico (together accounting for 83.6% of total imports from Mexico) (USDA, 2023f). Mexico also serves as the primary export destination for U.S. dairy products, apples, pears, and grapes (ITA, 2022).

Mexico and the USA are further interlinked by significant mutual direct investments as well as mutually agreed product quality standards and an increasingly integrated transport infrastructure between the two countries (USDA, 2023f). Agri-food import procedures, for example, are becoming more standardized, and retailers are developing advanced distribution systems (including cold chain technologies) to allow rapid and safe cross-border flows of produce. While Mexican consumers associate U.S. agri-food products with high quality, U.S. imported food products are more expensive than local produce, a differential exacerbated by exchange rate fluctuations, making U.S. products less affordable (FoodExport, 2023).

Despite these close links, Mexico has been seeking to diversify its agricultural sourcing activities to producers in the European Union, South America, and Asia, increasing the competition for imports from the USA (ITA, 2022). In addition, a greater emphasis has been placed on becoming self-sufficient in core agricultural products, including corn and milk, through the improvement of local, small-scale production, and consumption (ITA, 2022). A challenge that Mexico has been facing in agriculture is domestic food price inflation and food scarcity, with a governmental programme in place to maintain fair prices and support subsistence grain farmers (ITA, 2022).

According to a report by USDA's Foreign Agricultural Service and the Global Agricultural Information Network (Lozano, 2022), the southbound agri-food supply chain between the USA and Mexico heavily relies on the Laredo district in Mexico, due to the well-established logistics infrastructure in this district. Trade in this district is facilitated by several commercial border crossings close to major population centres. The primary southbound mode of transportation for bulk U.S. exports is rail (primarily via Texas border crossings), followed by ship (primarily through Gulf seaports), and a small percentage by truck (Lozano, 2022). Modes of transportation differ based on the commodity, e.g. with sorghum being primarily transported by truck. The export of meat and poultry products, as well as fresh fruit and vegetables, also heavily relies on trucks (Lozano, 2022).

The northbound agri-food supply chain between Mexico and the USA to also reliant on the Laredo region for border crossings, but not as much as the southbound supply chain. The northbound supply chain also uses ports of entry across the eastern part of the border. In contrast to the southbound supply chain, the northbound supply chain primarily uses land-based modes of transportation (Lozano, 2022).

The import and export of agricultural products creates a significant number of jobs. The import of fresh tomatoes from Mexico to both the USA and Canada, for example, supports almost 33 000 jobs in the USA and generates $1.8 billion in the U.S. economy from labour income (Duval et al., 2018). Data provided by the World Bank (World Bank, 2023) indicate the value added by Mexico's agriculture, forestry, and fishing sector as 3.9% of GDP in 2021, which is higher than the value added by that sector in Canada (1.7% of GDP in 2018) and the USA (1.0% of GDP in 2021) with the worldwide average being at 4.3% in 2021. In 2021, Mexico's exports of edible vegetables and certain roots and tubers ranked third in the world (10.3% of world exports), as were beverages, spirits, and vinegar (7.1% of world exports), with edible fruit and nuts, peel of citrus fruit and melons ranking fourth (5.7% of world exports) (ITC, 2021). In comparison, the USA's exports of cereals ranks first in the world (19.9% of world exports), as do edible fruit and nuts, peel of citrus fruit and melons (10.3% of world exports) (ITC, 2021).

Amongst the remaining Latin American countries, the OECD/Food and Agriculture Organization of the United Nations (2019) highlights Brazil, Argentina, and Chile as major exporters of agricultural products in terms of their value, with an average of 14.1% of the total labour force across all countries being employed in agriculture, and the agricultural sector in most countries accounting for more than 20% of the GDP. Latin America is the leading exporter of some specific agricultural products, including avocados (representing 75% of the world's production) and roses (representing 34% of the world's production) (Canales, 2023). Latin American countries export about a quarter of their agricultural production. However, this disadvantages small-scale producers who are less able to access export markets and compete with larger-scale producers, reinforcing or even exacerbating their income and food insecurity (Chicoma, 2023).

The second-largest share of agricultural imports to the USA comes from Canada. In 2021, the USA purchased 54.4% of Canada's $59.8 billion exports of agricultural products, with the USA supplying 57.0% of Canada's $42.6 billion imports of agricultural produce (USDA, 2023f). Canada accounts for 14.1% of the USA's agricultural exports and for 18.0% of its agricultural imports. Primary agricultural commodities moving from the USA to Canada include grains, fruit, vegetables, meat, and related products (65.2% of total exports to Canada in 2021), while the USA primarily imported Canadian meat and other animal products, grains and feeds, and oilseeds and oilseed products (61.5% of total imports from Canada). The leading agri-food imports in 2021 were biscuits and wafers, accounting for $3.4 billion. The two countries are further interlinked by significant mutual direct investments (USDA, 2023f). Canada's exports of edible vegetables and some roots and tubers ranks fifth in the world in 2021 (6.5% of world exports), as did preparations of cereals, flour, starch or milk, and pastrycooks' products (5.7% of world exports), and cereals ranking seventh (5.7% of world exports) (ITC, 2021). Characteristics that make Canada an agri-food leader include ample land and water resources, access to international markets, strong research and development, and the brand equity of products originating from Canada (Government of Canada, 2022, 2024).

The COVID-19 pandemic highlighted vulnerabilities in national and international agri-food supply chains. Boileau and Sydor (2020) characterized the agricultural sector in Canada, for example, as having a high vulnerability to supply disruptions but low vulnerability to demand disruptions. One industry that was especially vulnerable to supply disruptions was animal feed manufacture, due to high reliance on intermediate inputs imported primarily from the USA. Hui et al. (2022) also highlighted the fragility of agri-food supply chains, less in disruption of overall food availability but in the range of choices/ options and in higher prices, with rural areas being more vulnerable. This vulnerability was attributed to lean (just in time) supply chains with very low inventories, where minor disruptions can cause a ripple effect throughout the supply chain. Compounding factors during the COVID-19 pandemic included the inability of migrant workers to enter Canada due to the lockdown of borders (Hui et al., 2022). To address this challenge, the Agri-Food Pilot was launched in May 2020 to provide 'a pathway to permanent residence for experienced, non-seasonal workers in specific industries and occupations' (Government of Canada, 2020, 2021). Eligible industries include meat product manufacturing, and greenhouse and animal production, with the programme scheduled to run until May 2023 (Government of Canada, 2020). Additional challenges facing the agri-food supply chain in Canada include fluctuations in consumer demand and markets as well as compliance with regulatory frameworks affecting exports, requiring for example improving the regulatory cooperation with the USA (Tweed, 2013)

The Canadian government has been actively involved in supporting the agri-food supply chain, for example with investments to improve worker safety at food processing plants (Government of Canada, 2021a). This is also reflected at the provincial level. A policy brief by the Ontario Chamber of Commerce, for example, has provided guidance for building a more resilient food supply chain. Elements of this roadmap included (Kronfli, 2021):

- support for local food production and online sales;
- the improvement of the AgriStability programme (which stabilizes farm incomes in difficult economic periods);
- a reduction of red tape for farmers (such as interprovincial trade barriers for meat);
- support for the next generation of farmers (in the form of access to training, land, and capital);
- tackling food fraud; and
- addressing food insecurity amongst consumers.

Average food insecurity in Canada's ten provinces was reported as 15.9% in 2021, not including the territories where food insecurity has been very high (e.g. 46.1% in Nunavut in 2020) (Proof, 2022). To address these challenges, the Canadian Government has implemented a Food Policy for Canada, which creates 'a roadmap for a healthier and more sustainable food system in Canada'. Initiatives as part of this programme include (Government of Canada, 2019)

- the improvement of local food infrastructure;
- a buy-Canadian promotion campaign;
- a reduction of food waste; and
- tackling of food fraud.

The Canadian government is not only concerned with both national and international agri-food supply chains, but a greater coherence is also sought between national and international agendas, which include improved access to financing and control over resources by women smallholder farmers in countries exporting to Canada, and an emphasis on social and environmental sustainability in international agri-food supply chains (Government of Canada, 2021b).

7 Key challenges facing the U.S. agri-food supply chain

Agri-food supply chains are faced with a multitude of challenges, some of which are discussed here. A current assessment of the U.S. agricultural industry and the challenges it faces is provided by the Ag Economy Barometer, which is published monthly by Purdue University/CME Group. This barometer serves as a measure of the health of the U.S. agricultural economy. It produces an index based on a survey of 400 agricultural producers and their economic sentiments each month (each quarter the index is complemented by a survey of 100 agriculture and agribusiness leaders) (https://ag.purdue.edu/commercialag/ageconomybarometer/).

In addition to the main index value, finer-grained assessments are provided, such as reasons preventing respondents from making significant investments. While the primary reason for limiting new investment during early 2023 has been price increases for farm machinery and construction of farm buildings, this fell to second place in March 2023 (32% of respondents, compared to being consistently above 40% in previous months). The dominant reason for lack of new investment amongst 34% of respondents in the March 2023 survey was rising interest rates (Mintert and Langemeier, 2023). Other reasons included uncertainty about farm profitability (for 10% of respondents) and low farm machinery inventories at dealers limiting availability (for 4% of respondents).

We will be focusing on a key set of challenges in the following. Specifically, as already noted in the case of Canada, the COVID-19 pandemic has shown that the agri-food supply chain is more fragile than it may have appeared (Ali, 2022). Another critical challenge we will elaborate further on, is food safety which needs to be assured throughout the supply chain. This can be a complex undertaking, especially with international supply chains. A third challenge is food losses and waste which remains a significant weakness in the current supply chain. A further challenge is the shortage of labour, which significantly increased with pandemic restrictions on movement within and between countries, and has remained a significant problem even after the pandemic. Recent years have also shown the challenges created by droughts resulting from climate change, which have particularly impacted California as a major producer for the U.S. agri-food supply chain. A last challenge that will be covered in more detail in the following is the financial pressures the agri-food supply chain has had to manage.

8 Key challenges: the coronavirus disease 2019 pandemic

Like most supply chains, the agri-food supply has been challenged by the impact of the COVID-19 pandemic. Agri-food businesses, for example, faced significant shortages in key inputs, most notably seeds and pesticides (OECD, 2020). These constraints were created by limited mobility of workers involved in manufacture and distribution in the early phases of the pandemic (Barman et al., 2021), and the delays in shipments from countries that were on lockdown.

The problem was less availability of material inputs themselves but timely access to materials that might be produced in one group of countries but which could then not be transported to customers in other countries. This can be attributed to the complex global supply chains that have developed in recent years, characterized by an increasing division of labour and specialization between companies in different countries (OECD, 2020). What exacerbated the issue for the agri-food industry was that there are only certain times in a year when crops can be planted and/or fertilized and, if those windows were missed due to the lack of access to the input material, alternatives were limited.

Consumers of agri-food products also experienced associated supply shortages directly through the unavailability of products in supermarkets, especially during the early phases of the pandemic, creating an expectation amongst consumers of what the new reality (at least in the short-term) may look like. This led to unprecedented demand spikes sparked by panic buying. Managing the grocery supply chain thus became a challenge, with a demand planning/forecasting error of 59% (compared to 45% pre-pandemic) (Banker, 2021). A Statista (2023c) consumer poll (of March 15, 2020) revealed that most respondents (81%) believed food shortages to be caused by people stockpiling food, and not necessarily by food supply chain disruptions (19%). This suggests underlying consumer confidence in the agri-food supply chain's ability to ensure a stable provision of necessities.

While the unprecedented environment created by the COVID-19 pandemic exposed vulnerabilities and identified choke points in the supply chain, it provided a fruitful environment to create more resilient and robust agri-food supply chains. The U.S. Department of Agriculture (USDA) has invested more than $4 billion to improve agri-food supply chains as part of the Build Back Better initiative (Harper, 2021). Objectives include a strengthening of the food system, the creation of new market opportunities, addressing climate change, helping disadvantaged communities, and ensuring fair wages throughout the agri-food supply chain. The initiative seeks to address a food system considered to be "rigid, consolidated, and fragile", as stated by Agriculture Secretary Tom Vilsack (USDA, 2021). The initiative noticeably focuses on the entire supply chain, including changes in the following (USDA, 2021)

- food production (with a particular emphasis on essential workers);
- food processing (with a particular emphasis on expanding regional processing capacities to break up the current centralized system with, for example, a few companies in particular states, being primarily responsible for manufacturing particular products);
- food distribution and aggregation (with a particular emphasis on reducing shipping distances to put less strain on logistics and fostering local and regional production); and

- food marketing and sale to consumers (with a particular emphasis on supporting market access for small- and mid-sized producers, and enabling consumer access to healthy foods).

These initiatives seek to make the agri-food supply chain more resilient against shocks, enable it to deliver greater value to growers and workers, and provide consumers with an affordable selection of healthy food that is produced and sourced locally and regionally (Harper, 2021; USDA, 2021).

9 Key challenges: food safety

Despite decades of effort to ensure the safety of food products, food safety remains a significant concern. A key factor making this a persistent challenge has been agri-food supply chains becoming longer and more complex, providing a multitude of opportunities where food can become contaminated (Schoenherr et al., 2015). This was also highlighted as a primary challenge to food safety by the U.S. Centers for Disease Control and Prevention (CDC, 2023a).

Food can become contaminated at the time of production, for example, through contaminated soil (e.g. through improperly treated compost or manure), contaminated water being sprayed on food or drunk by animals, or livestock becoming infected with bacteria or diseases. Contaminated water can also taint food during processing (e.g. during washing), or food safety can be jeopardized by dirty work surfaces or poor worker hygiene practices. In the distribution stage, food safety can be challenged when refrigeration requirements are not met, even if this happens in just a small part of the product journey. In the food preparation stage, contamination can occur due to poor hygiene practices or a member of the restaurant kitchen staff being sick and transferring germs on the food product (CDC, 2023b). Additional challenges to food safety include environmental changes leading to food contamination, changes in consumer preferences and habits (e.g. for fresh or minimally processed food products which may not have been through a traditional decontamination step (e.g. application of heat)), and new and emerging bacteria and toxins, as well as the development of antibiotic resistance (CDC, 2023a).

The monitoring of foodborne illness in the U.S. is undertaken by the Centers for Disease Control (CDC). The CDC estimates that, on average, 48 million Americans get sick every year due to foodborne diseases, with 128 000 being hospitalized and 3000 dying each year (FDA, 2018). Oversight of food safety in the USA is divided between the Food and Drug Administration (FDA) and the USDA, with the former being responsible for 80% of the nation's food suppliers and the latter for 20% (Beach, 2023). A study by the Sedgwick Organization reported that the total number of units (e.g. number of packages or bags) recalled by the USDA increased by 700% in 2022 compared to 2021, amounting to 416.9 million units recalled across a total of 423 recalls. The number of recalls has remained steady in recent years, with the primary reason being undeclared allergens (Beach, 2023).

A recall can impact profitability significantly, with 77% of respondents to a survey by the Grocery Manufacturers Association estimating the cost of a recall up to $30 million (Holleman, 2018). This not only includes direct costs of the recall, but also indirect costs, such as penalties, lawsuits, lost sales, and brand damage (IBM, 2020). Chipotle was for instance fined $25 million for its role in norovirus outbreaks linked to its restaurants, with its stock declining to less than half of its value and leading to the founder and CEO having to step down (FSN, 2020). What makes managing and monitoring food safety so challenging is that, even when contamination has been detected, it may be difficult to trace it back to its source to eliminate the root cause. For example, the root cause for an *E. coli* contamination incident in Chipotle restaurants during October and December 2015 was never determined (Marves, 2022).

To address the challenge of food safety, the FDA developed the Food Safety Modernization Act (FSMA - enacted in 2011), which enables the FDA to be more proactive when it comes to improving food safety, including new enforcement authorities to ensure higher prevention compliance rates, as well as risk-based food safety standards. The law further establishes that imported food products are subject to the same standards as domestic products, and provides the foundation for "an integrated national food safety system in partnership with state and local authorities" (FDA, 2018). Preventive measures include mandatory preventive controls for food facilities, mandatory produce safety standards, and measures to prevent intentional contamination. Inspection and compliance measures include mandated inspection frequency, access to records, and testing by accredited laboratories. Response measures include the FDA's authority to mandate recalls, the suspension of registrations, the development of enhanced product tracing, and additional recordkeeping requirements for high-risk foods. Import measures include making the importer accountable for verifying that foreign suppliers have adequate preventive food safety controls in place, and establishing a third-party certification programme that assures foreign food facilities comply with U.S. food safety standards (CTNOFA, 2011; FDA, 2018).

Technology carries also great potential to address food safety, in particular blockchain technology. A blockchain is a decentralized and distributed digital ledger that records transactions across multiple organizations. IBM Food Trust™, for example, has leveraged this technology to create greater transparency across the agri-food supply chain from farmers to consumers. The application provides a 'complete history and current location of any food item, along with its accompanying information (i.e. certifications, test data, temperature data) readily available in seconds' (IBM, n.d.). Such solutions not only support improved food safety and quality, but can also generate supply chain efficiencies, reduce waste, improve sustainability, and prevent fraud. Walmart has been an early adopter of this technology, e.g. requiring all their fresh leafy green suppliers to adopt this technology to enable end-to-end traceability back to the farm within seconds (Redfield, et al., 2018). This initiative has expanded to other products and suppliers, and has also included partnerships with Dole, Kroger, McCormick, Nestlé, Tyson Foods, and Unilever (Sristy, 2021). Blockchain technology also has the potential to eliminate fraud and counterfeit agri-food products in the supply chain (Authentium, 2023), ensuring, for example, that the country of origin is indeed the place where the food was grown.

10 Key challenges: food losses and waste

One of the major challenges facing the U.S. and other agri-food supply chains is food losses and waste (FLW) (Prezkop et al., 2020). Food losses are generally related to food production and processing, while food waste is related to retail sale and consumer use. It has been estimated that the USA spends $218 billion a year (1.3% of GDP) on growing, processing, and transporting food that is never eaten (ReFED, 2016).

The organization ReFED (Rethink Food Waste Through Economics and Data) was established in 2016 as a non-profit organization to address the problem of FLW, bringing together a wide range of organizations, including food businesses, government and non-governmental organizations (NGOs). It has developed a *Roadmap to Reduce US Food Waste* highlighting where FLW is estimated to occur (ReFED, 2016):

- Farms: 16%;
- Food processors: 2%;
- Retailers and food service: 40%; and
- Consumers: 43%.

In addition, studies by the World Wildlife Fund (WWF) of the fresh produce sector have, for example, estimated post-harvest losses of (WWF, 2020)

- tomatoes: 40%;
- peaches: 39%; and
- lettuce: 56%.

Other studies have made similar estimates of losses (e.g. Johnson et al., 2018). Farmers growing fresh produce in particular face a wide range of challenges in matching supply with demand, including

- forecasting demand and likely yields (farmers will often deliberately produce a surplus in order to guarantee meeting contract requirements from processing or retail customers);
- timing harvesting to meet demand;
- ensuring sufficient labor for harvesting;
- strict quality standards (e.g. relating to appearance or anticipated shelf life) which means perfectly edible produce is rejected by processors or retailers; and
- the so-called walk-by field losses where a grower decides it is simply uneconomic to harvest a crop.

Solutions to this reality include better communication backwards through the supply chain to match supply with likely demand and improved, more flexible supplier contracts to allow farmers to match demand at less risk to themselves.

As an additional statistic, the USDA, for example, has estimated that over 30% of available food at the retail and consumer end of the agri-food supply chain is uneaten (Buzby et al., 2014). ReFED and the Food Waste Reduction Alliance (FWRA) estimate that over $18 billion is lost at the retail stage (ReFED, 2018). Causes of waste include

- poor stock management;
- poor storage conditions (e.g. inadequate temperature control);
- deliberate over-stocking at point of sale (in part to ensure customers have plenty of choice);
- removing edible produce with minor blemishes (to avoid potential customer complaints about quality);
- removal of edible produce that has just gone past a 'best before' or 'use by' date (to avoid potential customer complaints about freshness and safety); and
- packaging and promotional offers which encourage customers to buy in excess of what they actually need.

Solutions to this reality include, for example, Kroger's Zero Hunger/Zero Waste initiative which, for example, has developed better metrics to measure waste, pilot schemes in stores to reduce waste and campaigns to educate customers (e.g. about 'best before' or 'use by' dates) (https://www.thekrogerco.com/sustainability/zero-hunger-zero-waste/).

The U.S. Environmental Protection Agency (EPA) has also developed a Food Recovery Hierarchy which prioritizes minimizing the volume of surplus food in the first place, followed by measures to repurpose or re-use any surplus, e.g. by donating it to food banks, converting waste to animal feed, using waste as a feedstock (e.g. in anaerobic digestion to produce biofuel), or composting (for use as an organic soil amendment), with landfill or incineration as a last resort (Gunders, 2017). There are also a range of initiatives to reduce waste in other parts of the agri-food supply chain such as hospitality, restaurants, and to better educate consumers to reduce food waste at home (Prezkop et al., 2020).

11 Key challenges: labour shortages

Labour shortages have been plaguing the U.S. agri-food industry over many years for a number of reasons. Specifically, there has been a declining interest in agriculture, due to the strenuous work and the long hours involved as well as the sometimes poor financial returns. Other factors that make the industry unattractive are high real estate and land prices, significant up-front investments in machinery and technology, volatile commodity pricing, unpredictable weather changes, and unequal work-life balance (Agamerica Lending, 2019). This makes it especially challenging to attract younger generations. As such, only 9% of agri-food producers are 35 years of age or younger and the average principal operator age is almost 60 years (Agamerica Lending, 2019).

There are also often better opportunities available elsewhere, which has intensified the competition for skilled workers and the higher wages that need to be offered to attract talent. This situation is exacerbated by simultaneous labour shortages in trucking, machinery operations, and warehousing, which are sectors that can offer higher wages. Overall, agri-food businesses have only limited ability to compete, particularly in the current economic environment which is exacerbate further by higher seed and fuel prices (Gruber, 2022).

Analogous to most industries, agri-food businesses have also been challenged by labour shortages triggered by the COVID-19 -pandemic. While laying off personnel in the early phases of the pandemic was seen as a means to weather the crisis and remain in business, companies are now facing difficulties rehiring their staff (McCann, 2022). This has led to the number of posted job vacancies to be higher than ever, making the hiring and retaining of workers a critical challenge (Bacon, 2022). In addition, even when workers have successfully been hired, critical workers becoming ill can be a significant impediment to the proper functioning of the supply chain (Ali, 2022).

To address these labour shortages, farmers have been pushing for immigration reforms through the Farm Workforce Modernization Act. The objective of the act is to ensure a stable and reliable workforce by providing a path to citizenship for undocumented agricultural workers and undertaking reforms related to the seasonal farmworker visa programme (Ali, 2022). It is expected that this bill, through the addressing of labour shortages in the agri-food supply chain, will help reduce high food prices which have increased by more than 11% between 2021 and 2022 (Simpson, 2022). Part of the Farm

Workforce Modernization Act is also a reform of the H-2A temporary agricultural work programme (Congress, 2021), which will be described in more detail next.

For many decades, the U.S. agri-food supply chain has relied on migrant workers who move temporarily from other countries to the USA to work in the agri-food supply chain. Temporary employment opportunities generally occur during the planting and harvesting seasons, with migrant workers entering the USA as part of the H-2A temporary agricultural workers programme. The number of visas issued under this programme has increased significantly, reaching an all-time high in the 2021 fiscal year with 317 000 visas. In comparison, the number of visas issued in 2020 was 275 000, with a mere 100 000 issued in 2013 (Martin, 2022). In order for a visa to be issued, employers must demonstrate that they were not able to hire U.S. workers, with the growing numbers of visas indicating a significant domestic labour shortage. Employers of H-2A workers must pay for the workers' travel and housing, and provide a minimum wage as determined by the U.S. Department of Labor. This is done to avoid driving the wage rate for domestic workers. The hourly wage rates effective 1 January 2023, which are also known as the Adverse Effect Wage Rates, range from $13.67 in Alabama and Arkansas, to $18.65 in California (DOL, 2022).

A common criticism of the H-2A programme was that it only allowed migrant workers to stay for a maximum of 10 months before they had to return to their home country (Swenson, 2022). This disadvantages those agri-food sectors that are in need of year-round employments, such as the dairy or livestock industries. Updates of the Farm Workforce Modernization Act would alleviate this shortcoming and allow year-round employment of H-2A workers in those agri-food industries where year-long employment is warranted.

A study by Texas A&M International University (Maddan et al., 2022) concluded there to be a positive impact of more migrant and H-2A workers on lower inflation (given potentially lower food prices), higher average and minimum wages, and lower unemployment. In contrast, increased rejections of petitions for naturalization were noted to be associated with higher consumer prices and inflation. Overall, more petitions for naturalization have been associated with lower inflation (FGN, 2022).

12 Key challenges: droughts

Droughts can have a significant impact on agricultural production. They are ranked third, behind tropical cyclones and severe storms, as being responsible for 'billion-dollar weather disasters since 1980' (NIDIS, 2023). Prolonged droughts can have a significant impact on all aspects of agriculture, resulting in reduced crop production, livestock losses, loss of soil health (including erosion), and property damage. The magnitude of this impact is illustrated by the severe drought in 2012, which impacted 80% of U.S. agricultural land and necessitating the federal crop insurance programme to offer $14.5 billion in loss payments, or by the 2015 drought in California leading to the loss of more than 10 000 seasonal jobs (NIDIS, 2023). Another challenging drought year was 2022, when respondents to a survey by the American Farm Bureau Federation expected their farm-related revenue to be down by 38% due to drought conditions (Munch, 2022).

In addition to a direct impact on production losses, droughts can also fuel the prevalence of pests and diseases (enabled by high temperatures) and decrease water availability, which may be especially detrimental to specialty crops, such as fruits and vegetables, compared to field crops (NIDIS, 2023). As of April 2023, 1424 counties across the USA were experiencing droughts, impacting 198.3 million acres of crops and 18.3 million beef cattle (NIDIS, 2023). An up-to-date overview of current drought conditions is provided by the USA. Drought Monitor (https://droughtmonitor.unl.edu/), which determines drought severity based on soil moisture, streamflow, precipitation levels, and historical weather patterns. The National Drought Mitigation Center then overlays the data from the USA. Drought Monitor with maps for agricultural areas to compute the impacted acreage by agricultural commodity (https://agindrought.unl.edu/).

To address this significant challenge, the USDA established Climate Hubs, which are cross-agency collaborations led by the USDA Agricultural Research Service and the USDA Forest Service. The mission of these Climate Hubs is "to develop and deliver science-based, region-specific information and technologies, with USDA agencies and partners, to agricultural and natural resource managers that enable climate-informed decision-making, and to provide access to assistance to implement those decisions" (USDA, 2023d).

13 Key challenges: financial pressures

The agricultural sector is subject to significant financial pressures. Two of the most important are higher costs for inputs (such as fertilizers and feed) and rising interest rates. The rising costs of inputs was highlighted as a major concerns for example by producers in the March 2023 report of the Ag Economy Barometer. Thirty-four per cent of respondents indicated this to be their major concern, down from 53% during summer 2022 (Mintert and Langemeier, 2023).

Higher input costs can be driven by supply constraints, particularly in the availability of raw materials for fertilizer production. This led in 2021 to price spikes for fertilizers of more than 300%, driven by price increases for ammonia (up by 210%), liquid nitrogen (up by 159%), and urea (up by 155%) (Myers, 2021). While the USA is the third-largest producer of fertilizer worldwide, it still is dependent on these major raw materials which are often subject to significant import tariffs (Myers, 2021). With the USA producing a significant number of crops that remain dependent on synthetic fertilizers to maintain high yields (e.g. corn, wheat, and rice), farmers have been aiming to adjust what they are planting. This has been driven not only by the price of fertilizer, which farmers are not able to recoup by higher prices for their output, but also by its availability, which has been reduced in recent years (Doan, 2022). Farmers are therefore switching to crops that are less fertilizer-intensive, such as soybeans and peas (Doan, 2022). In addition, governmental initiatives have been put in place to increase production and access to domestic, innovative fertilizers for producers challenged by the price increases (Martin and Roden, 2023).

Another input that has been experiencing a significant increase in prices over the last 20 years are agricultural seeds. Conventional seed prices have experienced a 200% increase and genetically modified seed prices a 700% increase (Kovak, 2023). This increase can be attributed to industry consolidation amongst major seed producing firms, coupled with the intellectual property associated with genetically modified seeds. The multitude of patents already in place makes it difficult for smaller seed producers to enter the market, further restricting competition (Kovak, 2023). Rising interest rates were noted by 25% of respondents as being their primary concern in March 2023, a percentage that increased significantly from summer 2022 when it was just 14% of respondents (Mintert and Langemeier, 2023).

14 Recent developments: technological innovations

To respond to this range of challenges, agri-food supply chains have been making changes to their operations. For example, the industry has been transitioning to planting crops that are easier to manage and require less human labour, scaling back operations, leasing portions of their land, moving operations abroad, postponing expansion plans, and, if possible, raising wages and benefits to attract talent into farming (Agamerica Lending, 2019). While the opportunities are many, this section will highlight two recent developments that have come to the force, which are the leveraging of technological innovations (discussed in this section) and sustainable agriculture (discussed in the following section).

The leveraging of technological innovations has been one way to deal with the mounting challenges for agri-food supply chains, including the constrained labour environment. Some of the advanced technological solutions farmers are experimenting with include wireless sensors, predictive forecasting, advanced automation, robotics, and data analytics (Agamerica Lending, 2019).

For example, while there is great potential in the use of robots for picking produce, it is a challenging task, especially when one considers for instance the delicate picking needed to harvest a ripe strawberry. Key challenges and developments in robotics are reviewed in van Henten and Edan (2024). Inroads in this vein have, for example, been made by Bowery Farming based in in New York City, which acquired Traptic, a manufacturer of farming robots that use artificial intelligence and computer vision to direct robotic arms for the successful harvesting of strawberries and tomatoes (Oller, 2022). While the investment in these technologies is high, and while the technology may not be universally applicable, this strategy has the potential to increase productivity in a challenging labour market. Traptic plans for its robot to pick 100 000 strawberries per day. However, while a lot of attention has been paid to robotic automation especially in strawberry picking, adoption has been limited, partly because robots are only able to pick 70% of ripe strawberries, with the remainder still needing to be picked by humans (Johnson, 2022).

Vertical farms (also known as plant factories) provide a promising avenue for agri-food supply chains to deal with challenges such as increased weather variability due to climate change (Kozai and Hayashi, 2023). For example, Driscoll's, a California-based agri-food firm specializing in strawberries and other berries, has partnered with Plenty, a vertical farming company that is building the world's largest indoor vertical farming campus on 120 acres in the State of Virginia (Wang et al., 2022). The project, which includes 30-foot grow towers that are climate-controlled, pesticide-free, and maximize light exposure for the crops, is projected to produce up to 20 million pounds of produce on an annual basis. Robotic arms will be key in tending to the crops as they develop, with yields expected to be 150–350 times higher per acre than traditional farming, enabled by Plenty's expertise both in factory construction and operation using advanced data analytics for precise control (Oller, 2022).

Another area where information can come into play is precision agriculture, which has the potential to reduce costs by enabling farmers to tailor operating decisions based on identifying site-specific characteristics (Stafford, 2018). A range of remote and proximal sensors can track variations, for example, in soil type and quality, in crop growth and health, in harvested yield, and other variables for small parcels of a field, providing guidance for what actions are needed for each parcel (e.g. more or less fertilizer, pesticide or irrigation required). These differentiated treatments can also be made automatically by automatically guided equipment using global positioning system (GPS) tracking technologies. As an example, use of precision agriculture has made it possible to reduce the area of corn requiring irrigation from 15% to 11% between 1996 and 2016 (Effland et al., 2022). The development of precision livestock farming technologies (Berckmans, 2022), which have significantly improved monitoring of livestock, has, for example, reduced the amount of feed needed for a dairy cow to produce 100 pounds of milk by about 40% over the last 40 years (American Farm Bureau Federation, 2023).

Another promising technology has been implementation of digital twins, which mirror the physical supply chain from farmers to end consumers (Denis et al., 2020). The myriad of decisions that need to be made along the supply chain include planning for the next growing season, determining the best sequence to fill empty silos, and transporting produce to markets that will ensure greatest profitability. This complex and interlinked system and set of decisions is suitable for advanced simulation to optimize system design and decision-making (Denis et al., 2020).

15 Recent developments: sustainable agriculture

Sustainability has become increasingly important to consumers, with 63% of Americans indicating that they are willing to pay more for sustainably sourced products (Amway, 2020). This section introduces regenerative agriculture as a promising means to address sustainability, and discusses the sustainability impact of food packaging.

15.1 Regenerative agriculture

One way to address sustainability concerns is regenerative agriculture, which focus on understanding and optimizing the physical, chemical, and biological health of soils that have been damaged by conventional, intensive agricultural practices such as use of heavy machinery, extensive use of fertilizers and/or pesticides (Masterson, 2022). This form of agriculture aims to improve soil health through stopping or substantially reducing ploughing (no-till or min-till), rotating crops, and using manure and compost as fertilizers to optimize soil organic matter (SOM) (Masterson, 2022). With current industrial food systems contributing 44–57% to global greenhouse gas (GHG) emissions (Grain, 2011), regenerative agriculture has the potential to not only reduce GHG emissions (through fewer inputs and a greater capacity to sequester carbon in soil), but to also to stabilize or increase yields in cases of extreme weather and climate change (through more resilient soils), create drought-resistant soil with improved water-holding capacity (e.g. through higher SOM and use of cover crops to lock in moisture), foster biodiversity, improve nutritional quality of crops, and restore grasslands (70% of which have already been degraded) (Regeneration International, 2019).

One company that has been pioneering this approach is Unilever, which issued a new set of regenerative agriculture principles in 2021. These principles focus on 'nourishing the soil, increasing biodiversity, improving water quality and climate resilience, capturing carbon and restoring and regenerating the land' (Unilever, 2021), and complement the company's Sustainable Agriculture Code (Unilever, 2017). This code has been a fundamental pillar of Unilever's sustainable sourcing programme since 2010, impacting thousands of its suppliers and hundreds of farmers (Unilever, 2023). The company is now also providing help and assistance to farmers to implement regenerative agricultural practices.

15.2 Food packaging

While packaging can maintain the safety and freshness of food, it can also have a significant impact on the environment, especially when designed for single use. About two-thirds of solid waste generated in the USA consists of packing (both for food and other products) (FoodPrint, 2020). The environmental impact, however, starts with production of packaging, whether it be made from glass, aluminium, paper, or plastic. Significant resources are needed in the form of energy and water to manufacture packaging while, at the same time, generating emissions and wastewater. Once the packaging has served its purpose of transporting and protecting the food, it is frequently discarded and ends up in landfills.

A significant challenge has also been created by plastic debris in the world's oceans, which Oceana (2023) estimates to amount to 33 billion pounds on a yearly basis. Most of this packaging does not degrade quickly, with plastic bottles, for example, needing 450 years to degrade (FoodPrint, 2020). When discarded in the oceans, the plastic debris also creates significant dangers for animals, including fish, seabirds, sea turtles, and marine mammals, who can ingest the debris or become entangled in it. This can also have an impact on humans through consumption of marine food products such as fish (Reddy, 2018).

Innovative food packaging solutions to address this challenge include edible packaging. Stonyfield Farms, for example, has collaborated with WikiFoods to encase its frozen yoghurt in pearls. While individual servings do not need any packaging, the entire bag still does (Brown, 2014). Another approach is to make packaging biodegradable, e.g. using algae and seaweed that function like plastic once wet, but degrade completely within 2 months (Patel, 2020). Companies have also started to offer packaging made of folded fibre that contains herb seeds - once the packaging has fulfilled its function, it can be planted (Packaging Digest, 2014).

16 Conclusion

This chapter focused on the U.S. agri-food supply chain and traced its developments, discussed various types of agri-food supply chains, including local, national, and international supply chain flows, highlighted some of the key challenges facing the U.S. agri-food supply chain, and showcased some of the recent developments in the sector. In doing so, the chapter illustrated the importance of agri-food supply chains and highlighted it as a promising research avenue. It is hoped that this chapter will serve as a starting point for researchers to investigate this domain further.

Perhaps, the most important underlying theme arising from this overview is the pressure to improve agri-food supply chains beyond the traditional focus on optimizing product safety and quality, choice, pricing, and availability. Agri-food supply chains are now under pressure to meet a new set of environmental and social objectives, whether it is reducing GHG emissions, minimizing FLW or developing greater resilience to the effects of climate change and other shocks (such as the COVID-19 pandemic). Agri-food supply chains are also under greater pressure to provide more equitable returns for farmers and farm workers (e.g. to offset labour shortages), and healthier, more accessible and affordable food for more disadvantaged communities.

Advances in technologies such as precision agriculture, robotics, and automation and blockchain technology will help improve supply chain efficiency in some of these areas, e.g. by reducing the volume of inputs needed (and therefore reducing GHG emissions and the pollution and biodiversity risks related to large-scale application of fertilizers and pesticides), substituting some labour requirements and improving data flow and decision-making. However, more profound, structural changes may also be necessary, such as moving from intensive to more regenerative methods of farming, rebalancing risks in the system which can seem to fall disproportionately on farmers (e.g. by providing more financial and technical support in developing more sustainable farming methods), and developing more circular supply chains where waste and co-products are recycled back into the system (e.g. by repurposing manure and crop residues as organic fertilizer).

It is also clear that some of these changes will involve trade-offs between competing priorities. As an example, regenerative agriculture promises to both reduce agriculture's contribution on climate change and improve resilience to climate change impacts, but potentially at the expense of the high yields achieved by more intensive methods of production. Similarly, the interest in promoting SFSCs has the potential to improve resilience (by decentralizing risks away from a primary dependence on highly complex, national supply chains which are vulnerable to shocks such as the COVID-19 pandemic) as well as strengthening local communities, but possibly at the expense of a higher overall environmental footprint. Managing these trade-offs will possibly be the greatest challenge of all.

17 Where to look for further information

- A whitepaper by the Association of American Railroads (AAR) positions U.S. freight railroads as a vital component of the agri-food industry, transporting everything from agri-food products to farm equipment and fertilizers. Freight railroads also enable the nation to be a top grain exporter. The whitepaper can be accessed at: https://www.aar.org/wp-content/uploads/2021/09/AAR-Grain-Fact-Sheet.pdf.

- The importance of inland waterways for the U.S. Agriculture was highlighted in a whitepaper by Agribusiness Consulting, which can be accessed at: https://www.ams.usda.gov/sites/default/files/media/ImportanceofInlandWaterwaystoUSAgricultureFullReport.pdf.
- Agriculture supply chain management is described by Authentium (2023) at: https://authentium.io/agriculture-supply-chain.html. The article covers agriculture warehouse management systems, agriculture supply chain software, and blockchain technology in farming.
- An interesting investigative report highlighting food monopolies is Lakhani et al (2021).

17.1 Associations

- The American Farm Bureau Federation is the national advocate for farmers, ranchers and rural communities. On its website (https://www.fb.org/) it provides a range of resources, including news, market intelligence, viewpoints, and information about events (such as conferences), leadership development through its Farm Bureau University, initiatives (such as its Ag Innovation Challenge) and programs (e.g. Young Farmers and Ranchers, and Women's Leadership Development).

17.2 Websites

- The website Food Safety News (https://www.foodsafetynews.com/) reports on food safety issues, including outbreaks and recalls. Beach, C. (2023) for example finds a significant increase in the number of food items recalled in 2022.
- The U.S. Bureau of Labor Statistics reports on employment projections for major industries, including agriculture, forestry, fishing, and hunting (https://www.bls.gov/emp/tables/employment-by-major-industry-sector.htm).
- The American Farm Bureau Federation (https://www.fb.org/) provides impressive statistics about agriculture and food on the following website: https://www.fb.org/newsroom/fast-facts.
- The Economic Research Service (ERS) of the US Department of Agriculture (USDA) (https://www.ers.usda.gov/) provides a range of impressive statistics and insights on the US ag and food industry, For example, general statistics are provided at https://www.ers.usda.gov/data-products/ag-and-food-statistics-charting-the-essentials/ag-and-food-sectors-and-the-economy (general statistics) and information on the concentration in the

US seed industry is provided athttps://www.ers.usda.gov/data-products/chart-gallery/gallery/chart-detail/?chartId=107516#:~:text=In%20recent%20decades%2C%20the%20U.S.,seed%20and%20agricultural%20chemical%20sales. In addition, a report posted on that website reports on the *importance of inland waterways to U.S. agriculture* (Agribusiness Consulting, 2019).

- The USDA also maintains a website with Agricultural Marketing Services (https://www.ams.usda.gov/), including rules and regulations, grades and standards, and market news.
- The Association of American Railroads (AAR, https://www.aar.org/) offers a wide range of information on the U.S. railroad industry, including rail traffic data, rail cost indexes, and publications and reports, such as a report on the criticality of f*reight railroads for the agri-food supply chain* (https://www.aar.org/wp-content/uploads/2021/09/AAR-Grain-Fact-Sheet.pdf).
- The U.S. Centers for Disease Control and Prevention (CDC, https://www.cdc.gov/) provide information on food safety challenges CDC (2023a) as well as how food can get contaminated along the supply chain CDC (2023b).
- A current assessment of the U.S. agricultural industry and the challenges it faces is provided by the Ag Economy Barometer, which is published monthly by Purdue University/CME Group. This barometer serves as a measure of the health of the U.S. agricultural economy. It produces an index based on a survey of 400 agricultural producers and their economic sentiments each month (each quarter the index is complemented by a survey of 100 agriculture and agribusiness leaders) (https://ag.purdue.edu/commercialag/ageconomybarometer/).

18 References

AAR (2023). *Freight Railroads & Grain*. Association of American Railroads, April 2023. Available at: https://www.aar.org/wp-content/uploads/2021/09/AAR-Grain-Fact-Sheet.pdf.

Agribusiness Consulting (2019). *Importance of Inland Waterways to U.S. Agriculture*. Agribusiness Consulting, August 2019. Available at: https://www.ams.usda.gov/sites/default/files/media/ImportanceofInlandWaterwaystoUSAgricultureFullReport.pdf.

American Farm Bureau Federation (2023). *Fast Facts about Agriculture and Food*. American Farm Bureau Federation. Available at: https://www.fb.org/newsroom/fast-facts.

Amway (2020). *Consumers Want Traceability, Transparency in Their Products*. Amway Connections, January 9, 2020. Available at: https://amwayconnections.com/healthy-living/consumers-want-traceability-transparency-in-their-products/.

Authentium (2023). *Agriculture Supply Chain Management*. Available at: https://authentium.io/agriculture-supply-chain.html.

Beach, C. (2023). *Report Finds an Enormous Increase in the Number of Food Items Recalled in 2022*. Food Safety News, March 15, 2023. Available at: https://www.foodsafetynews.com/2023/03/report-finds-enormous-increase-in-number-of-food-items-recalled-in-2022/.

Berckmans, D. (Eds) (2022). *Advances in Precision Livestock Farming*. Burleigh Dodds Science Publishing, Cambridge.

BLS (2023). *Employment Projections*. U.S. Bureau of Labor Statistics. Available at: https://www.bls.gov/emp/tables/employment-by-major-industry-sector.htm.

Boileau, D. and Sydor, A. (2020). *Vulnerability of Canadian Industries to Disruptions in Global Supply Chains*. Government of Canada / Gouvernement du Canada, June 2020. Available at: https://www.international.gc.ca/trade-commerce/economist-economiste/analysis-analyse/supply-chain-vulnerability.aspx?lang=eng.

Brown, A. (2014). *WikiFoods Develops Edible Packaging for Stonyfield Frozen Yogurt*. Food Dive, March 21, 2014. Available at: https://www.fooddive.com/news/wikifoods-develops-edible-packaging-for-stonyfield-frozen-yogurt/241901/.

Buzby, J. C., Hodan, F.-W. and Jeffrey, H. (2014). *The Estimated Amount, Value and Calories of Postharvest Losses at the Retail and Consumer Levels in the United States*. United States Department of Agriculture – Economic Research Service (USDA-ERS) Report EIB-21.

Cafer, A., Ramachandran, S., Mann, G. and Kaiser, M. (2019). *National Food Affordability: A County-Level Analysis*. Public Health Post, March 4, 2019. Available at: https://www.publichealthpost.org/research/national-food-affordability-a-county-level-analysis/.

Canales, M. (2023). *Latin America Has Become a Leading Exporter of These 2 Hugely Popular Products*. World Economic Forum, March 28, 2023. Available at: https://www.weforum.org/agenda/2023/03/avocados-roses-latin-america-exports/.

Carbone, A. (2017). Food supply chains: coordination governance and other shaping forces. *Agricultural and Food Economics* 5(1): 3.

CDC (2023a). *Challenges in Food Safety*. U.S. Centers for Disease Control and Prevention. Available at: https://www.cdc.gov/foodsafety/challenges/index.html.

CDC (2023b). *How Food Gets Contaminated - The Food Production Chain*. U.S. Centers for Disease Control and Prevention. Available at: https://www.cdc.gov/foodsafety/production-chain.html.

Chicoma, J. L. (2023). *Latin America's Food Paradox*. Food Tank. Available at: https://foodtank.com/news/2023/02/latin-americas-food-paradox/.

Chiffoleau, Y. and Dourian, T. (2020). Sustainable food supply chains: is shortening the answer? A literature review for a research and innovation agenda. *Sustainability* 12(23): 9831.

Cicatiello, C. (2020). Alternative food shoppers and the "quantity dilemma": a study on the determinants of their purchases at alternative markets. *Agricultural and Food Economics* 8(1): 15.

Congress (2021). *H.R.1603 - Farm Workforce Modernization Act of 2021*. H.R.1603 – 117th Congress (2021-2022). Available at: https://www.congress.gov/bill/117th-congress/house-bill/1603.

Conti, C., Hall, A., Orr, A., Hambloch, C. and Mausch, R. D. Embracing Complexity at Play. How Food and Agriculture Interventions Could Better Navigate the Unpredictable Dynamics of Agri-Food Systems. *How Food and Agriculture Interventions Could Better Navigate the Unpredictable Dynamics of Agri-Food Systems*. Available at: https://papers.ssrn.com/sol3/papers.cfm?abstract_id=4673009.

CTNOFA (2011). *Food Safety Modernization Act*. The Northeast Organic.

Denis, N., Dilda, V., Kalouche, R. and Sabah, R. (2020). *Agriculture Supply-Chain Optimization and Value Creation*. McKinsey white paper, May 12, 2020, Available at: https://www.mckinsey.com/industries/agriculture/our-insights/agriculture-supply-chain-optimization-and-value-creation#/.

Doan, L. (2022). *Russia's Ukraine War Is Stressing Farmers in U.S.* CBS News, April 7, 2022. Available at: https://www.cbsnews.com/news/russia-ukraine-war-farmers-economic-impact/.

Doernberg, A., Piorr, A., Zasada, I., Wascher, D. and Schmutz, U. (2022). Sustainability assessment of short food supply chains (SFSCs): developing and testing a rapid assessment tool in one African and three European city regions. *Agriculture and Human Values* 39(3): 885–904.

Douglas, L. (2023). *U.S. Farm Agency Announces Working Group on Seed Industry Consolidation*. Reuters, March 6, 2023. Available at: https://www.reuters.com/markets/commodities/us-farm-agency-announces-working-group-seed-industry-consolidation-2023-03-06/.

Duckworth, C. (2023). *Everything You Need to Know Before Signing Up for a CSA*. Tasting Table, February 14, 2023. Available at: https://www.tastingtable.com/1198167/everything-you-need-to-know-before-signing-up-for-a-csa/.

Dutia, S. G. (2014). *AgTech: Challenges and Opportunities for Sustainable Growth*. Ewing Marion Kauffman Foundation, Kansas City, MO.

Duval, D., Bickel, A. K. and Frisvold, G. (2018). *Mexican Fresh Tomatoes Agribusiness Value Chain Contributions to the U.S. Economy*. College of Agriculture & Live Sciences, The University of Arizona, November 2018. Available at: https://economics.arizona.edu/mexican-fresh-tomatoes-agribusiness-value-chain-contributions-us-economy.

Effland, A., Saavoss, M., Capehart, T., McBride, W. D. and Boline, A. (2022). *Innovations in Seed and Farming Technologies Drive Productivity Gains and Costs on Corn Farms*. United States Department of Agriculture, Economic Research Service, Amber Waves, April 4, 2022. Available at: https://www.ers.usda.gov/amber-waves/2022/april/innovations-in-seed-and-farming-technologies-drive-productivity-gains-and-costs-on-corn-farms/.

Farming Association of Connecticut (June 21, 2011). Available at: https://ctnofa1982.blogspot.com/2011/06/food-safety-modernization-act.html.

FDA (2018). *Background on the FDA Food Safety Modernization Act (FSMA)*. U.S. Food & Drug Administration, January 30, 2018. Available at: https://www.fda.gov/food/food-safety-modernization-act-fsma/background-fda-food-safety-modernization-act-fsma.

Feeding America (2022). *Charitable Food Assistance Participation in 2021*. Feeding America, June 2022. Available at: https://www.feedingamerica.org/sites/default/files/2022-06/Charitable%20Food%20Assistance%20Participation%20in%202021.pdf.

FGN (2022). *Representatives, Industry Members Implore Senate to Pass ag Labor Bill*. Fruit Growers News, July 13, 2022. Available at: https://fruitgrowersnews.com/news/representatives-industry-members-implore-senate-to-pass-ag-labor-bill/.

FoodExport (2023). *Mexico Country Profile*. Food Export. Available at: https://www.foodexport.org/export-insights/market-and-country-profiles/mexico-country-profile.

FoodPrint (2020). *The Environmental Impact of Food Packaging*. FoodPrint, November 13, 2020. Available at: https://foodprint.org/issues/the-environmental-impact-of-food-packaging.

FSN (2020). *Chipotle Agrees to Pay $25 Million Federal Fine for Role in Some Outbreaks*. Food Safety News, April 22, 2020. Available at: https://www.foodsafetynews.com/2020/04/chipotle-agrees-to-pay-25-million-federal-fine-for-role-in-some-outbreaks/.

Government of Canada (2019). *Food Policy of Canada*. Government of Canada/Gouvernement du Canada, June 17, 2019. https://www.canada.ca/en/agriculture-agri-food/news/2019/06/food-policy-for-canada--backgrounder.html.

Government of Canada (2020). *Agri-Food Pilot*. Government of Canada/Gouvernement du Canada, August 7, 2020. Available at: https://www.canada.ca/en/immigration-refugees-citizenship/services/immigrate-canada/agri-food-pilot.html.

Government of Canada (2021a). *Government of Canada Invests $16.9 Million to Improve Worker Safety and Strengthen the Food Supply Chain in British Columbia*. Government of Canada/Gouvernement du Canada, March 5, 2021. Available at: https://www.canada.ca/en/agriculture-agri-food/news/2021/03/government-of-canada-invests-169-million-to-improve-worker-safety-and-strengthen-the-food-supply-chain-in-british-columbia.html.

Government of Canada (2021b). *Sustainable Agri-Food Value Chains*. Government of Canada/Gouvernement du Canada, April 23, 2021. Available at: https://agriculture.canada.ca/en/department/initiatives/food-policy/food-systems-summit/sustainable-agri-food-value-chains.

Government of Canada (2021c). *Longitudinal Immigration Database (IMDB) Technical Report*, Available at: www150.statcan.gc.ca/n1/pub/11-633-x/11-633-x2022009-eng.pdf.

Government of Canada (2022). *Overview of Canada's Agriculture and Agri-Food Sector*, Government of Canada/Gouvernement du Canada, July 18, 2022. Available at: https://agriculture.canada.ca/en/sector/overview.

Government of Canada (2024). *Ministerial Transition Book: Overview Documents, 2024*. Available at: https://agriculture.canada.ca/en/department/transparency/briefing-documents/ministers-transition-books/ministerial-transition-book-overview-documents.

Grain (2011). *Food and Climate Change: The Forgotten Link*. Grain, September 28, 2011. Available at: https://grain.org/article/entries/4357-food-and-climate-change-the-forgotten-link.

Gunders, D. (2017). *Wasted: How America Is Losing up to 40 Per Cent of Its Food from Farm to Fork to Landfill*. Natural Resources Defence Council - NRRC Issue Paper R: 17-05-A. Available at: https://www.nrdc.org/sites/default/files/wasted-food-IP.pdf.

Harper, J. M. (2021). *USDA to Invest More Than $4 Billion to Strengthen Food System*. UCCE Lake Country, University of California, Agriculture and Natural Resources, June 8, 2021. Available at: https://celake.ucanr.edu/?blogasset=21880&blogpost=47777.

Hemphill, T. A. (2023). Managing food waste in the United States: a proposed public-private governance and sustainability model. *Technological Sustainability* 2(1): 67–88.

Holleman, M. (2018). *Evaluating the Real Costs of a Food Product Recall*. Snack Food and & Wholesale Bakery, September 14, 2018. Available at: https://www.snackandbakery.com/articles/92105-evaluating-the-real-costs-of-a-food-product-recall.

Hui, A., Robertson, S. K. and Atkins, E. (2022). *How Canada's Fragile Food Supply Chain is Being Disrupted*. The Globe and Mail, January 20, 2022. Available at: https://www.theglobeandmail.com/business/article-how-canadas-fragile-food-supply-chain-is-being-disrupted/.

IBM (2020). *Focus on Food Safety*. IBM Food Trust. Available at: https://www.ibm.com/downloads/cas/KWNWD3ZM.

IBM (n.d.). *About IBM Food Trust*. IBM Food Trust. Available at: https://www.ibm.com/downloads/cas/EX1MA1OX.

ITA (2022). *Mexico - Country Commercial Guide, Agriculture*. International Trade Administration, U.S. Department of Commerce, September 23, 2022. Available at: https://www.trade.gov/country-commercial-guides/mexico-agriculture.

ITC (2021). *Trade Map*. International Trade Centre. Available at: https://www.trademap.org/Product_SelProductCountry.aspx.

Johnson, L. K., Dunning, R. D., Gunter, C. C., Dara Bloom, J., Boyette, M. D. and Creamer, N. G. (2018). Field measurement in vegetable crops indicates re-evaluation of on-farm food loss estimates in North America. *Agricultural Systems* 167: 136–142.

Kapadia, S. (2022). *Food Suppliers Wade into Logistics as Trucking Rates Soar*. Supply Chain Dive, February 9, 2022. Available at: https://www.supplychaindive.com/news/food-supplier-distributor-ltl-trucking-freight/617062/.

Kapala, A. (2022). Legal status of farmers involved in short food supply chains, a comparative study. *Review of Central and East European Law*, 48: 43.

Kenner, B., Jiang, H., Russel, D. and Kaufman, J. (2023). *Outlook for U.S. agricultural trade: February 2023, AES-123*. USDA, Economic Research Service and USDA, Foreign Agricultural Service, February 23, 2023.

KMPG (2013). *The Agricultural and Food Value Chain: Entering a New Era of Cooperation*. KPMG International. Available at: https://assets.kpmg.com/content/dam/kpmg/pdf/2013/06/agricultural-and-food-value-chain-v2.pdf.

Knorr, D. and Watzke, H. (2019). Food processing at a crossroad. *Frontiers in Nutrition* 6(85): 1–8.

Konar, M. (2019). *We Mapped How Food Gets from Farms to Your Home*. The Conversation, October 25, 2019. Available at: https://theconversation.com/we-mapped-how-food-gets-from-farms-to-your-home-125475.

Kovak, E. (2023). *Seed Prices Have Soared–Is Intellectual Property the Problem?* The Breakthrough Institute, January 17, 2023. Available at: https://thebreakthrough.org/issues/food-agriculture-environment/seed-prices-have-soared-is-intellectual-property-the-problem.

Kozai, T. and Hayashi, E. (Eds) (2023). *Advances in Plant Factories: New Technologies in Indoor Vertical Farming*. Burleigh Dodds Science Publishing, Cambridge.

Kronfli, C. (2021). *Growing a More Resilient Food Supply Chain in Ontario*. Ontario Chamber of Commerce, ISBN: 978-1-928052-75-3. Available at: https://occ.ca/wp-content/uploads/COVID19-Policy-Brief-Food-Supply-Chain-final.pdf.

Lakhani, N., Uteuova, A. and Chang, A. (2021). *Revealed: The True Extend of America's Food Monopolies, and Who Pays the Price*. The Guardian, July 14, 2021. Available at: https://www.theguardian.com/environment/ng-interactive/2021/jul/14/food-monopoly-meals-profits-data-investigation.

Lin, X., Ruess, P. J., Marston, L. and Konar, M. (2019). Food flows between counties in the United States. *Environmental Research Letters* 14(8): 084011.

Lobao, L. and Meyer, K. (2001). The great agricultural transition: crisis, change, and social consequences of twentieth century US farming. *Annual Review of Sociology* 27(1): 103–124.

Lozano, E. (2022). *United States-Mexico Agricultural Trade Logistics Review*. United States Department of Agriculture, Foreign Agricultural Service, Global Agricultural Information Network, Report Number: MX2022-0066, December 23, 2022. Available at: https://apps.fas.usda.gov/newgainapi/api/Report/DownloadReportByFileName?fileName=United%20States-Mexico%20Agricultural%20Trade%20Logistics%20Review_Monterrey%20ATO_Mexico_MX2022-0066.pdf.

Lucas, A. (2019). *Dean Foods, America's Biggest Milk Producer, Files for Bankruptcy*. NBC News, November 12, 2019. Available at: https://www.nbcnews.com/business/business-news/dean-foods-america-s-biggest-milk-producer-files-bankruptcy-n1080586.

Majewski, E., Komerska, A., Kwiatkowski, J., Malak-Rawlikowska, A., Wąs, A., Sulewski, P., Gołaś, M., Pogodzińska, K., Lecoeur, J., Tocco, B., Török, Á., Donati, M. and Vittersø, G. (2020). Are short food supply chains more environmentally sustainable than long chains? A life cycle assessment (LCA) of the eco-efficiency of food chains in selected EU countries. *Energies* 13(18): 4853.

Malak-Rawlilowska, A., Majewski, E., Wąs, A., Borgen, S. O., Csillag, P., Donati, M., Freeman, R., Hoàng, V., Lecoeur, J., Mancini, M. C., Nguyen, A., Saïdi, M., Tocco, B., Török, Á., Veneziani, M., Vittersø, G. and Wavresky, P. (2019). Measuring the economic, environmental and social sustainability of short food supply chains. *Sustainability* 11(15): 4004.

Martin, A. and Roden, A. (2023). *Higher Cost and Inflation Weigh on the Mind of Growers in 2023*. Seed World, February 9, 2023. Available at: https://www.seedworld.com/higher-cost-and-inflation-weigh-on-the-mind-of-growers-in-2023/.

Martinez, S. W. and Park, T. (2021). *Marketing Practices and Financial Performance of Local Food Producers: A Comparison of Beginning and Experienced Farmers*. United States Department of Agriculture, Economic Research Service, Economic Information Bulletin Number 225, August 2021, Available at: https://www.ers.usda.gov/webdocs/publications/101786/eib-225.pdf?v=2970.4.

Marves, C. (2022). *Food Poisoning at Chipotle: A History of Food Safety Issues*. Food Poisoning News, March 12, 2022. Available at: https://www.foodpoisoningnews.com/food-poisoning-at-chipotle-a-history-of-food-safety-issues/.

Masterson, V. (2022). *What Is Regenerative Agriculture?* World Economic Forum, October 11, 2022. Available at: https://www.weforum.org/agenda/2022/10/what-is-regenerative-agriculture/.

Mintert, J. and Langemeier, M. (2023). *Commodity Price Outlook and Interest Rate Concerns Cloud Farmer Sentiment*. AG Economy Barometer, Purdue University, CME Group, April 4, 2023. Available at: https://ag.purdue.edu/commercialag/ageconomybarometer/commodity-price-outlook-and-interest-rate-concerns-cloud-farmer-sentiment/.

Munch, D. (2022). *New AFBF Survey Shows Drought's Increasing Toll on Farmers and Ranchers*. American Farm Bureau Federation, August 14, 2022. Available at: https://www.fb.org/market-intel/new-afbf-survey-shows-droughts-increasing-toll-on-farmers-and-ranchers.

Myers, S. (2021). *Too Many to Count: Factors Driving Fertilizer Prices Higher and Higher*. American Farm Bureau Federation, December 31, 2011. Available at: https://www

.fb.org/market-intel/too-many-to-count-factors-driving-fertilizer-prices-higher-and-higher.

National Research Council (2015). *A Framework for Assessing Effects of the Food System*. The National Academies Press, Washington, DC.

NIDS (2023). *Agriculture, National Integrated Drought Information System*. Available at: https://www.drought.gov/sectors/agriculture.

Oceana (2023). *Plastic Pollution: Tacking the Plastics Crisis at the Source*. Oceana. Available at: https://usa.oceana.org/our-campaigns/plastic/.

OECD/Food and Agriculture Organization of the United Nations (2019). Latin American agriculture: prospects and challenges. In: *OECD-FAO Agricultural Outlook 2019–2028*. OECD Publishing, Paris/Food and Agriculture Organization of the United Nations, Rome.

Packaging Digest (2014). *Biodegradable Packaging for Pangea Organics Incorporates Seeds That Can Be Planted*. Packaging Digest, January 29, 2014. Available at: https://www.packagingdigest.com/biodegradable-packaging-pangea-organics-incorporates-seeds-can-be-planted.

Patel, P. (2020). *The Time Is Now for Edible Packaging*. c&en, January 26, 2020. Available at: https://cen.acs.org/food/food-science/time-edible-packaging/98/i4.

PPI (2023). *Producer Price Index by Industry: General Freight Trucking, Long-Distance Truckload (PCU484121484121)*. FRED. Available at: https://fred.stlouisfed.org/series/PCU484121484121.

Prezkop, L., Stone, A., Baker G. A., Johson, L. K. and Deutsch, J. (2020). Challenges and initiatives in reducing food losses and waste: United States. In: Yahia, E. M. (Ed.) *Preventing Food Losses and Waste to Achieve Food Security and Sustainability* (pp. 629–652). Burleigh Dodds Science Publishing, Cambridge.

Proof (2022). *Household Food Insecurity in Canada*. Proof, University of Toronto. Available at: https://proof.utoronto.ca/food-insecurity/.

Reddy, S. (2018). *Plastic Pollution Affects Sea Life Throughout the Ocean*. Pew, September 24, 2018. Available at: https://www.pewtrusts.org/en/research-and-analysis/articles/2018/09/24/plastic-pollution-affects-sea-life-throughout-the-ocean.

Redfield, C., Mounce, R., Mundo, M. and Yiannas, F. (2018). *Food Traceability Initiative: Fresh Leafy Greens*. Walmart, Letter to Leafy Green Suppliers, September 24, 2018. Available at: https://corporate.walmart.com/media-library/document/blockchain-supplier-letter-september-2018/_proxyDocument?id=00000166-088d-dc77-a7ff-4dff689f0001.

ReFED (2016). *Rethink Food Waste: A Roadmap to Reduce Food Waste by Twenty Percent*. Available at: https://refed.org/downloads/ReFED_Report_2016.pdf.

ReFED (2018). *ReFED Analysis: Grocery Sector*. Available at: https://refed.org/articles/refed-analysis-reveals-food-waste-represents-18-2-billion-profit-opportunity-for-grocery-retailers/.

Regeneration International (2019). *Why Regenerative Agriculture?* Regeneration International. Available at: https://regenerationinternational.org/why-regenerative-agriculture/.

Schoenherr, T., Narasimhan, R. and Bandyopadhyay, P. (2015). The assurance of food safety in supply chains via relational networking: A social network perspective. *International Journal of Operations and Production Management* 35(12): 1662–1687.

Semuels, A. (2019). *They're Trying to Wipe Us Off the Map: Small American Farmers Are Nearing Extinction*. Time, November 27, 2019. Available at: https://time.com/5736789/small-american-farmers-debt-crisis-extinction/.

Sristy, A. (2021). *Blockchain in the Food Supply Chain - What Does the Future Look Like?* Walmart Global Tech, November 30, 2021. Available at: https://tech.walmart.com/content/walmart-global-tech/en_us/news/articles/blockchain-in-the-food-supply-chain.html.

Stafford, J. (Eds) (2018). *Precision Agriculture for Sustainability*. Burleigh Dodds Science Publishing, Cambridge.

Statista (2023a). *U.S. Agriculture - Statistics & Facts*. Statista. Available at: https://www.statista.com/topics/1126/us-agriculture/.

Statista (2023b). *Leading Food and Beverage Companies in North America in 2021/2022, Based on Sales*. Statista. Available at: https://www.statista.com/statistics/459276/leading-food-and-beverage-processors-in-north-america/.

Statista (2023c). *Most Likely Perceived Cause of Food and Supply Shortages in Local Grocery Stores Due to Coronavirus (COVID-19) in the United States as of March 15, 2020*. Statista. Available at: https://www.statista.com/statistics/1105594/coronavirus-cause-grocery-store-food-and-supply-shortages-us/.

Taylor, R. (2023). *Farm to Table: A Movement for Local and Organic Food. Webstaurantstore Blog, Foodservice Trends*. Available at: https://www.webstaurantstore.com/blog/4336/farm-to-table-meaning.html.

The Global Food Banking Network (2023). *Our Role in Food Systems*. Available at: https://www.foodbanking.org/our-role-in-food-systems/.

Tweed, M. (2013). *Toward a Common Goal: Canada's Food Supply Chain–Part 1. Report of the Standing Committee on Agriculture and Agri-Food*. 41st Parliament, First Session, June 2013. Available at: https://www.ourcommons.ca/Content/Committee/411/AGRI/Reports/RP6226525/agrirp10/agrirp10-e.pdf.

Unilever (2017). *Sustainable Agriculture*. Unilever. Available at: https://assets.unilever.com/files/92ui5egz/production/555201fc976d5389fbdf9845099c15b61d1ceb53.pdf/sustainable-agriculture-code--sac---2017.pdf.

Unilever (2021). *The Unilever Regenerative Agriculture Principles, with Implementation Guides 2021*. Unilever. Available at: https://www.unilever.com/files/92ui5egz/production/489410442380812907bc3d97be02ccda1a44ab4b.pdf.

Unilever (2023). *Regenerating Nature*. Unilever. Available at: https://www.unilever.com/planet-and-society/protect-and-regenerate-nature/regenerating-nature/.

USDA (2021). *USDA to Invest More Than $4 Billion to Strengthen Food System*. Available at: https://www.usda.gov/media/press-releases/2021/06/08/usda-invest-more-4-billion-strengthen-food-system.

USDA (2022). *Growth in the Number of U.S. Farmers Markets Slows in Recent Years*. United States Department of Agriculture, Economic Research Service, August 10, 2022. Available at: https://www.ers.usda.gov/data-products/chart-gallery/gallery/chart-detail/?chartId=104402.

USDA (2023a). *Farming and Farm Income*. United States Department of Agriculture, Economic Research Service. Available at: https://www.ers.usda.gov/data-products/ag-and-food-statistics-charting-the-essentials/farming-and-farm-income/.

USDA (2023b). *U.S. Agricultural Trade at a Glance*. United States Department of Agriculture, Economic Research Service. Available at: https://www.ers.usda.gov/

topics/international-markets-u-s-trade/u-s-agricultural-trade/u-s-agricultural-trade-at-a-glance/.

USDA (2023c). *Farm Structure and Contracting*. United States Department of Agriculture, Economic Research Service. Available at: https://www.ers.usda.gov/topics/farm-economy/farm-structure-and-organization/farm-structure-and-contracting/.

USDA (2023d). *About us--What are the USDA Climate Hubs?* United States Department of Agriculture, Climate Hubs. Available at: https://www.climatehubs.usda.gov/about-us.

USDA (2023e). *Food Environment Atlas*. United States Department of Agriculture, Economic Research Service. Available at: https://www.ers.usda.gov/data-products/food-environment-atlas/.

USDA (2023f). *Mexico: Trade & FDI*. United States Department of Agriculture, Economic Research Service. Available at: https://www.ers.usda.gov/topics/international-markets-u-s-trade/countries-regions/usmca-canada-mexico/mexico-trade-fdi/.

USGS (2023). *California's Central Valley*. U.S. Geological Survey. Available at: https://ca.water.usgs.gov/projects/central-valley/about-central-valley.html.

van Henten, E. and Edan, Y. (Eds) (2024). *Advances in Agri-Food Robotics*. Burleigh Dodds Science Publishing, Cambridge.

Vittersø, G., Torjusen, H., Laitala, K., Tocco, B., Biasini, B., Csillag, P., de Labarre, M. D., Lecoeur, J., Maj, A., Majewski, E., Malak-Rawlikowska, A., Menozzi, D., Török, Á. and Wavresky, P. (2019). Short food supply chains and their contributions to sustainability. *Sustainability* 11(17): 4800.

Wang, S. L., Hoppe, R. A. and Hertz, T. (2022). *Increases in Labor Quality Contributed to Growth in US Agricultural Output*. Amber Waves: The Economics of Food, Farming, Natural Resources, and Rural America (1490-2022-477).

WWF (World Wildlife Fund) (2020). *Maximizing Farm Resources and Edible Food Rescue: Specialty Crop Loss Report*. Available at: https://www.worldwildlife.org/publications/maximizing-farm-resources-and-edible-food-rescue-specialty-crop-loss-report.

Yih, P. T., Oria, M., and Nesheim, M. C. (Eds.). (2015). *A Framework for Assessing Effects of the Food System*. Washington, DC: National Academies Press.

York (2023). *Complex Supply Chains*. Food Policy in Canada, York University. Available at: https://foodpolicyforcanada.info.yorku.ca/backgrounder/problems/complex-supply-chains/.

Chapter 3

Agri-food supply chains in Western and Northern Europe

Michael Bourlakis and Emmanuel Sawyerr, Cranfield University, UK

1 Introduction

A complex network of related supply chain members, including producers, logistics companies (providing transport and warehousing services), manufacturers, retailers, packaging enterprises and agricultural machinery manufacturers and service providers, work together synergistically to form the food system that provides food to consumers in Europe. This system relies primarily on about 10.3 million farms, the majority of which are small- to medium-sized farms, with only a few very large ones (Eurostat, 2021a). More than half of these are specialist crop farms (general field cropping, cereals, oilseed and protein crops, olives, fruits and citrus fruits, vineyards and horticulture) while almost a quarter specialises in livestock (sheep, goats and other grazing livestock, dairying, poultry, pigs etc.). The rest consists of mixed farming and non-classified farms. In 2019, the European Union's (EU) total agricultural output stood at a little below €400 billion, most of which were processed into food and beverages by hundreds of thousands of enterprises (Rossi, 2020).

http://dx.doi.org/10.19103/AS.2023.0122.05

Western and Northern Europe make a strong contribution to food production. The joint classification of these regions in the chapter is based on selected countries in the two regions as defined by the United Nations Statistical Division (UNSD, 2022). Significant focus is given to Ireland, the United Kingdom (UK), France, Belgium, the Netherlands, Germany, Denmark, Sweden, Norway, Finland and Iceland. Consideration was given to the size of the country (land and population) as well as membership in the EU due to data availability and accessibility. France (17.8%) accounts for the largest utilised agricultural area in the EU, while the Netherlands is the biggest international trader (importer and exporter) of agricultural, fisheries and food and beverage products. The countries in this region also account for the highest levels of fish and aquaculture production, value-added agricultural production and food and beverage processing.

This chapter takes a closer look at agri-food supply chains (AFSC) in this region that facilitate the production and delivery of agri-food products to consumers within and outside of the region. First, a discussion on the types of supply chains is presented, followed by the strategies adopted to procure and deliver food to consumers. Despite the high levels of efficiency of these AFSCs and food affordability (compared to average wages), various key challenges require urgent attention. Some of these are discussed and the current state of the supply chains in this region is presented. Here, insights about the application of blockchain technology in some major retailers in the region are discussed and this is followed by a brief presentation of the opportunities for further research and policy interventions, and a summary of the chapter. The reader is finally directed to sources for further information.

2 The agricultural sector in Western and Northern Europe

In 2020, the agricultural industry's output value, which encompasses the value of crops, animals, agricultural services as well as goods and services, was an estimated €414.1 billion (Eurostat, 2021b). About 53% of this came from crops (with vegetables, horticultural plants and cereals being the most valuable crops) and 38.4% from animals and animal products – most of which were from milk and pigs (Eurostat, 2021b). The remaining 8.6% were from agricultural services and inseparable non-agricultural activities. The biggest contributors, who were jointly responsible for 58.7% of the total output value were France (€76.3 billion), Germany (€57.6 billion), Italy (€56.9 billion) and Spain (€52.3 billion). Along with the Netherlands, Poland and Romania, about 76% of the EU's overall agricultural output came from these countries. Of these, France, Germany and Netherlands – countries in the Northern and Western regions, contributed €162.1 billion worth of agricultural output value. Overall harvested cereal production was 286.5 million tonnes, with France and Germany accounting for 35.2% of these. Of its 100.1 million tonnes of sugar beets produced in 2020, Germany, France and the Netherlands were responsible for 61.5%. Germany (21.7%), France (16.1%) and the Netherlands (13%) were the first, second and fourth highest contributors to the EU's 54.0 million tonnes of potatoes produced in 2020 (Eurostat, 2021c).

Western and Northern European regions are also responsible for a significant amount of livestock production in Europe. France and Germany are two of the top three producers of livestock (including bovine animals, sheep, goats and pigs) in Europe (Eurostat, 2021d). For poultry, France and Germany feature in the top five as third and fourth, respectively while Denmark, France and the Netherlands are the top three in dairy production (European Commission, 2022a; Eurostat, 2021a). Table 1 shows at least one country in this region represented in the top five productions of cereals, root crops, vegetables, fruits, meat, livestock and dairy.

Iceland and Norway combined for 3.13 million tonnes of fish from the Northeast Atlantic Ocean in 2019. The fish products include herring, sprat, blue whiting and mackerel. France and Denmark had a fishing catch of 525 000 tonnes and 400 000 tonnes of live weight, respectively. Western and Northern Europe is thus also responsible for the supply of most of the fish consumed in Europe (Eurostat, 2021a, 2022a).

3 Types of agri-food supply chains in Western and Northern Europe

A variety of AFSCs is at the core of the cardinal role that Western and Northern European regions play in food production and agricultural value addition in Europe. These AFSCs may be classified either as local or commodity AFSCs while the products provided within either may be fresh, conserved or manufactured (see Fig. 1).

3.1 Local agri-food supply chains

Most of the food produced in this region is consumed locally. In the UK, for example, 54% of all food consumed in the country is produced locally (DEFRA, 2021b). Local AFSCs in this chapter encompass all supply chains that supply agri-food products within a country so that it ranges from short, rural supply chains to relatively longer ones whereby food may be produced in one part of a country and consumed in another part of it. Thus, these local AFSCs may be face to face, spatially proximate or spatially extended (Marsden et al., 2000). Face-to-face AFSCs are short supply chains where the consumer purchases directly from the producer or processor in person or a variant of face-to-face interaction through the internet. With spatial proximity, consumers are made

Table 1 The dominance of Western and Northern European countries in the EU's agricultural industry in 2020

Food production category	Top five producers 1	2	3	4	5
Cereals					
Common wheat and spelt	France	Germany	Poland	UK	Spain
Durum wheat	Italy	France	Greece	Spain	Germany
Grain maize and corn-cob-mix	France	Romania	Hungary	Poland	Italy
Barley	Spain	Germany	France	UK	Denmark
Rye and winter cereal mixtures	Germany	Poland	Denmark	Spain	Austria
Oats	Poland	Spain	Finland	UK	Sweden
Rice	Italy	Spain	Greece	Portugal	France
Grapes for wine	Italy	Spain	France	Germany	Romania
Root crops					
Sugar beets	Germany	France	Poland	Netherlands	UK
Potatoes	Germany	France	Poland	Netherlands	UK
Vegetables					
Carrots	Germany	UK	Poland	Netherlands	France
Onions	Netherlands	Spain	Poland	France	Germany
Fresh peas	France	UK	Spain	Hungary	Italy
Cauliflowers and broccoli	Spain	Italy	France	Poland	UK
Lettuce	Spain	Italy	Germany	France	UK
Tomatoes	Italy	Spain	Portugal	Netherlands	Greece
Cucumbers	Spain	Netherlands	Poland	France	Greece

Fruits					
Apples	Poland	Italy	France	Germany	Romania
Pears	Italy	Netherlands	Belgium	Spain	France
Peaches	Spain	Greece	Italy	France	Portugal
Plums	Romania	France	Italy	Spain	Poland
Kiwis	Italy	Greece	France	Portugal	Spain
Nuts	Spain	Italy	Greece	Portugal	France
Citrus	Spain	Italy	Greece	Portugal	France
Livestock population					
Pigs	Spain	Germany	France	Denmark	Poland
Bovine animals	France	Germany	Spain	Ireland	Italy
Sheep	Spain	Romania	Greece	France	Italy
Goats	Greece	Spain	Romania	France	Italy
Meat production					
Sheep meat	Spain	France	Ireland	Greece	Italy
Goat meat	Greece	Spain	France	Netherlands	Cyrus
Veal	Netherlands	Spain	France	Italy	Belgium
Poultry	Poland	Spain	France	Germany	Italy
Pigs	Germany	Spain	France	Poland	Netherlands
Bovine meat	France	Germany	Italy	Spain	Ireland
Dairy products					
Drinking milk	Germany	Spain	France	Italy	Poland
Whey	Germany	Netherlands	Poland	Italy	Ireland
Butter	Germany	France	Ireland	Poland	Netherlands
Cheese	Germany	France	Italy	Netherlands	Poland

NB: Shaded countries are in Western and Northern Europe.
Sources: DEFRA (2021a) and Eurostat (2021a,c,d).

<table>
<tr><td colspan="2"></td><td colspan="3">Products</td></tr>
<tr><td rowspan="2">Supply Chain</td><td>local</td><td rowspan="2">fresh</td><td rowspan="2">conserved</td><td rowspan="2">manufactured</td></tr>
<tr><td>commodity</td></tr>
</table>

Figure 1 Typology of AFSCs. Source: Adapted from: Smith (2008).

aware of the local nature of the products at the point of retail. These products are produced and retailed in the specific region of production. The spatially extended AFSC is where value-laden information about the place of production is translated to consumers who may have no personal experience of and/or are outside of the region of production (Pérez-Mesa et al., 2021; Renting et al., 2003). Regardless of the type of local AFSC, they are the supply chains that provide the fresh, conserved and manufactured food products consumed within a country.

Local AFSCs are typically considered more sustainable as they tend to be shorter and involve lower emissions because of their relatively lower food miles (Jones et al., 2008; Pretty et al., 2005; Smith, 2008). In recent years, there has been a push for more local AFSCs due to a myriad of reasons including sustainability, consumer confidence in food integrity and traceability. Additionally, these AFSCs help generate rural enterprises and regenerate rural communities (Smith, 2008). For these reasons, the EU's Farm to Fork Strategy aimed to engender sustainable and resilient farming where consumers ate healthier, locally farmed and sustainably sourced food. It sought to reward primary food producers and other actors in the supply chain who had transitioned into sustainable practices (Eurostat, 2021a).

Face-to-face AFSCs are very popular for fresh food products, where farmers sell their produce through producers' markets, basket delivery systems or community-supported agriculture. Farmers benefit from fairer pricing while consumers access fresh and seasonal food products. In its rural development policy 2014–20, the EU created opportunities for producers wishing to involve themselves in local food systems to receive benefits from European Agricultural Fund for Rural Development's co-financed interventions (Augère-Granier, 2016). Despite these incentives and the local sustainability benefits that are rife in supporting short supply chains, only about 2% of the fresh food market produce is directly sold between farmers and consumers (Augère-Granier, 2016). Significantly more fresh food produce goes through large retailers. Further discussion on the dominance of food retailers in this region is presented later in the chapter.

Conserved food products include stable food products preserved by old methods such as drying, smoking and salting, as well as modern methods like pasteurisation, chilling and freezing to allow for the transportation of perishable food from typically rural or coastal production sites to urban populations (Smith, 2008). Food conservation is particularly critical to prevent food waste as it helps reduce food degradation for perishable food products such as dairy, meat, fish, fruits and vegetables which may either be fresh or processed. The supply chains that supply these products tend to be longer than the face-to-face supply chains for fresh food products sold by primary producers. There are a few more actors in these chains. Ireland and Sweden are highly specialised in the production of fresh or chilled cuts of beef and veal (Eurostat, 2021a) which not only serve their citizens but are also exported to the rest of Europe. In the UK, over 50% of domestically consumed vegetables, 67% of sugar beet, 71% of potatoes, 79% of oilseeds and 16% of fruits (DEFRA, 2021c) are supplied by these supply chains, where the products are preserved (to keep their freshness) by various forms of packaging, temperature-controlled containers, chilled vehicles, etc.

The supply chains for manufactured food products tend to be the longest of the local supply chains that serve a country's population. The Western and Northern parts of Europe contribute significantly to value addition to food products. Outside of Italy, which contributes 18.3% of the total in the EU, France and Germany combine for 29.2% of value addition in food production (Eurostat, 2021a). Even though some locally manufactured foods are consumed locally, a lot more are exported to the rest of Europe through commodity AFSCs.

3.2 Commodity agri-food supply chains

Commodity AFSCs are long supply chains that may source raw materials internationally or deliver their products to an international market. In this chapter, commodities refer to agri-food products (whether fresh, conserved or manufactured) that are traded internationally, often using long-distance land (road or rail), air or sea transport. These supply chains tend to have multiple layers of producers, intermediaries (agents, brokers), large purchasing centres, international logistics firms and retailer distribution chains before products get to the final consumer (Pérez-Mesa et al., 2021). These supply chains are therefore more complex and involve more processing steps (Smith, 2008). These products include meat products, fish and other seafood, beverages, bakery and starch-based products, dairy products, fruit and vegetables as well as other manufactured food products. They may range from staple food products to luxury, sometimes even high-value items.

The highest manufactured product is non-processed cheese (with products such as Brie, Edam, Feta, and Gorgonzola, among others) accounting for €30.5

billion. France and Germany are the second and third biggest manufacturers behind Italy. Germany is also the principal producer of fresh bread and beer, which represent the second and third-highest product values in the EU (Eurostat, 2021a). Tomato is an essential ingredient in North and Western Europe. In Norway, it is the most sold fresh vegetable product with an annual consumption (including processed tomato products) of 16.3 kg per capita, a third of which is locally produced (Løvdal et al., 2019) while in Belgium, the consumption is 27.5 kg (Pérez-Mesa et al., 2021).

Commodity AFSCs facilitate international trade within Europe and elsewhere. More than 67% of total trade in the EU in agricultural, fisheries and food and beverage products were within the EU in 2020 (Eurostat, 2021a). Tables 2, 3 and 4 provide the details of international trade for meat (including edible meat offal), fish (including crustaceans, molluscs and other aquatic invertebrates), dairy produce (including birds' eggs, edible products of animal origin, not elsewhere specified or included), edible vegetables (including certain roots and tubers), edible fruit (nuts, peel of citrus fruit or melons) and cereals in 2020. For Tables 2 and 3, the values represent the quantities of products that are imported and exported by the countries indicated. For Table 4, the imports represent the values of the food products the EU imports from the UK, Iceland and Norway while the export values indicate how much the EU exports to these three nations.

France leads in the export of dairy products and cereals outside the EU but is also the highest importer of vegetables. Of the countries in Western and Northern Europe that are not part of the EU, the highest importer of meat from the EU is the UK while Norway is the highest exporter of fish to the EU. Of the selected products, the highest export from Iceland to the EU is fish. The UK is a net importer of fish. The types of fish typically consumed locally are largely non-native and so most of its fish catch (like mackerel and herring) is exported while most of its domestically consumed fish (such as tuna, haddock, cod, shrimps and prawns) are imported (DEFRA, 2021c). About 23.4% (€42.0 billion) of the EU's total agricultural, fisheries and food and beverage products in 2020 were exported to the UK. It is the main export destination for crop and crop products; almost a quarter of these products are exported to the UK.

Due to the Netherlands' favourable location as a logistics hub (Rotterdam has the largest sea freight port in Europe), the country plays a crucial role in the import and export of agricultural, fisheries and food and beverage products (Eurostat, 2021a). The Netherlands had the largest intra-EU trade surplus in 2020 (€30.9 billion) as it imports from the world over and re-exports to other EU states.

Table 2 Values of international trade (in €'000 000) outside of the EU for selected food commodities by EU countries in Western and Northern Europe in 2020

	Meat		Fish		Dairy		Vegetables		Fruits and nuts		Cereals	
Country	Imports	Exports	Imports	Exports	Imports	Exports	Imports	Exports	Imports	Exports	Imports	Exports
Belgium	204.57	345.26	648.37	20.46	180.68	1108.48	268.81	463.28	1845.90	249.61	402.70	53.55
Denmark	71.50	2379.80	2282.61	828.33	72.03	1092.60	26.59	61.55	124.71	25.51	19.66	79.77
Finland	4.94	98.81	221.97	37.84	1.58	175.38	5.29	1.05	90.74	2.10	6.29	31.35
France	423.86	1110.28	2009.84	315.76	227.37	2606.99	850.35	455.81	1561.74	521.01	282.61	3372.29
Germany	743.23	2280.05	1246.49	281.37	595.18	2154.17	424.12	185.46	3006.00	256.72	209.40	1545.29
Ireland	312.28	1893.20	111.36	178.95	543.89	1802.40	146.62	134.86	228.02	31.48	303.78	38.35
Netherlands	1424.63	3205.93	2126.47	1023.74	215.50	2499.00	746.42	2077.27	6407.37	1086.73	1041.12	57.25
Sweden	52.61	41.12	3879.67	329.21	20.35	136.62	34.68	10.88	297.61	35.65	38.65	112.31

Source: Eurostat (2022b).

Table 3 Values of international trade (in €'000 000) within the EU for selected food commodities by EU countries in Western and Northern Europe in 2020

	Meat		Fish		Dairy		Vegetables		Fruits and nuts		Cereals	
Country	Imports	Exports	Imports	Exports	Imports	Exports	Imports	Exports	Imports	Exports	Imports	Exports
Belgium	1344.61	2704.04	773.39	705.49	3351.96	2616.25	1734.36	1803.02	1377.34	2013.75	1534.70	520.01
Denmark	873.31	1220.96	283.13	2222.68	670.79	1197.97	469.00	108.08	620.92	85.21	154.47	251.37
Finland	177.57	64.79	130.47	143.00	362.94	266.48	268.95	8.87	426.29	26.41	29.04	109.00
France	3696.87	1984.09	2389.48	787.33	3613.94	3737.49	2390.09	1759.26	4035.12	984.22	656.67	3453.53
Germany	5448.36	5032.75	2555.07	1053.22	7656.98	7179.39	6246.45	944.09	8060.31	1744.84	2749.69	1094.38
Ireland	274.37	1299.77	66.80	309.45	309.71	1157.82	221.66	0.94	246.87	5.45	87.11	1.26
Netherlands	2489.60	5614.26	858.72	2748.67	3760.69	5819.59	1890.61	5064.57	2061.00	6950.31	1843.33	560.41
Sweden	840.71	98.63	252.35	3383.83	953.69	212.67	670.46	76.52	724.33	83.31	95.24	239.54

Source: Eurostat (2022c).

Table 4 Values of international trade (in €'000 000) for selected food commodities between the EU and non-EU countries (United Kingdom, Iceland, Norway) in Western and Northern Europe in 2020

	EU/United Kingdom		EU/Iceland		EU/Norway	
Selected products	Imports	Exports	Imports	Exports	Imports	Exports
Meat	1493.90	4271.12	1.45	13.05	18.04	88.72
Fish	1463.89	883.05	939.80	18.09	6128.04	120.90
Dairy	1401.67	3310.78	1.23	6.56	15.79	144.87
Vegetables	400.14	2761.83	0.03	22.85	1.39	324.75
Fruits and nuts	514.90	2461.05	0.03	33.99	1.81	429.98
Cereals	407.45	721.56	0.03	10.40	0.41	85.45

Source: Eurostat (2022d).

4 Agri-food supply chain strategies in Western and Northern Europe

As technological advancements have progressed and populations in cities and other urban areas have increased significantly, the markets for food products have also grown astronomically. Not only do AFSCs seek to meet their in-country demands, but there is also the attractiveness of the international market and hence, the need to level up the capabilities of the supply chains to meet these higher levels of demand. Conventional AFSCs, in response to these, have developed industrialised processes to reconfigure their supply chain network (production facilities, warehouses, distribution centres and stores) for more efficient handling of large product volumes (Dreyer et al., 2015; Hübner et al., 2013).

With increasing complexity as supplier numbers and geographical dispersion increase, organisations in the agri-food chains - particularly manufacturers and retailers - have implemented systems (such as multi-criteria supplier selection decision systems) that not only alleviate but also optimise operational and logistical processes (Lyu et al., 2010). Strategic network design decisions (such as the use of central versus regional warehouses) aimed at optimising capital and operational costs as well as the adoption of management technologies for efficient inventory management have resulted in highly effective handling of perishable, ambient, chilled and frozen food products (Hübner et al., 2013). Based on supplier location and product characteristics, retailers tend to use a distribution mix whereby warehouses may serve storage, picking and cross-docking purposes (Melo et al., 2009).

To increase the value proposition to customers and fine-tune distribution from warehouses to retail outlets, organisations continually make and review outlet planning decisions. Considerations include store location (which may affect market density, geographical expansion and accessibility from warehouses) and store sizes (depending on market size and also to influence customer purchasing behaviour) (Cachon and Kök, 2007; Gupta et al., 2009; Hübner et al., 2013). The layout and infrastructure of the in-store showrooms aim to reflect a retailer's brand, ease in-store logistical processes, reduce consumer search costs and influence their buying decisions as well as backroom design (which affects offloading) and outlet storage capacity are also important factors (Ge et al., 2009; Kotzab and Teller, 2005).

With the rapidly changing market environment in Western and Northern Europe, supply chains have sought increased flexibility and agility to better align supply to demand (Hübner et al., 2013; Thatte et al., 2013). One of the common strategies to mitigate against demand variability is using extra (safety) stock. Food manufacturers produce and store in high volumes at a variety of locations for retail distributors to source and make available to consumers

(Dreyer et al., 2015). This strategy is utilised for products with long shelf lives but it is not sustainable for highly perishable products where a more responsive supply chain is required (Blackburn and Scudder, 2009).

4.1 The dominant role of retail and manufacturing

Multiple food retailers dominate the food retailing industry in Western and Northern Europe (Bourlakis and Weightman, 2004; Fagotto, 2014; Maglaras et al., 2015). They have been described as holding the 'gatekeeper role' in these supply chains (Fox and Vorley, 2004, p. 7) because, unlike in countries where retailers are fragmented and thus manufacturers hold greater influence (Clay, 2005), food retailers in these regions are consolidated and thus hold greater power. The top ten retailers in Western Europe control about 71% of the continent's market (Von Redaktion LZ, 2021). In the UK, only four large retail companies make up about 75% of the market (Beitzen-Heineke et al., 2017). Similarly, four retailers control close to 71% of the German food market (Deloitte, 2021).

Notwithstanding, retailers in Western Europe continue to grow bigger. Tesco and Carrefour are reported to have created a joint purchasing platform while Asda and Sainsbury are pursuing similar agreements (Pérez-Mesa et al., 2021). Amazon plans to open 260 'Go'-like checkout-free stores in the UK alone and many more in Germany and Spain in the next few years (Sabbe, 2021). Several retailers act as dominant channels for food marketing. They also build alliances with food manufacturers to develop and market their own branded/private label products, thereby providing economic alternatives to brands of national and global manufacturers (Yakovleva, 2007). This is consumer-driven, as they are increasingly choosing cheaper retailer-owned labels over more acclaimed branded products. In 2020, private label share of retailers' sales volume in the UK was 46.8%, 44% in Belgium, 43.1% in Germany and 30.7% in Finland, with fresh, frozen and dairy products being the top food products in this category (PLMA, 2021). For retailers, this allows for cost-effectiveness while also enabling effective planning for the increasing levels of competitive promotional activities (Kumar et al., 2013).

Resulting from these, there is a shift of the balance of power within the supply chain towards retailers, thereby allowing them on one hand, to offer potential shortcuts and access to positive change, but also providing them with a mechanism to shift responsibility to the less powerful actors in the supply chain - such as farmers (Fox and Vorley, 2004; Hingley, 2005; Smith, 2008). As an example, some retailers push product ownership and responsibility to supply chain actors like primary producers and logistics companies so that they only have direct ownership for the period the products are held in their retail stores. Thus, the retailers minimise the direct responsibility for dealing with the consequences of damaged or surplus products (Alexander and Smaje, 2008). Retailers are able to control prices and this adversely impacts supplier prices across various sectors of the industry, sometimes even resulting in a deficit in the infrastructural investment made by farmers on their farms (Kumar et al., 2013; Manikas and Terry, 2009). Meanwhile, customers often perceive that the price reduction at the farmgate does not translate into cheaper food products for them (Elghannam et al., 2019).

Large retailers can impose quality and safety standards that cripple aspects of producers' management control and the flexibility for business continuity (McDonagh and Commins, 1999; Thompson and Lockie, 2013). This is not the case only for local farmers, but also for primary food producers in developing countries looking to export to developed countries. The challenge in the case of exporting developing country producers is with regards to European private standards requiring these exporters to invest heavily (e.g. in training requirements, third party audits, infrastructural and technological updates) to standardise (often well above) the safety and quality of their products with European regulatory requirements and consumer expectations (Lin, 2014; Rao et al., 2021).

A central feature of AFSCs in Western and Northern Europe is vertical integration. Food manufacturers produce products according to retailers' specifications and are contractually required to undertake value-adding activities such as quality control and packaging (Cotterill, 1997; Hingley, 2005; Kumar et al., 2013; Manikas and Terry, 2009; Yakovleva, 2007). Furthermore, downstream actors in the food supply chain are constantly pushing for product tracking through the supply chain as this is critical for the prevailing 'Just-In-Time' delivery structure in the sector (Fernie et al., 2010; Jones et al., 2005). Regardless of the already high levels of vertical integration in these chains, these factors dictate closer relationships across these AFSCs, particularly as retailers are increasingly inclined to use preferred suppliers (Kumar et al., 2013; Yakovleva, 2007). These relationships, though they enhance the overall output of AFSCs, and in some aspects, are beneficial to producers, usually favour retailers (Hingley, 2005) as they are able to avoid taking on the added risk of ownership that comes with vertical integration (Rao et al., 2021). Consequently, there has been an increasing consolidation of upstream actors (producers and processors) - albeit at varying levels, across all food sectors in this region (Kumar et al., 2013). Even among competitors, there is increased cooperation to ensure they survive their relationships with multiple large retailers (Walley and Custance, 2010) - a phenomenon known as coopetition (Beckeman et al., 2013).

Notwithstanding, the growing demand for food provenance has presented local producers with opportunities to directly integrate with retailers on a store-by-store basis, thereby opening major new channels to consumers and access to the customer bases of these retailers (Kumar et al., 2013). On the part of the retailers, this direct integration with local producers allows them to meet consumer demand for greater choice and local provenance, thereby providing them with a competitive advantage. Furthermore, in response to retailers' policies, farmers' markets have sprung up in many towns and cities in the countries in these regions over the past couple of decades (Walley and Custance, 2010). This allows farmers to get closer to their customers and to offer them a greater variety of produce while being able to retain a greater share of the profits.

4.2 Sustainability in agri-food supply chains

With the growing global population, concerns about food security began to go beyond immediate food provision to include the ability to preserve and regenerate resources for future generations. Thus, concerns about the effects of agricultural practices on the environment have gathered significant attention, with increasing efforts aimed at making processes from farm to fork more eco-friendly (Bourlakis et al., 2014). Organisations have consequently shown great commitment to reporting their energy consumption, greenhouse gas (GHG) emissions, water consumption, waste, packaging, raw material use, recycling and the use of chemicals (Jones et al., 2005).

In addition to these, consumers have had growing interest over the past couple of decades in how especially, the most powerful players in the agri-food chain – retailers (and large manufacturers), can ensure fairness in economic benefits across the chain, fair wages, healthier food options, greater accessibility and affordability for all and the overall social impact and well-being of the communities within which they operate. For example, the UK Government's Sustainable Development Commission in its 'Green, Healthy and Fair' report indicated that '*As gatekeepers of the food system, supermarkets are in a powerful position to create a greener, healthier, fairer food system through their influence on supply chains, consumer behaviour and their own operations*' (SDC, 2008, p. 6).

Governmental pressure through legislation, intra- and international regulations and policy directives by regional bodies (such as the EU) all serve as major drivers for sustainability in the food chain (Chkanikova and Mont, 2015). Nevertheless, there continues to be reasonable scepticism about the apparent interest in sustainability by retailers, with some suggesting they are more driven by marketing and public relations initiatives, cost savings and the pre-emption of unwelcome statutory regulations (Jones et al., 2008). There is increasing evidence that interest in sustainability by large retailers in Western and Northern Europe may also be because of increasing stakeholder demands, increased investor appeal and better brand reputation – all of which lead to increased competitive advantage and economic sustainability (Chkanikova and Mont, 2015; Jones et al., 2005).

A lot of convenience food products (such as ready meals and sandwiches) are manufactured for specific supermarket brands. Due to demand volatility and the short product shelf lives of these products, retailers give their manufacturers very short order lead times. To compensate for these, convenience food product manufacturers actively overproduce, resulting in about 20–40% of material waste (Darlington et al., 2009). The last-minute cancellation of orders along with sanctions placed on food manufacturers by retailers when order volumes are unfulfilled also contribute to food waste. Additionally, manufacturing processes such as product change in production lines or interruptions in production lines and cold chains also contribute to food loss and waste in the form of product residues, loss of raw materials and compromised product quality which may either become unusable for later processing or get rejected by customers (Dora et al., 2020).

Furthermore, the threat of legal recourse as a result of irresponsible social behaviour, the threat of negative press, consumer demand for healthier food options, ethical consumerism, increasingly sustainable industrial process benchmarks, sound scientific inputs for improved sustainability and pressure from civil society and advocacy groups continue to drive sustainability higher up the priority lists for agri-food chain actors - especially retailers and large manufacturers (Boström et al., 2005; Chkanikova and Mont, 2015; Hoffman, 2000; SDC, 2008; Smith, 2008).

Another key aspect of sustainability in AFSCs in Western and Northern Europe has been the push for shorter food supply chains. This was partly due to the need to increase consumer confidence in food integrity and improve traceability, and to reduce food miles within supply chains (Bourlakis and Bourlakis, 2004; Renting et al., 2003). A key characteristic was the ability of these AFSCs to re-socialise or re-spatialise food to allow the consumer to make better ethical and sustainability judgements (Marsden et al., 2000). There has been particular emphasis on the origin of food products in these supply chains in terms of the locality or even the particular farms where they are produced as producers seek to emphasise these as indications of quality. The benefits of short food supply chains are evident in all three dimensions of sustainability.

Economically, arguments have been made for an improvement in producer margins by reconnecting farmers directly to consumers, showing support for sustainable and healthy farming methods and helping generate resilient farm-based livelihoods through the generation of local employment, reduction in production costs and an improvement in synergies with other sectors (Elghannam et al., 2019; Jarzębowski et al., 2020; Oglethorpe and Heron, 2013). From an environmental perspective, benefits include the promotion of less polluting production methods as well as the overall reduction in GHG emissions, food miles, food waste, energy use and resource usage. Socially, consumers get to connect better with producers thereby fostering trust, social inclusion, consumer empowerment and a sense of community (Jarzębowski et al., 2020).

Notwithstanding, the development of short AFSCs presents various challenges, particularly to institutions responsible for ensuring the economic and social regeneration of rural communities (Marsden et al., 2000). Firstly, the use of short AFSCs still remains marginal. In all of Europe, only about 15% of farms sell more than 50% of their production directly to consumers (Augère-Granier, 2016). Also, competition among local producers can cripple them from cooperating and coordinating with each other to be able to compete in major export markets (Marsden et al., 2000). In certain contexts, particularly where logistical operations are not adequately optimised or the chains are overstrained, prices paid by consumers could be higher (Elghannam et al., 2019). Therefore, even though local sustainable benefits are rife with these supply chains, there is some required balance between these benefits and the strategic considerations (such as greater yield volumes, bigger markets, economies of scale, currency exchange benefits, optimised production, supply chain processes, etc.) that inform the establishment of longer supply chains as local economies need to be able to compete in bigger and more major markets at national and regional levels.

5 Key challenges to agri-food supply chains

Challenges in food systems vary quite significantly depending on the type of product, location of production, intended market, prevailing legislation and regulations, among other variables. Notwithstanding, there are key challenges that are common in AFSCs, especially in Western and Northern Europe. A closer look at some of these is presented in the following sections.

5.1 Primary production

With the great power retailers wield in AFSCs, the adoption of category management is prevalent in vertical food channels, especially for fresh foods, which are usually unbranded. Here, the supplier base tends to shrink to guarantee consistency, resulting therefore in preferred lead suppliers dominating particular products or product lines. Furthermore, with the limited bargaining power as one moves upstream in the AFSC, there is growing evidence for increasingly poor overall well-being for food producers with lower levels of negotiating power (Estevam and Santos, 2018).

Additionally, the thin operating margins upstream of AFSCs have fuelled harsh working conditions and exploitation of agri-food workers in both crop and animal production in Germany, Sweden, the UK and the Netherlands, among others. At the governmental level, there have been various forms of interventions to tackle the power imbalances. In commodity AFSCs, the increasing globalisation and complexity to satisfy customer demand for cheaper products present challenges in ensuring sustainability across the whole chain. Having thousands of suppliers geographically dispersed across the globe, it proves difficult for organisations to have good oversight of working conditions, remuneration rates and modern slavery - especially when in some cases there is insufficient or total lack of enforcement of local labour and environmental legislations (Chkanikova and Mont, 2015; Jones et al., 2005).

There are ongoing civil society-led initiatives aimed at ensuring Fairtrade, to drive improved livelihoods and working conditions of producers, particularly in developing countries. Retailers such as Sainsbury's, Tesco and Marks and Spencer in the UK have shown commitment towards this by dedicating whole product lines to Fairtrade (SDC, 2008). whilst producers in developing countries continue to supply significant tonnage of various agri-food products to the region.

5.2 Food safety

The management of food safety ensures the physical safety-related attributes of the food products as well as aspects such as occupational health and safety, technical training, chemical safety infrastructure and record-keeping (Rao et al., 2021). Following a series of food crises in the late 1990s such as the Bovine spongiform encephalopathy (BSE) crisis (popularly known as the mad cow diseases), Salmonella and dioxin contaminations, the European Commission published a policy paper that led to the formation of the European Food Safety Authority (EFSA) which was set up in 2002 to be a source of scientific advice and communication on food chain-related risks (EFSA, 2022; European Commission, 2012). Food safety has consequently become high on the agenda of food retailers, with many reporting it in their Corporate Social Responsibility (CSR) reports. Organisations continue to show high commitment to retaining customer confidence in response to concerns about genetically modified foods, and the use of pesticides and heavy metals.

Over the years, there have been discussions about the impact of food safety regulations on producers in developing countries. Some argue they only benefit retailers and consumers in developed countries while others insist that there are spillover benefits for the producers and consumers in developing countries as well (Fuchs et al., 2011; Rao et al., 2021). Regardless, regulatory uncertainty regarding the EU's standards on food safety has made private standards attractive to food retailers and manufacturers (Charlier and Valceschini, 2010). This has led to the morphing of originally strict food standards to now incorporate issues that are not entirely food safety-related, leading to producers in developing countries accepting standards that producers in developed countries like USA and Japan who also export to Europe do not adhere to (Burrell, 2011; Naiki, 2014; Rao et al., 2021). Resultantly, there is the stifling of the businesses of primary producers in developing countries due to being subjected to stricter regulatory requirements that go beyond food safety and quality (Jones et al., 2005).

5.3 Carbon emissions, food waste and other sustainability challenges

Many challenges impede the expeditiousness of the implementation of sustainability across the AFSC. These include increased competition in the retail sector (Smith, 2008) as well as inconsistent and unclear policy directions by governmental and regulatory bodies within countries. Internationally, there is a lack of harmonised regulations on food safety, quality and labelling requirements across Western and Northern European countries (Chkanikova and Mont, 2015; Jones et al., 2008). The challenges created for smaller producers and producers from developing countries through the implementation of private standards have also been identified as inhibiting the sustainability of food production and suppressing the economic viability of small food producers in relatively smaller markets (Chen et al., 2008; Fagotto, 2014; Fuchs et al., 2009).

Some retailers have highlighted the restrictiveness of governmental legislation and how it impedes their innovation. Cost and the lack of economic incentives, along with insufficient expertise have also been identified as barriers to pursuing sustainability, especially upstream of the food chain (Seuring and Müller, 2008; Smith, 2008). The limited availability of sustainable products (especially top-quality organic and fair-trade options) in the supplier market drives prices higher, which stifles demand levels and consequently hinders widespread sustainable sourcing (Chkanikova and Mont, 2015).

Environmentally, sustainability issues revolve around the availability of water, discharge of wastewater, solid water disposal, product disposal or utilisation, food packing material, chemical residues, transportation and air emissions (Avermaete and Viaene, 2002). Globally, the food supply chain creates approximately 13.7 billion metric tons of carbon dioxide equivalents (CO_2eq) and 26% of anthropogenic GHG emissions (Poore and Nemecek, 2018). Contrary to concerns that the transportation and distribution of food products are major sources of emissions in AFSCs (Validi et al., 2014), the majority of emissions from average diets in Europe are from land-use change (30%) or emissions at farm-level (manure management and methane from cattle, 22% each) (Ritchie and Roser, 2020; Sandström et al., 2018). Transport only accounts for 6%. Thus, the farm stage by far dominates emissions for food production and supply (Poore and Nemecek, 2018). As shown in Fig. 2, countries

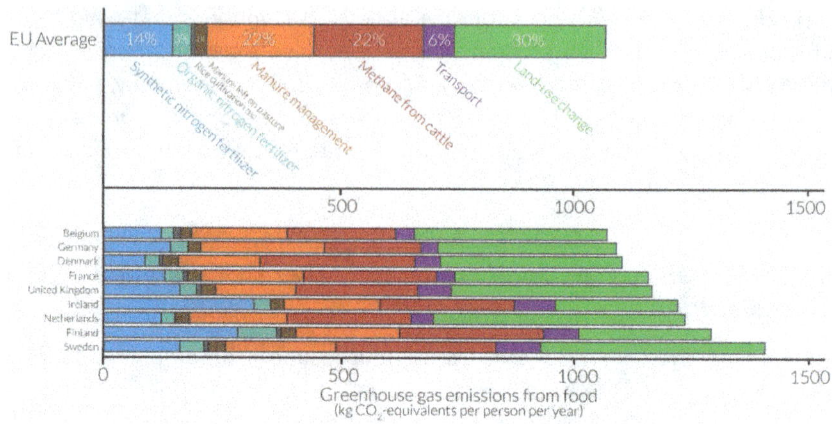

Figure 2 Supply chain-related (production and trade) dietary emissions of the average diets in Western and Northern European countries. Source: Adapted from: Ritchie and Roser (2020) and Sandström et al. (2018).

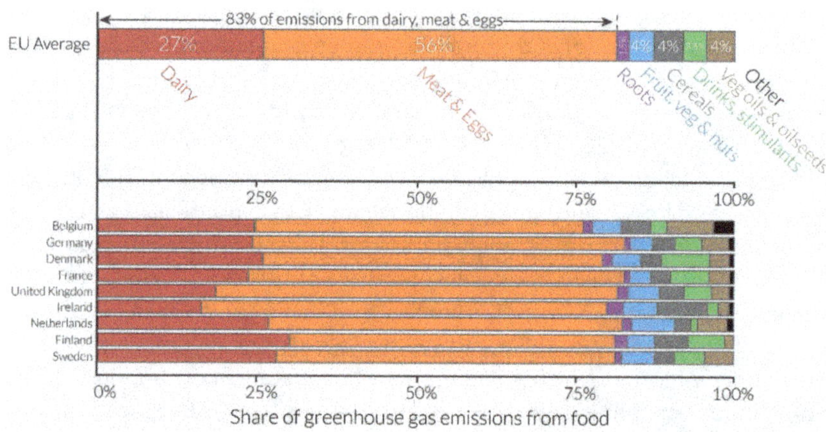

Figure 3 Dietary emissions of selected food items groups in Western and Northern European countries. Source: Adapted from: Ritchie and Roser (2020) and Sandström et al. (2018).

in Western and Northern Europe produce above the continent's average, with Sweden producing the highest emissions in the region.

Dairy products (27%) and meat and eggs (56%) account for 83% of total emissions from the average European diet, with plant-based diets accounting for only 17% (see Fig. 3). The differences observed in the relevant Western and Northern countries largely depend on the level of meat consumption. Consequently, despite the heightened drive for shorter supply chains due to, inter alia, environmental benefits, diet modification holds greater potential to reduce the adverse environmental impact of food production and supply.

Poore and Nemecek (2018) provide a framework to pursue emission reduction in food production. For Western and Northern Europe where traceability of meat products is stringent, a combination of producer incentivisation through taxes or subsidies that reflect environmental costs in product prices along with consumer education on the true cost of food could help drive a reduction in carbon emissions resulting from meat and egg production.

Another important sustainability challenge with AFSCs in these regions is food waste. The third target of the 12th United Nations Sustainable Development Goal is to ensure that by 2030, per capita global food waste at the retail and consumer levels is halved and food losses along production and supply chains (including post-harvest losses) are reduced (United Nations, 2021). This is particularly crucial, seeing that a third of all food produced globally (estimated to be around 1.3 billion tonnes of food, worth about $1 trillion) goes to waste (United Nations, 2021). In the EU (the UK included), approximately 88 million tonnes of food waste are generated annually with associated costs estimated at around €143 billion (Stenmarck et al., 2016).

The annual average household food waste in Western and Northern Europe is 139 kg per capita (Forbes et al., 2021). Germany leads the region with 6 263 775 metric tonnes of household food waste, followed by France and the United Kingdom with 5 522 358 metric tonnes and 5 199 825 metric tonnes, respectively (Forbes et al., 2021). As may be seen from Fig. 4, countries

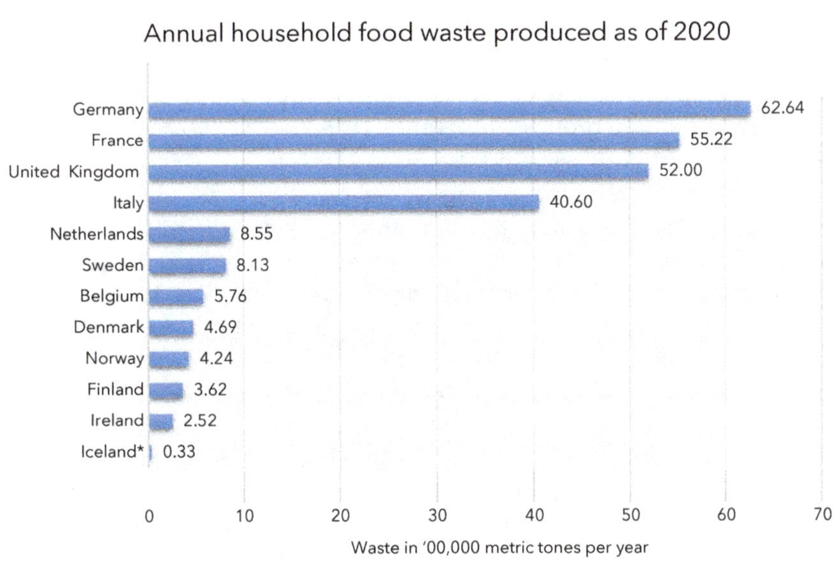

Figure 4 Annual household food waste produced ('00 000 metric tonnes per year) in Western and Northern Europe countries as of 2020. Sources: EPA, 2021; Forbes et al., 2021; Stefánsdóttir, 2020; Tiseo, 2021). NB: Data for Iceland are from 2019.

in Western Europe produce significant food waste, while Finland, Norway and Denmark - all in Northern Europe, account for the least levels of food waste. In the UK, 10 million tonnes of food and drink waste arises post-farmgate annually; 60% of which is avoidable, that is, it could be eaten (Environment Food and Rural Affairs Committee, 2017).

There are a variety of causes of food waste in AFSCs. Weather, food production methods, retailer and supplier strategies, and governance of the food value chain have all been identified as factors that influence food waste (Rohm et al., 2017). The large amounts of food waste generated upstream of AFSCs are a result of problematic relationships between producers and retailers (Food and Agriculture Organization, 2013). With the power imbalance within AFSCs, last-minute cancellation or modification of forecast orders, as well as unrealistic quality standards (tight cosmetic standards which result in about 30% of fruits and vegetables not being harvested), are major contributing factors to food waste upstream of supply chains (Facchini et al., 2018; Stuart, 2009). There are also significant amounts of food waste downstream of the supply chain due to consumer behaviour (Chauhan et al., 2021; Gustavsson et al., 2011). A widespread lack of knowledge about the difference between "best-before-date" and "use-by-date" causes a lot of food that is still good to eat to be thrown away while leftover foods also contribute to household food wastage. Quality standards by retailers, especially as relating to aesthetic defects as well as poor temperature management increases waste of food downstream.

It is also worth mentioning that in 2019, about 6.8% of people in the EU could not afford a meal with chicken, fish, meat or a vegetarian equivalent every other day (Eurostat, 2021a). Children (under 16 years) were slightly less likely to experience this categorisation of food poverty while among adults, the most vulnerable age groups were 16-24 years and 45-54 years with elderly people (65 years or over) being the worst affected group. Food poverty was also identified to be higher among women aged over 16 than among men (Eurostat, 2021a). Even though overall food poverty in the EU has been on the decline since 2011, Germany and France, which both lead this region with the highest levels of annual household food waste (see Fig. 4), also have the highest prevalence of food poverty in the region with 11.6% and 12.1% of the population, respectively, in 2019 (Eurostat, 2022e).

Food insecurity has also increased due to the Coronavirus pandemic, and in the UK, this has been exacerbated by Brexit as well. This has gotten worse with the Russian-Ukrainian war. The two countries jointly account for 64% of sunflower oil, 23% of wheat, 19% of barley and 18% of maize exports globally (Ritchie, 2022). Western and Northern Europe rely on these two countries for some of their supply of these food products and food prices are therefore adversely affected and food insecurity is worsening as a result.

Seeking to attend to some of these sustainability issues, the EU through its European Green Deal, part of which is the Farm to Fork Strategy, among other things, seeks to thoroughly deal with the challenges impeding the attainment of sustainable food systems economically, environmentally and socially (Eurostat, 2020). The Deal seeks to transform the EU into a modern, resource-efficient and competitive economy, ensuring that there will be no net emissions of GHG by 2050, economic growth is decoupled from resource use and that no person or place is left behind (European Commission, 2022b). Efforts towards avoiding food waste have been through information and capacity building, retail and supply chain alteration and food redistribution (Rohm et al., 2017). Even though the data on food redistribution is limited, the European Food Banks Federation (FEBA) reports redistributing 860 000 tonnes of food in 2020 (FEBA, 2021) - a 209% increase since 2014 (Stenmarck et al., 2016). In the UK, 92 015 tonnes of food were redistributed in 2020 from the retail, manufacturing, farming, hospitality and food service sectors through both commercial and charitable channels to disadvantaged consumers (WRAP, 2021).

6 Current state of agri-food supply chains

AFSCs have not have not been excluded from the increasing digitisation of supply chains to improve efficiency, integration, performance and the value proposition to customers. E-commerce has offered firms in the AFSC opportunities to provide online vendor catalogues, 24/7 service provision, responsive and efficient processing of customer queries and complaints, prediction of customer preferences and real-time tracking of shipments (Bourlakis and Bourlakis, 2004). E-commerce also affords for small and medium-scale enterprises in the food sector to exploit innovative modes, thereby reaching out to a greater market (Falguieres et al., 2015). In Belgium, the e-commerce sector witnessed an annual turnover growth to €8 billion in 2019 - 25% of which was achieved in neighbouring countries (van Rompaey, 2020).

Online food retailing has enjoyed a growing interest in Western and Northern European countries over the past decade. The UK, Netherlands, and Germany by far have had the highest percentage of customers who purchase food and groceries online (Eurostat, 2022f) as shown in Fig. 5. In 2019, the Netherlands led the region with 36% of customers shopping online. After the continent was hit by the coronavirus pandemic in 2020, overall retail sales growth declined across Europe. However, food and grocery sales increased by 7.5% during the period, with the rate of the online grocery sector growth almost tripling from 19.5% in 2019 to 56.1% in 2020 (Savills, 2021). In Western Europe, the online share of food and drink retail increased on average from 3.4% to 5.3%, with expectations that this could reach 12.6% by 2025. Nevertheless,

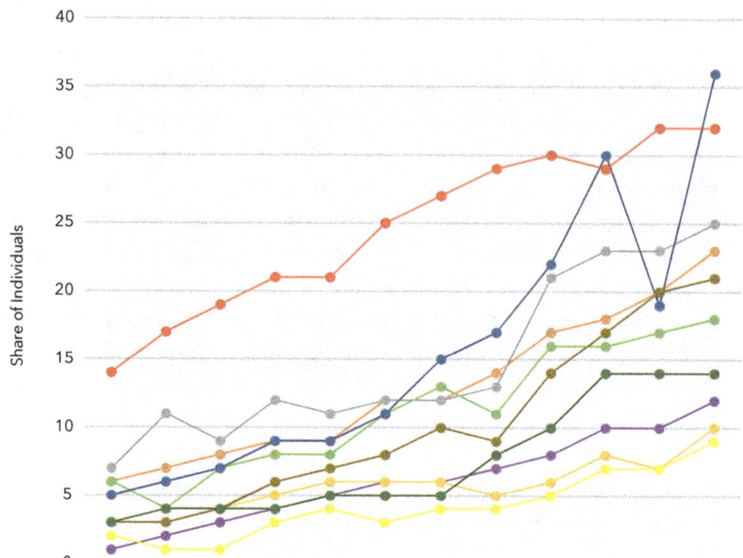

	2008	2009	2010	2011	2012	2013	2014	2015	2016	2017	2018	2019
Belgium	1	2	3	4	5	6	6	7	8	10	10	12
Denmark	6	7	8	9	9	12	12	14	17	18	20	23
Germany	7	11	9	12	11	12	12	13	21	23	23	25
Ireland	3	3	4	5	6	6	6	5	6	8	7	10
France	6	4	7	8	8	11	13	11	16	16	17	18
Netherlands	5	6	7	9	9	11	15	17	22	30	19	36
Finland	2	1	1	3	4	3	4	4	5	7	7	9
Sweden	3	3	4	6	7	8	10	9	14	17	20	21
Norway	3	4	4	4	5	5	5	8	10	14	14	14
United Kingdom	14	17	19	21	21	25	27	29	30	29	32	32

Year

Figure 5 Percentage of consumers who purchased food or groceries online in Western and Northern Europe from 2008 to 2019. Source: Adapted from: Eurostat (2022f).

customers continue to have concerns about the cost and quality of delivery, limited product assortment, poor food quality and the overall cost of online groceries (Galante et al., 2013).

As organisations have sought to reduce last-mile distribution costs, large food retailers (such as Tesco and Carrefour) have adopted elaborate omnichannel strategies such as dark stores and micro-fulfilment centres to ensure consistent customer experience across channels and expedite their service offerings. Customers now have options to buy in-store, buy online pickup in-store (click and collect) or have online purchases delivered to them. The pandemic expanded the rollout of these strategies significantly (Özbekler, 2022). Dark stores are traditional retail stores that have been converted to fulfilment centres not opened to the public whereas micro-fulfilment centres integrate the efficiencies provided by robotics fulfilment in large, automated warehouse facilities with the speed of local delivery usually associated with in-store picking (Hübner et al., 2022; Ladd, 2020). Tesco is collaborating with Gorillas (an on-demand grocery delivery company) while Asda is partnering with Just Eat to serve their customers using the micro-fulfilment model (Tatum, 2022).

With increasing economic difficulties due to the Coronavirus pandemic, consumers' shift towards value-orientated retailers (such as Aldi, Lidl and Asda) increased. In the UK, these types of brands have accounted for more than half of all new grocery openings and about 75% of all supermarket acquisitions in 2020 (Savills, 2021). As the effects of the pandemic, Brexit and the Russian-Ukrainian war continue to unravel, and the expenditure laxity of customers continues to tighten due to economic challenges, it is expected that more and more UK customers will turn to value-orientated brands for their groceries. Food retailers are now more likely to create alliances with suppliers that favour their planning processes and have shorter lead times due to proximity, implement independently developed protocols for addressing food safety and quality challenges and to commit to newer technologies that facilitate better visibility, information sharing and collaboration along the agri-food chain (Pérez-Mesa et al., 2021).

On this front, there has been an introduction and adoption of blockchain technology in the management of AFSCs with 40% of all blockchain technology applications being in the agri-food sector (Vadgama and Tasca, 2021).

The main features of blockchain technology, namely, decentralisation, immutability, security and smart contract, bring significant benefits that are invaluable in AFSCs (Duan et al., 2020). As may be seen from Fig. 6, blockchain technology has been applied significantly in pursuit of traceability in AFSCs (see Caro et al., 2018; Kamilaris et al., 2019; Rogerson and Parry, 2020; Tian, 2016). Additionally, it can improve transparency and ensure information

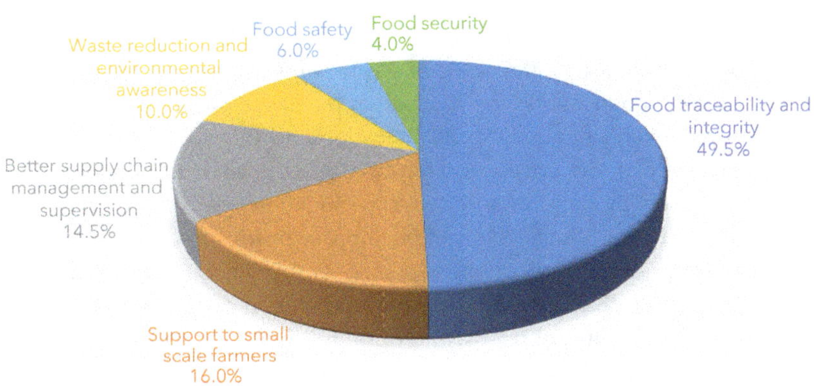

Figure 6 Application of blockchain technology in agri-food supply chains. Source: Adapted from: Kamilaris et al. (2019).

authenticity and efficiency for better management of AFSCs, provide support for small-scale farmers and improve sustainability through food security, food waste management and food safety (Duan et al., 2020; Kamilaris et al., 2019; Shahid et al., 2020).

Leading retailers in Western Europe have adopted blockchain technology. Auchan Retail, a French retail giant, has rolled out blockchain technology in France (and in Spain, Portugal, Italy and Senegal) to improve the traceability of products along its supply chains after initially testing the technology in Vietnam alongside the German blockchain start-up, TE-Food (Auchan Retail, 2018). Customers now have information on all stages of the product's life from farm to fork. Another French retailer, Carrefour has an acclaimed record of blockchain application in their operations. Using IBM's Food Trust platform, Carrefour provides a variety of data (including the farmer and location of the farm, hygiene certifications, nutritional information and even preparation tips) which can be accessed by scanning a Quick Response (QR) code on the labels of a select group of products (Carrefour, 2018; Win, 2020). In its chicken product line, the company confirms the comprehensive and detailed information provided through blockchain technology differentiates its value offering to customers and gives them a competitive edge (Thomasson, 2019).

7 Future trends for research

Over the past decade, the volume of international trade (imports and exports) of agri-food products from this region has continually risen, thereby evincing the rise of globalisation as well as the excellent application of new supply chain management strategies for optimised performance. This has however compromised the resilience of these supply chains (Sawyerr and Harrison, 2019). For example, due to insufficient capacity in domestic food production, Brexit-related labour market and just-in-time supply chains, the UK's agri-food system has been exposed for its vulnerability (Garnett et al., 2020). Food supplies in both France and Germany, two of the most developed food-producing countries in the EU, suffered disruptions during the COVID-19 pandemic, owing not only to hikes in demand but also to a reduction in the production of food products (Liu, 2020). Disruptions in agri-food chains were also caused by nationally imposed restrictions which resulted in limited movement, and hence, a shortage of seasonal farm workers, blocked transport routes and long queues at border checks for commodity transport in the region (Rossi, 2020). With the potential that the application of Industry 4.0 in AFSC management has, exploring how increasing vulnerabilities of the agri-food chains in these regions can be addressed is needed.

At the height of the pandemic, having an online presence became critical for retailers as online grocery shopping demand spiked. Studies into the pandemic as a driver for omnichannel food retailing as well as further studies into the various omnichannel strategies adopted by retailers are needed. As global food supplies have been adversely hit by the Russian–Ukrainian conflict, research into the impact of resultant disruptions on AFSCs in Western and Northern Europe as well as how organisations addressed them ought to be pursued.

Furthermore, even with commitment of retail giants - such as Carrefour - to applying blockchain technology to enhance traceability, there are still many products where the technology is not used and there are many companies which could exploit it. Besides traceability, more research is also needed in exploring how blockchains can be applied in other aspects of the AFSC including food waste reduction, food redistribution and resilience. Challenges with the application of blockchain technology in the agri-food sector include a widespread lack of understanding and knowledge of the technology and there are many concerns with scalability, stakeholder cooperation, manipulation of raw data and outdated legislation (Duan et al., 2020). Each of these presents great opportunities for further research, considering country and/or product-specific issues and prescriptive insights.

Moreover, the application of robotics in warehousing is becoming increasingly common especially as warehousing costs (considering its labour intensiveness) significantly affect logistics costs. Robotics will be critical to the successful integration of Industry 4.0 technology into the management of supply chains (Voss, 2021). With the pickers-to-goods system being the most widespread order-picking system in Western Europe (van Gils et al., 2018), exploring Gialos and Zeimpekis' (2021) proposed approach to the more efficient vision-picking technology in the order-picking process in the region's AFSCs and the implications on affordability and the value proposition to customers will be insightful. Similarly, there are research opportunities for exploring collaborative robotics in AFSCs as e-commerce continues to grow in food retailing. Studies can be conducted into the key opportunities and the major barriers within the food system, not only seeking process efficiency and supply chain optimisation but also meeting the sustainability challenges around food waste, surplus food redistribution, food poverty, insecurity and inequalities.

There is also much research to be done on omnichannel distribution strategies such as customer fulfilment centres, in-store picking, dark stores and micro-fulfilment centres. Efforts at identifying the best strategy or combination of strategies depending on various sets of variables (such as population density, store locations, sustainability considerations, cost efficiencies, etc.) are needed as food retailers continue to evolve their delivery systems to enhance service offerings and value proposition to customers.

Also, despite the significant attention given to making the logistics of agri-food more sustainable, it is evident that the greater source of emissions in the agri-food sector is land-use change. More attention must be given to exploring ways by which farmers can be incentivised to reduce emissions. There are opportunities to explore what role Industry 4.0 (through artificial intelligence, robotics, drone technology, the Internet of Things, Big Data, etc.) can play in addressing land use challenges and food waste on primary food production sites. Considering dairy, meat and egg production account for 83% of carbon emissions and these form a significant part of meals in this region, a radical dietary change does not seem realistic in the short term. Notwithstanding, exploring how to make the production of these products more sustainable while educating consumers on more sustainable alternatives may prove useful.

Household food waste continues to be an elusive challenge in the overall food waste management campaign. Although educating consumers on when food is no longer good for consumption has been sufficiently made, this does not reflect in records on household food waste. Both research and policy interventions on addressing household food waste are very much needed.

8 Conclusion

This chapter has taken a closer look at AFSCs in Western and Northern Europe. The supply chains were classified as either local or commodity and some strategies adopted in AFSCs in the region were discussed. The dominant role of retailers and, to some extent, manufacturers was highlighted and the implications of this on other actors in the food chain, as well as on food prices were discussed. Powerful retailers were indicated to enjoy the benefits of vertical integration (which is a central feature in AFSCs in the region) while avoiding some of its added risks. We have presented the complex sustainability issues in the region and the role retailers play in pursuing sustainability goals.

The dominance of retailers was also identified to adversely impact the operating margins of primary producers locally and in developing countries. Pressures to ensure food safety and regulatory uncertainty continue to result in private standards which stifle primary production in developing countries due to stricter requirements that suppliers from other developing countries do not adhere to. A variety of sustainability challenges, including food production carbon emissions, food waste and food poverty are highlighted. Presenting the current state of AFSCs in the region, the growing role of e-commerce, especially due to the COVID-19 pandemic is presented, and this led to discussions around the use of blockchain technology in the sector. Opportunities for further research are then presented to conclude the chapter.

9 Where to look for further information

9.1 Books

a. Bourlakis, M. and Weightman, P. W. H. (Eds.) (2004). *Food Supply Chain Management*, Blackwell Publishing Ltd.
b. Paloviita, A. and Järvelä, M. (Eds.) (2016). *Climate Change Adaptation and Food Supply Chain Management*. New York: Routledge.
c. Bourlakis, M., Vlachos, I. and Zeimpekis, V. (eds.) (2011). *Intelligent Agri-food Chains and Networks,* Blackwell-Wiley.
d. Aktas, E. and Bourlakis, M. (Eds.) (2020). *Food Supply Chains in Cities: Modern Tools for Circularity and Sustainability*, Palgrave Macmillan.

9.2 Journals

a. *British Food Journal.*
b. *Journal of Chain and Network Science.*
c. *Agribusiness Journal.*
d. *Conferences: The Annual International European Forum* (IGLS Forum series).

9.3 Research centres

a. Centre for Procurement, Logistics and Supply Chain Management, School of Management, Cranfield University.

10 References

Alexander, C. and Smaje, C. (2008). Surplus retail food redistribution: An analysis of a third sector model, *Resources, Conservation and Recycling* 52(11), 1290–1298.

Auchan Retail. (2018). *Food Traceability: Auchan Retail Is Launching Blockchain Technology Internationally*, Auchan Retail.

Augère-Granier, M.-L. (2016). *Short Food Supply Chains and Local Food Systems in the EU, European Parliamentary Research Service*. Available at: https://www.europarl.europa.eu/RegData/etudes/BRIE/2016/586650/EPRS_BRI(2016)586650_EN.pdf.

Avermaete, T. and Viaene, J. (2002). On innovation and meeting regulation: The case of the Belgian food industry. In: *Industrial Dynamics of the New and Old Economy - Who Is Embracing Whom?, DRUID Summer Conference*, Copenhagen, 1–22.

Beckeman, M., Bourlakis, M. and Olsson, A. (2013). The role of manufacturers in food innovations in Sweden, *British Food Journal* 115(7), 953-974.

Beitzen-Heineke, E. F., Balta-Ozkan, N. and Reefke, H. (2017). The prospects of zero-packaging grocery stores to improve the social and environmental impacts of the food supply chain, *Journal of Cleaner Production* 140, 1528–1541.

Blackburn, J. and Scudder, G. (2009). Supply chain strategies for perishable products: The case of fresh produce, *Production and Operations Management* 18(2), 129–137.

Boström, M., Føllesdal, A., Klintman, M., Micheletti, M. and Sørensen., M. P. (2005). Studying political consumerism. In: Boström, M., Føllesdal, A., Klintman, M., Micheletti, M. and Sørensen., M. P. (Eds). *Political Consumerism: Its Motivations, Power, and Conditions in the Nordic Countries and Elsewhere*, Nordisk Ministerråd, København, Oslo, 145–164.

Bourlakis, C. and Bourlakis, M. (2004). The future of food supply chain management. In: Bourlakis, M. A. and Weightman, P. W. H. (Eds). *Food Supply Chain Management*, Blackwell Publishing Ltd, Oxford, UK, 221-230.

Bourlakis, M., Maglaras, G., Gallear, D. and Fotopoulos, C. (2014). Examining sustainability performance in the supply chain: The case of the Greek dairy sector, *Industrial Marketing Management* 43(1), 56–66.

Bourlakis, M. and Weightman, P. W. H. (2004). Introduction to the UK food supply chain. In: Bourlakis, M. A. and Weightman, P. W. H. (Eds). *Food Supply Chain Management*, Blackwell Publishing Ltd, Oxford, UK, 2–10.

Burrell, A. (2011). 'Good agricultural practices' in the agri-food supply chain, *Environmental Law Review* 13(4), 251-270.

Cachon, G. P. and Kök, A. G. (2007). Category management and coordination in retail assortment planning in the presence of basket shopping consumers, *Management Science* 53(6), 934–951.

Caro, M. P., Ali, M. S., Vecchio, M. and Giaffreda, R. (2018). Blockchain-based traceability in agri-food supply chain management: A practical implementation. In: *2018 IoT Vertical and Topical Summit on Agriculture - Tuscany, IOT Tuscany 2018*, IEEE, Tuscany, Italy, 1–4.

Carrefour. (2018). Carrefour launches Europe's first food blockchain, *Quality*, 6 March. Available at: https://www.carrefour.com/en/newsroom/carrefour-launches-europes-first-food-blockchain.

Charlier, C. and Valceschini, E. (2010). Food safety, market power and private standards: An analysis of the emerging strategies of food operators, *International Journal on Food System Dynamics* 2, 103–110.

Chauhan, C., Dhir, A., Akram, M. U. and Salo, J. (2021). Food loss and waste in food supply chains: A systematic literature review and framework development approach, *Journal of Cleaner Production* 295, 126438.

Chen, C., Yang, J. and Findlay, C. (2008). Measuring the effect of food safety standards on China's agricultural exports, *Review of World Economics* 144(1), 83–106.

Chkanikova, O. and Mont, O. (2015). Corporate supply chain responsibility: Drivers and barriers for sustainable food retailing, *Corporate Social Responsibility and Environmental Management* 22(2), 65–82.

Clay, J. (2005). *Exploring the Links between International Business and Poverty Reduction: A Case Study of Unilever in Indonesia*, Oxfam, GB, Oxford, UK

Cotterill, R. W. (1997). The food distribution system of the future: Convergence towards the US or UK model? *Agribusiness* 13(2), 123–135.

Darlington, R., Staikos, T. and Rahimifard, S. (2009). Analytical methods for waste minimisation in the convenience food industry, *Waste Management* 29(4), 1274–1281.

DEFRA. (2021a). Horticultural statistics. Available at: https://view.officeapps.live.com/op/view.aspx?src=https%3A%2F%2Fassets.publishing.service.gov.uk%2Fgovernment%2Fuploads%2Fsystem%2Fuploads%2Fattachment_data%2Ffile%2F1010176%2Fhort-dataset-11aug21.xlsx&wdOrigin=BROWSELINK.

DEFRA. (2021b). *Agriculture in the UK Evidence Pack.*

DEFRA. (2021c). *UK Food Security Report 2021*, HH Associates Ltd. Available at: https://assets.publishing.service.gov.uk/government/uploads/system/uploads/attachment_data/file/1041623/United_Kingdom_Food_Security_Report_2021_16dec2021b.pdf.

Deloitte. (2021). *Global Powers of Retailing 2021*. Available at: http://www2.deloitte.com/global/en/pages/consumer-business/articles/global-powers-of-retailing-2014.html.

Dora, M., Wesana, J., Gellynck, X., Seth, N., Dey, B. and De Steur, H. (2020). Importance of sustainable operations in food loss: Evidence from the Belgian food processing industry, *Annals of Operations Research* 290(1–2), 47–72.

Dreyer, H. C., Swahn, N., Kiil, K., Strandhagen, J. O. and Romsdal, A. (2015). The responsiveness of food retail supply chains: A Norwegian case study. In Umeda, S., Nakano, M., Mizuyama, H., Hibino, N., Kiritsis, D. and von Cieminski, G. (Eds). *Advances in Production Management Systems: Innovative Production Management Towards Sustainable Growth* (Vol. 459), Springer International Publishing, Cham, 152–160.

Duan, J., Zhang, C., Gong, Y., Brown, S. and Li, Z. (2020). A content-analysis based literature review in blockchain adoption within food supply chain, *International Journal of Environmental Research and Public Health* 17(5), 1784.

EFSA. (2022). About us: Who we are and what we do. Available at: https://www.efsa.europa.eu/en/aboutefsa (accessed 23 March 2022).

Elghannam, A., Mesias, F. J., Escribano, M., Fouad, L., Horrillo, A. and Escribano, A. J. (2019). Consumers' perspectives on alternative short food supply chains based on social media: A focus group study in Spain, *Foods* 9(1), 22.

Environment Food and Rural Affairs Committee. (2017). *Food Waste in England, Eigth Report of Session 2016–17*. Available at: https://publications.parliament.uk/pa/cm201617/cmselect/cmenvfru/429/429.pdf.

EPA (2021). *How Much Food Do We Waste in Ireland?, EPA Website*. Available at: https://www.epa.ie/publications/circular-economy/resources/nature-and-extent-of-food-waste-in-ireland.php.

Estevam, D. de O. and Santos, M. J. (2018). The challenges of concentration in the agricultural product supply chain: A study of Producer Organizations in Portugal, *International Journal of Advanced Engineering Research and Science* 5(10), 208–218.

European Commission. (2012), *FAQ: European Food Safety Authority (EFSA)*, Brussels. Available at: https://ec.europa.eu/commission/presscorner/detail/en/MEMO_12_869.

European Commission. (2022a). Market data on national and European agriculture. Available at: https://agridata.ec.europa.eu/extensions/DataPortal/production.html (accessed 7 March 2022).

European Commission. (2022b). A European green deal: Striving to be the first climate-neutral continent. Available at: https://ec.europa.eu/info/strategy/priorities-2019-2024/european-green-deal_en (accessed 29 March 2022).

Eurostat. (2020). *Agriculture, Forestry and Fishery Statistics: 2020 Edition*, edited by Cook, E., Publications Office of the European Union, Luxembourg. Available at: https://doi.org/10.2785/143455.

Eurostat. (2021a). *Key Figures on the European Food Chain: 2021 Edition*, edited by Cook, E., Imprimerie Bietlot, Luxembourg. Available at: https://doi.org/10.2785/180958.

Eurostat. (2021b). Performance of the agricultural sector, *Eurostat*. Available at: http://agris.fao.org/agris-search/search.do?recordID=GB2013201757 (accessed 11 March 2022).

Eurostat. (2021c). Agricultural production - crops Agrometeorological review. Available at: https://ec.europa.eu/eurostat/statistics-explained/index.php?title=Agricultural_production_-_crops.

Eurostat. (2021d). Agricultural production - livestock and meat, *Eurostat*. Available at: https://ec.europa.eu/eurostat/statistics-explained/index.php?title=Agricultural_production_-_livestock_and_meat&oldid=549389#Livestock_population%0Ahttps://ec. Available at: europa.eu/eurostat/statistics-explained/index.php?title=Agricultural_production_-_livestock_and_.

Eurostat. (2022a). Fish production in Europe, Luxembourg. Available at: https://ec.europa.eu/eurostat/databrowser/view/FISH_CA_MAIN__custom_1188880/default/table.

Eurostat (2022b). Share of extra-EU trade in agricultural, fisheries and F&B products, Luxembourg. Available at: https://appsso.eurostat.ec.europa.eu/nui/show.do?query=BOOKMARK_DS-645593_QID_-16FF0DE4_UID_-3F171EB0&layout=PERIOD,L,X,0;REPORTER,L,X,1;FLOW,L,Y,0;PRODUCT,L,Y,1;PARTNER,C,Z,0;INDICATORS,C,Z,1;&zSelection=DS-645593INDICATORS,VALUE_IN_EUROS;DS-645593PARTNE.

Eurostat (2022c). *Share of Intra-EU Trade in Agricultural, Fisheries and F&B Products*, Luxembourg. Available at: https://doi.org/10.4324/9781315033853-30.

Eurostat (2022d). International trade partners, Luxembourg. Available at: https://appsso.eurostat.ec.europa.eu/nui/show.do?query=BOOKMARK_DS-645593_QID_-4F9B022A_UID_-3F171EB0&layout=PERIOD,B,X,0;PRODUCT,B,X,1;FLOW,B,Y,0;PARTNER,B,Y,1;REPORTER,B,Z,0;INDICATORS,C,Z,1;&zSelection=DS-645593INDICATORS,VALUE_IN_EUROS;DS-645593REPORT.

Eurostat (2022e). Inability to afford a meal with meat, chicken, fish (or vegetarian equivalent) every second day by level of activity limitation, sex and age, Available at: https://appsso.eurostat.ec.europa.eu/nui/show.do?dataset=hlth_dm030&lang=en (accessed 29 March 2022).

Eurostat (2022f). Internet purchases by individuals (until 2019), *Isoc_ec_ibuy*. Available at: https://appsso.eurostat.ec.europa.eu/nui/submitViewTableAction.do (accessed 30 March 2022).

Facchini, E., Iacovidou, E., Gronow, J. and Voulvoulis, N. (2018). Food flows in the United Kingdom: The potential of surplus food redistribution to reduce waste, *Journal of the Air and Waste Management Association* 68(9), 887–899.

Fagotto, E. (2014). Private roles in food safety provision: The law and economics of private food safety, *European Journal of Law and Economics* 37(1), 83–109.

Falguieres, M., Kumar, V., Garza-Reyes, J. A., Kumari, A., Lim, M. K. and Rocha-Lona, L. (2015). Investigating the impact of short food supply chain on emigration: A study of Valencia community in Spain, *IFAC-PapersOnLine* 48(3), 2226–2232.

Fernie, J., Sparks, L. and McKinnon, A. C. (2010). Retail logistics in the UK: Past, present and future, *International Journal of Retail and Distribution Management* 38(11/12), 894–914.

Food and Agriculture Organization. (2013). *Food Wastage Footprint: Impacts on Natural Resources*, Rome, Italy.

Forbes, H., Quested, T. and O'Connor, C. (2021). *Food Waste Index Report 2021*, UNEP, Nairobi. Available at: https://www.unep.org/resources/report/unep-food-waste-index-report-2021.

Fox, T. and Vorley, B. (2004). *Stakeholder Accountability in the UK Supermarket Sector Final, Report of the 'Race to the Top' Project*. Available at: www.racetothetop.org.

Fuchs, D., Kalfagianni, A. and Arentsen, M. (2009). Retail power, private standards, and sustainability in the global food system. In: Clapp, J. and Fuchs, D. (Eds). *Corporate Power in Global Agrifood Governance*, MIT Press, Cambridge, 29-59.

Fuchs, D., Kalfagianni, A. and Havinga, T. (2011). Actors in private food governance: The legitimacy of retail standards and multistakeholder initiatives with civil society participation, *Agriculture and Human Values* 28(3), 353-367.

Galante, N., López, E. G. and Monroe, S. (2013). *The Future of Online Grocery in Europe, Perspectives on Retail and Consumer Goods*, McKinsey & Company, London, UK

Garnett, P., Doherty, B. and Heron, T. (2020). Vulnerability of the United Kingdom's food supply chains exposed by COVID-19, *Nature Food* 1(6), 315-318.

Ge, X., Messinger, P. R. and Li, J. (2009). Influence of soldout products on consumer choice, *Journal of Retailing* 85(3), 274-287.

Gialos, A. and Zeimpekis, V. (2021). Defining and testing system parameters for enhancing vision picking technology in warehouse operations. In: Aktas, E., Bourlakis, M., Minis, I. and Zeimpekis, V. (Eds). *Supply Chain 4.0: Improving Supply Chains with Analytics and Industry 4.0 Technologies*, Kogan Page Limited, London, 113-149.

Gupta, S., Heng, X. and Sahu, V. (2009). Impact of store size on impulse purchase, *IUP Journal of Marketing Management* 8(1), 7-22.

Gustavsson, J., Christel, C. and Ulf, S. (2011). *Global Food Losses and Food Waste, Save Food Congress*. Available at: https://www.madr.ro/docs/ind-alimentara/risipa_alimentara/presentation_food_waste.pdf.

Hingley, M. K. (2005). Power imbalance in UK Agri-food supply channels: Learning to live with the supermarkets? *Journal of Marketing Management* 21(1-2), 63-88.

Hoffman, A. J. (2000). *Competitive Environmental Strategy. A Guide to the Changing Business Landscapes*, Island Press, Washington, DC.

Hübner, A., Hense, J. and Dethlefs, C. (2022). The revival of retail stores via omnichannel operations: A literature review and research framework, *European Journal of Operational Research* 302(3), 799-818. https://doi.org/10.1016/j.ejor.2021.12.021.

Hübner, A. H., Kuhn, H. and Sternbeck, M. G. (2013). Demand and supply chain planning in grocery retail: An operations planning framework, *International Journal of Retail and Distribution Management* 41(7), 512-530.

Jarzębowski, S., Bourlakis, M. and Bezat-Jarzębowska, A. (2020). Short food supply chains (SFSC) as local and sustainable systems, *Sustainability* 12(11), 1-13.

Jones, P., Clarke-Hill, C., Comfort, D., Hillier, D. and Shears, P. (2005). Radio frequency identification and food retailing in the UK, *British Food Journal* 107(6), 356-360.

Jones, P., Comfort, D. and Hillier, D. (2005). Corporate social responsibility and the UK's top ten retailers, *International Journal of Retail and Distribution Management* 33(12), 882-892.

Jones, P., Comfort, D. and Hillier, D. (2008). Moving towards sustainable food retailing? *International Journal of Retail and Distribution Management* 36(12), 995-1001.

Kamilaris, A., Fonts, A. and Prenafeta-Boldú, F. X. (2019). The rise of blockchain technology in agriculture and food supply chains, *Trends in Food Science and Technology* 91(July), 640-652.

Kotzab, H. and Teller, C. (2005). Development and empirical test of a grocery retail instore logistics model, *British Food Journal* 107(8), 594-605.

Kumar, M., Srai, J., Pattinson, L. and Gregory, M. (2013). Mapping of the UK food supply chains: Capturing trends and structural changes, *Journal of Advances in Management Research* 10(2), 299–326.

Ladd, B. (2020). A look at the micro-fulfillment model and the future of grocery retail, *Forbes*. Available at: https://www.forbes.com/sites/forbescommunicationscouncil/2020/10/05/a-look-at-the-micro-fulfillment-model-and-the-future-of-grocery-retail/?sh=5633d7e662bf (accessed 5 October).

Lin, C. F. (2014). Public-private interactions in global food safety governance, *Food and Drug Law Journal* 69(2), 143–160.

Liu, S. (2020). Food supply pressure in France and Germany during COVID-19: Causes from manufacturing, *Journal of Agriculture, Food Systems, and Community Development* 9(4), 1–4.

Løvdal, T., Droogenbroeck, B. V., Eroglu, E. C., Kaniszewski, S., Agati, G., Verheul, M. and Skipnes, D. (2019). Valorization of tomato surplus and waste fractions: A case study using Norway, Belgium, Poland, and Turkey as examples, *Foods* 8(7), 229.

Lyu, J., Ding, J.-H. and Chen, P.-S. (2010). Coordinating replenishment mechanisms in supply chain: From the collaborative supplier and store-level retailer perspective, *International Journal of Production Economics* 123(1), 221–234.

Maglaras, G., Bourlakis, M. and Fotopoulos, C. (2015). Power-imbalanced relationships in the dyadic food chain: An empirical investigation of retailers' commercial practices with suppliers, *Industrial Marketing Management* 48, 187–201.

Manikas, I. and Terry, L. A. (2009). A case study assessment of the operational performance of a multiple fresh produce distribution centre in the UK, *British Food Journal* 111(5), 421–435.

Marsden, T., Banks, J. and Bristow, G. (2000). Food supply chain approaches: Exploring their role in Rural Development, *Sociologia Ruralis* 40(4), 424–438.

McDonagh, P. and Commins, P. (1999). Food chains, small-scale food enterprises and rural development: Illustrations from Ireland, *International Planning Studies* 4(3), 349–371.

Melo, M. T., Nickel, S. and Saldanha-da-Gama, F. (2009). Facility location and supply chain management – a review, *European Journal of Operational Research* 196(2), 401–412.

Naiki, Y. (2014). The dynamics of private food safety standards: A case study on the regulatory diffusion of global G.A.P., *International and Comparative Law Quarterly* 63(1), 137–166.

Oglethorpe, D. and Heron, G. (2013). Testing the theory of constraints in UK local food supply chains, *International Journal of Operations and Production Management* 33(10), 1346–1367.

Özbekler, T. M. (2022). Insights from supply chain disruptions in the COVID-19 era. In: Lopes, A. P. (Ed). *Cases on Supply Chain Management and Lessons Learned From COVID-19*, IGI Global, Hershey, USA, 62–85.

Pérez-Mesa, J. C., Piedra-Muñoz, L., Galdeano-Gómez, E. and Giagnocavo, C. (2021). Management strategies and collaborative relationships for sustainability in the agrifood supply chain, *Sustainability* 13(2), 749.

PLMA. (2021). *Private Label Maintains Strong Position across Europe*, Private Label Member Association International Council, Amsterdam, April. Available at: https://www.plmainternational.com/industry-news/private-label-today.

Poore, J. and Nemecek, T. (2018). Reducing food's environmental impacts through producers and consumers, *Science* 360(6392), 987–992.

Rao, M., Bast, A. and de Boer, A. (2021). European private food safety standards in global agri-food supply chains: A systematic review, *International Food and Agribusiness Management Review* 24(5), 739–754.

Renting, H., Marsden, T. K. and Banks, J. (2003). Understanding alternative food networks: Exploring the role of short food supply chains in rural development, *Environment and Planning A* 35(3), 393–411.

Ritchie, H. (2022). *How Could the War in Ukraine Impact Global Food Supplies?* Available at: https://ourworldindata.org/ukraine-russia-food.

Ritchie, H. and Roser, M. (2020). Environmental impacts of food production, *OurworldInData.Org*. Available at: https://ourworldindata.org/environmental-impacts-of-food (accessed 25 March 2022).

Rogerson, M. and Parry, G. C. (2020). Blockchain: Case studies in food supply chain visibility, *Supply Chain Management: An International Journal*, 25(5), 601–614.

Rohm, H., Oostindjer, M., Aschemann-Witzel, J., Symmank, C., L Almli, V., de Hooge, I. E., Normann, A. and Karantininis, K. (2017). Consumers in a sustainable food supply chain (COSUS): Understanding consumer behavior to encourage food waste reduction, *Foods* 6(12), 104.

Rossi, R. (2020). *Protecting the EU Agri-Food Supply Chain in the Face of COVID-19, European Parliamentary Research Service*. Available at: https://www.europarl.europa.eu/RegData/etudes/BRIE/2020/649360/EPRS_BRI(2020)649360_EN.pdf.

Sabbe, P. (2021). Amazon wants hundreds of Go supermarkets in Europe, *Retail Detail*. Available at: https://www.retaildetail.eu/en/news/food/amazon-wants-hundreds-go-supermarkets-europe (accessed 18 March 2022).

Sandström, V., Valin, H., Krisztin, T., Havlík, P., Herrero, M. and Kastner, T. (2018). The role of trade in the greenhouse gas footprints of EU diets, *Global Food Security* 19(May), 48–55.

Savills. (2021). European food and groceries sector wider consumer trends. Available at: https://pdf.euro.savills.co.uk/european/europe-retail-markets/spotlight---european-food-and-groceries-sector---2021.pdf.

Sawyerr, E. A. and Harrison, C. (2019). Developing resilient supply chains: Lessons from high-reliability organisations, *Supply Chain Management: An International Journal*, 25(1), 77–100.

SDC. (2008). Green, healthy and fair: A review of government's role in supporting sustainable supermarket food. Available at: http://www.sd-commission.org.uk/data/files/publications/GreenHealthyAndFair.pdf.

Seuring, S. and Müller, M. (2008). From a literature review to a conceptual framework for sustainable supply chain management, *Journal of Cleaner Production* 16(15), 1699–1710.

Shahid, A., Almogren, A., Javaid, N., Al-Zahrani, F. A., Zuair, M. and Alam, M. (2020). Blockchain-based agri-food supply chain: A complete solution, IEEE Access (Vol. 8), 69230–69243.

Smith, B. G. (2008). Developing sustainable food supply chains, *Philosophical Transactions of the Royal Society B: Biological Sciences* 363(1492), 849–861.

Stefánsdóttir, B. (2020). *Food Waste Statistics for Iceland in 2019*, Department for Climate and Green Communities, Reykjavík, Iceland.

Stenmarck, Å., Jensen, C., Quested, T. and Moates, G. (2016). *Estimates of European Food Waste Levels, Fusions*, Stockholm. Available at: https://www.eu-fusions.org

/phocadownload/publications/estimatesofeuropeanfoodwastelevels.pdf%5cn; https://phys.org/news/2016-12-quarter-million-tonnes-food-logistics.html#nrlv.

Stuart, T. (2009). *Waste: Uncovering the Global Food Scandal*, Penguin Books Ltd., London, UK.

Tatum, M. (2022). How rapid delivery has led a space race in the UK's cities, *The Grocer*, 8 March. Available at: https://www.thegrocer.co.uk/property-and-planning/how-rapid-delivery-has-led-a-space-race-in-the-uks-cities/665263.article.

Thatte, A. A., Rao, S. S. and Ragu-Nathan, T. S. (2013). Impact of supply chain management practices of a firm on supply chain responsiveness and competitive advantage of a firm, *Journal of Applied Business Research* 29(2), 499–530.

Thomasson, E. (2019). Carrefour says blockchain tracking boosting sales of some products, Reuters, 3 June. Available at: https://www.reuters.com/article/us-carrefour-blockchain/carrefour-says-blockchain-tracking-boosting-sales-of-some-products-idUSKCN1T42A5?mc_cid=a41c15e70c&mc_eid=295ead2126.

Thompson, L.-J. and Lockie, S. (2013). Private standards, grower networks, and power in a food supply system, *Agriculture and Human Values* 30(3), 379–388.

Tian, F. (2016). An agri-food supply chain traceability system for China based on RFID & blockchain technology. In: *2016 13th International Conference on Service Systems and Service Management (ICSSSM)*, IEEE, Kunming, China, 1–6.

Tiseo, I. (2021). Annual household food waste produced in selected countries in Europe as of 2020. Available at: https://www.statista.com/statistics/1219815/food-waste-of-selected-countries-in-europe/.

United Nations. (2021). Goal 12: Ensure sustainable consumption and production patterns, *United Nations Sustainable Development Goals*. Available at: https://www.un.org/sustainabledevelopment/sustainable-consumption-production/ (accessed 9 July 2021).

UNSD. (2022). Standard country or area codes for statistical use. Available at: https://unstats.un.org/unsd/methodology/m49/.

Vadgama, N. and Tasca, P. (2021). An analysis of blockchain adoption in supply chains between 2010 and 2020, *Frontiers in Blockchain* 4(March), 1–14.

Validi, S., Bhattacharya, A. and Byrne, P. J. (2014). A case analysis of a sustainable food supply chain distribution system - A multi-objective approach, *International Journal of Production Economics* 152, 71–87.

van Gils, T., Ramaekers, K., Caris, A. and de Koster, R. B. M. (2018). Designing efficient order picking systems by combining planning problems: State-of-the-art classification and review, *European Journal of Operational Research* 267(1), 1–15.

van Rompaey, S. (2020). Belgian webshops get 25 % of turnover from neighbours, *Retail Detail*. Available at: https://www.retaildetail.eu/en/news/general/belgian-webshops-get-25-turnover-neighbours (accessed 18 March 2022).

Von Redaktion LZ. (2021). Top 50 food retail Europe 2021, *Lebensmittel Zeitung*, 26 August. Available at: https://www.lebensmittelzeitung.net/handel/rankings/ranking-top-50-lebensmittelhandel-europa-2021-161132?crefresh=1.

Voss, M. (2021). Collaborative robotics. In: Aktas, E., Bourlakis, M., Minis, I. and Zeimpekis, V. (Eds). *Supply Chain 4.0: Improving Supply Chains with Analytics and Industry 4.0 Technologies*, Kogan Page Limited, London, 150–170.

Walley, K. and Custance, P. (2010). Coopetition: Insights from the agri-food supply chain, *Journal on Chain and Network Science* 10(3), 185–192.

Win, T. L. (2020). Apps and blockchain help European supermarkets lure climate-conscious consumers, Reuters, 26 February. Available at: https://www.reuters.com/article/us-europe-food-climate-change/apps-and-blockchain-help-european-supermarkets-lure-climate-conscious-consumers-idUSKCN20K0VN%0Ahttps://www.reuters.com/article/us-europe-food-climate-change/apps-and-blockchain-help-european-s.

WRAP. (2021). Surplus food redistribution in the UK 2015–2020, Banbury. Available at: https://wrap.org.uk/sites/default/files/2021-06/WRAP-Surplus-food-redistribution-in-the-UK-2015-2020.pdf.

Yakovleva, N. (2007). Measuring the sustainability of the food supply chain: A case study of the UK, *Journal of Environmental Policy and Planning* 9(1), 75–100.

Chapter 4

Agri-food supply chains in Southern and Eastern Europe

F. J. Peña Rodríguez, University of Granada, Spain; C. Yacamán Ochoa, Autonomous University of Madrid, Spain; and A. Matarán Ruiz, University of Granada, Spain

1 Introduction

One of the most remarkable changes in modern societies has occurred in the agri-food sector. Since the middle of the last century, the way in which food is produced, distributed and consumed has been transformed. After the Second World War, industrailized countries prioritised increasing the volume and efficiency of agricultural production and largely supported the growth of a global market in food production and supply (Woods, 2005; Marsden and Symes, 1987). Food production has become an increasingly globalised activity characterised by greater intensification in production, increasing specialisation and complexity in supply chains and greater concentration of ownership (Woods, 2005).

One of the main features of this model is the predominance of longer, more complex food supply chains in which farmers as producers are separated from consumers by a series of intermediaries such as wholesalers, logistics companies, food manufacturers, packagers and retailers such as supermarkets, hypermarkets, self-service and discount shops. This separation has been accentuated by the shift to global sourcing of agricultural raw materials, enabled by the work of organisations such as the World Trade Organization (WTO) in breaking down trade barriers and liberalising trade between countries. This

http://dx.doi.org/10.19103/AS.2023.0122.06

neoliberal model has come under increasing criticism as allowing rich countries and a relatively small number of multinational companies in the Global North effectively to exploit cheap resources and labour in the Global South (Rosset and Martínez, 2014; Vivas Esteve, 2015; Yacamán Ochoa, 2022).

This transformation in agri-food supply chains has occurred later in Southern and Eastern Europe compared to other regions such as Northern Europe or North America, in part due to the political context in these countries. Southern European countries such as Spain faced a legacy of authoritarian regimes characterised by policies favouring autarchy, including economic policies based on economic self-sufficiency and state intervention. After World War II, Eastern European countries were part of a Communist bloc dominated by the Soviet Union which limited trade with Western countries and imposed a model dominated by state-owned trading companies operating under a system of centralised state control (Hruška et al., 2020).

It was not until these authoritarian regimes ended (e.g. with the death of General Franco in Spain in 1975 and the collapse of the Soviet Union in the early 1990s) that many Southern and Eastern European countries such as Spain and Poland started a new phase of economic development and integration with Western European and global markets. This can be seen in the later accession of these countries to the European Union (EU) which, among its core principles, promotes free trade within its member countries. Greece joined the European Economic Community in 1981, while Spain and Portugal joined in 1986. Former Eastern bloc countries such as Poland did not join the EU until as late as 2004. Another indicator is the later emergence of supermarkets, a characteristic feature of modern food supply chains, in some countries in these regions. Whilst the first supermarket in Spain (Caprabo) opened in 1959, they were not introduced in Poland until the 1990s (Rudawska and Bilinska-Reformat, 2018).

Within this context, this chapter reviews current trends and opportunities in agri-food supply chains in Southern and Eastern Europe, with a focus on Spain and Poland as examples.

2 Current agri-food supply chains in Southern and Eastern Europe

This section focuses on Spain and Poland as examples of wider trends in Southern and Eastern Europe. Despite a late start, countries such as Spain have rapidly adopted many of the characteristics of longer, complex food supply chains that typify many countries in Western Europe. Evidence suggests significant concentration in the current food retail sector in Spain (Medina and Oubiña, 1999; MAGRAMA, 2008; MERCASA, 2013). Figures from the Spanish Ministry of Agriculture now suggest that over 70% of food purchased by consumers comes from supermarkets and other large retailers (Ministry of Agriculture, Fisheries and Food, 2019). A recent analysis suggested the following breakdown (Savills, 2022):

- Supermarkets: 47%;
- Hypermarkets: 13%;
- Discount stores: 13%; and
- Online shopping: 2%.

According to this analysis, independent retailers accounted for only 18% of food sales.

Figures also show a high degree of consolidation within the supermarket sector itself. Figure 1 shows the market share of the main supermarket chains in Spain in 2021. It shows the Mercadona supermarket chain as the sector leader with double the market share of its main French competitor, Carrefour. Mercadona had a 28% market share with sales of €27 million in 2019. Over 80% of fresh produce in a major city like Cordoba, for example, is now supplied by Mercadona (Reina-Usuga et al., 2023).

Spanish consumers have become increasingly habituated to this new retail environment. While they understand the potential benefits of shorter food supply chains (in which produce is bought direct from local producers), consumers in countries such as Spain still place a high priority on price, quality, choice and convenience which are key benefits offered by supermarket-dominated long food chains (Chang et al., 2022). Even more environmentally conscious consumers, who might be expected to favour more traditional, local

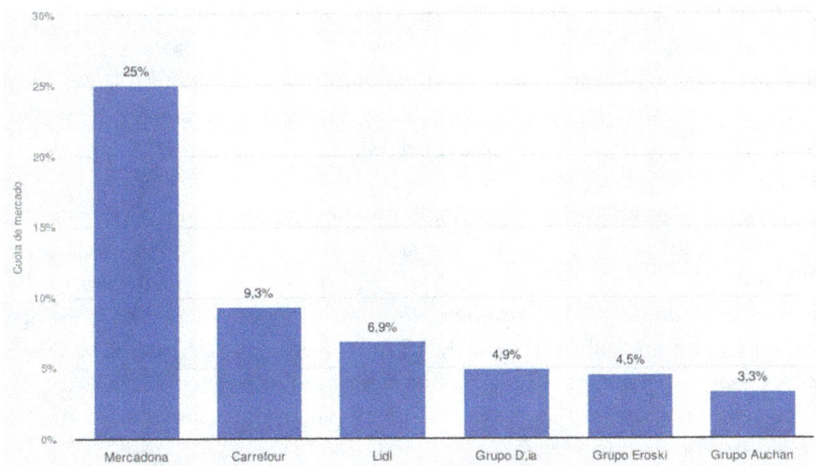

Figure 1 Market share of the main supermarket chains in Spain in 2021. Source: Adapted from: Statista (2021).

sources of food, have become more reliant on a retail environment dominated by supermarkets. Only 29% of consumers of organic foods purchase them in an independent, specialist outlet (Escobar-Lopez 2021). This picture contrasts with other Southern European countries such as Italy which retain a more fragmented and diversified retail sector where supermarkets (with an 18% share of grocery sales in 2021) and discount stores (a 5% share) co-exist with a strong local network of independent retail outlets (a 27% share) and traditional corner grocery stores (a 9% share), particularly in central and southern Italy (USDA, 2022b).

However, these figures on the share of food sales represented by supermarkets in Spain conceal a distinctive characteristic of the Spanish retail sector (that can also be seen in other Southern European countries such as Italy): the continuing strength of regional supermarket chains which compete with national chains such as Mercadona and have an estimated 25% share of the overall food retail market. These regional chains include Ahorramas and Sánchez Romero in the Madrid area, Consum in Valencia, BM in the Basque Country, Condis in Catalonia, Gadisa y Froiz in Galicia and Coviran and El Jamón in Andalusia (USDA, 2020).

Complementing trends in the retail sector, there is also evidence of growing concentration in agricultural production in Spain linked to increasing integration into global food supply chains. Whilst some countries such as Portugal and Greece import more than they export, others such as Spain have emerged as a major food exporter. In 2020, the Spanish agri-food and fisheries sector exported produce worth almost €54 million, 4.1% more than in 2019 (Ministry of Agriculture, Fisheries and Food, 2022), and 63% went to countries in the EU, mainly France, Germany and Italy. The remaining 37% went to other countries, mainly the United Kingdom, China and the United States.

The effects of this export-driven growth on the agricultural sector can be seen, for example, in Spanish pig production. Meat production is the largest sector in Spanish agriculture (accounting for 25% of overall sales) and is dominated by pig production for export (Giménez et al., 2021). This growth has been accompanied by increasing consolidation in pig farming. The number of pig farms is estimated to have fallen by over 66% between 1999 and 2009, equivalent to losing over 100 000 farms. A further 18 000 disappeared between 2009 and 2013. An estimated 65% of pig meat is now produced by large national or multinational companies, with farmer cooperatives accounting for 18% and independent farmers only 15% of production (Win, 2021). Looking more broadly, according to the Spanish farmers' organisation COAG, 7% of agricultural companies now own 42% of the value of fresh produce grown in Spain (http://www.coag.org/).

Turning to Eastern Europe, while some Eastern European countries within the EU are net importers of food (e.g. Bulgaria, the Czech Republic, Slovakia and Romania), others (e.g. Poland and Hungary) have, like Spain, become net exporters of food products (World International Trade Solution, 2019). In countries such as the Czech Republic or Hungary, the population growing at least some of their own food is between 40% and 48% while, in countries such as Germany, the Netherlands or the United Kingdom, this percentage is between 14% and 38%. This has sustained a tradition of food self-sufficiency in rural areas in countries such as the Czech Republic (Fonte, 2008; Spilková and Perlín, 2013; Vávra et al., 2018; Hruška et al., 2020).

In contrast, opening up of export opportunities after joining the EU has had a significant impact in transforming the agri-food sector in countries such as Poland (Firlej et al., 2017). After accession to the EU in 2004, Polish agri-food exports increased seven-fold between 2004 and 2021, reaching almost 38 billion€. Agri-food exports now represent 13% of overall exports by volume, making Poland the fifth largest food producer in the EU with a 10% share of overall food production and the largest food producer in Central and Eastern Europe (Pawlak and Smutka, 2022; USDA, 2022a).

With over 1 million farms, Poland retains a sizeable smallholder sector with over 50% of farms under 5 hectares in size. Many of these holdings only supply food for the families cultivating them with any surplus provided to relatives or neighbours or sold locally (Mincyte, 2011; Bilewicz and Śpiewak, 2019). However, the overall number of farms in the agricultural sector decreased by 190 000 (a 13% fall) between 2010 and 2021 and just under 9% of farms are now responsible for over 60% of overall agricultural production. Analyses of fruit and vegetable production in Poland suggest that, while is still dominated by small-scale producers, the trend is towards greater concentration of production on fewer, larger farms. Since the 1990s, multinational companies such as Heinz, Danone, Unilever and Nestle have also invested heavily in food production in Poland, now controlling up to 80% of production in some sectors (Figurska et al., 2023). As a result, a smaller number of large food processors and retailers are becoming more dominant and are able to impose strict quality requirements on their suppliers (Klepacki, 2019; Klepacki et al., 2020).

There has been a similar trend towards concentration in the retail sector. With market liberalisation in the 1990s, the number of supermarkets quadrupled between 1993 and 2005 through privatisation also prompted a rapid initial increase in small, independent shops (Hermann et al., 2009). It has been estimated that the 20 largest grocery retailers in Poland now have a combined market share of 75% of grocery sales (https://www.lloydsbanktrade.com/en/market-potential/poland/distribution). Of the top ten food retailers, five are foreign owned with Biedronka (Portugal) and Lidl (Germany) as the two leading retailers (USDA, 2022a). Research suggests that Polish consumers were initially attracted by the novelty of supermarkets and their association with more affluent, Western European lifestyles as well as a demand for greater choice, including for imported food products (Spilková et al., 2013; Bilewicz and Śpiewak, 2019; USDA, 2022a). These developments have been complemented by significant improvements in logistics in areas such as refrigerated transport and storage to create more efficient and flexible supply chains (Klepacki, 2018).

It has been estimated that the number of independent, local food stores has continued to decline by 5% year, though independent (often family-owned and operated) shops and local markets (typically supplied from neighbouring farms) remain common in many areas (Hruška et al., 2020; USDA, 2022a). This characteristic has been seen as contributing to greater resilience in maintaining food supplies during the coronavirus disease 2019 (COVID-19) pandemic as well as providing a foundation for promoting urban agriculture (Spilková and Perlín, 2013; Rudawska and Bilinska-Reformat, 2018; Skaniska and Zaluski, 2021). Research suggests both the resilience of local food cultures and a more recent shift in consumer attitudes back towards favouring local food supply chains as more ethical and environmentally sustainable (Gorlach et al., 2006; Horska et al., 2011; Spilková et al., 2013). Recent research suggests that, although they may have limited awareness of alternative, localised means of food supply such as farmers' markets, urban consumers in Poland do value local food produce, suggesting a potential demand to be tapped (USDA, 2022a; Solarz et al., 2023). These trends suggest that the more recent integration of Eastern European countries like Poland into the global economy means that regional agri-food chains have undergone less transformation than, e.g., those in countries such as Spain. This can be seen, e.g. for example, in the case of countries like Romania where there remains a strong tradition of local food production and consumption, and where the agricultural, food production, logistics and retail sectors remain relatively fragmented (USDA, 2018).

3 Transforming agri-food supply chains

There is now growing awareness of the problems associated with modern long agri-food supply chains and the importance of developing alternative, shorter food networks (AFNs) (Yacamán Ochoa et al., 2020). An alternative vision is that food systems should not be designed to fit a paradigm prioritising growth and based on extraction, land grabbing and profit maximisation and controlled by a small number of multinational corporations. Instead they should be designed according to principles of cooperation, sufficiency, justice, equity, regeneration and understanding food as a common right (González de Molina and López García, 2021; McGreevey et al., 2022).

An example of this shift in values can be seen in changing policy in the EU. Although EU policies such as the Common Agricultural Policy (CAP) have been criticised as continuing to favour the status quo, in 2019 the EU adopted the 'Farm to Fork' strategy in line with the European Green Deal trying to superficially integrate these values. This strategy is a non-binding roadmap to promote the transformation of the agri-food system. A key aspect is the need to increase the resilience of local and regional agri-food systems by increasing support for shorter, localised supply chains based on family farming using regenerative (e.g. organic) practices and selling to local consumers.

In line with the EU 'Farm to Fork' strategy, some cities have made significant progress in fulfilling the Milan Urban Food Policy Pact, an agreement committed to developing local urban food systems that are sustainable, inclusive, resilient, safe, climate friendly and that provide healthy and affordable food for all (https://www.milanurbanfoodpolicypact.org/). The Pact sets out over 30 initiatives in six categories that cities can adopt to achieve these goals: governance, sustainable diets and nutrition, social and economic equity, food production, food supply and distribution, and food waste. In Spain, the Pact has been adopted by 16 municipalities (Barcelona, Bilbao, Cadiz, Denia, Fuenlabrada, Godella, Granollers, Madrid, Malaga, Rivas Vaciamadrid, San Sebastian, Seville, Valencia, Valladolid, Vitoria-Gastiez and Zaragoza). The adoption of the Pact in Madrid and Valencia is, e.g., reviewed in Martin and del Fuente (2022).

As this suggests, the drive to develop alternative food systems is not just based on grounds of economic fairness and social justice. It is also driven by concerns about the sustainability of global agri-food supply chains which are seen as both contributing and vulnerable to a range of political and environmental shocks (Fig. 2). These include the reliance of such systems on steadily dwindling sources of energy and raw materials and the contribution of energy and input-intensive methods of production and distribution to resource depletion and climate change which, in turn, increases supply chain vulnerabilities (Ellen MacArthur Foundation, 2019). There has also been a

Figure2 Reasons to strengthen short food supply chains. Source: Authors.

focus on the high levels of food losses and waste (30–40% by some estimates) associated with modern food production and distribution (Jurgilevich et al., 2016).

This has led to a focus on designing new, more 'circular' systems of food production and distribution based on more flexible and responsive local food supply chains requiring fewer inputs and with less waste, with a focus on more sustainable methods of production, closing nutrient loops and recycling waste products back into the system (Ellen MacArthur Foundation, 2015). This requires improving cooperation between producers within agriculture, between producers in different sectors and between producers and local consumers (Fanfani and Matarán Ruiz, 2020). Examples range from using renewable energy sources and biodegradable packaging to cooperative food hubs to ensure efficient waste-free food distribution and the processing of urban waste into compost for growing crops.

4 Defining and developing short food supply chains

The EU Rural Development Regulation 1305/2013 defines short food supply chains (SFSCs) as 'a supply chain involving a limited number of economic operators, committed to cooperation, local economic development, and close geographical and social relations between food producers, processors, and consumers'. This recognises the potential role of SFSCs in re-localising agricultural production (e.g. by supporting local farmers) and, more broadly, in improving social cohesion at a local-regional scale. The definition of the EIP-Agri Focus Group (final report 2015) defines SFSCs as a means to 'restructure food chains in order to support sustainable and healthy farming methods, generate resilient farm-based livelihoods (in rural, peri-urban and urban areas) and re-localise control of food economies'.

SFSCs favour relations of proximity and trust between local producers, retailers, restaurants and consumers with the aim of ensuring food safety, food sovereignty, environmental sustainability, social capital and economic efficiency (Aubry and Kebir, 2013). Among other objectives, SFSCs also support recovery of economic, social and geographical links between food products and rural knowledge, local varieties, local culture, as well as the agroecological and climatic conditions of bioregions (Fanfani and Matarán Ruiz, 2020). A comprehensive definition of SFSCs highlights not just economic and social impacts but also addresses environmental and ethical problems from a bioregional perspective, in order that they become a real strategy for achieving truly resilient food systems.

However, despite the potential benefits of developing SFSCs, there are formidable barriers to implementing them in practice. These are summarised in Table 1.

Table 1 Short food supply chains: barriers and opportunities

Barriers
Atomisation of the primary sector
Predominance of conventional extensive crops
Low production capacity of farms with organic certification
No organisation of production between holdings
Lack of crop planning according to season and seasonal demand
Low economic capacity of farmers to address farm improvements
High fixed costs and intermediate consumption per production
Poor efficiency in direct distribution services to retail stores
Low adaptation to local market demands and proximity
Food supply predominates through large shopping malls
Scarce direct marketing between local producers and agro-food industries
Lack of specialisation in short food supply chains
Little commercial cooperation in the agricultural sector
Lack of public policies to support short food supply chains and small-scale production
Little diversification of marketing channels
Lack of innovation and added value of products
Strong competition from conventional food systems
Low level of associationism and individualistic culture
Lack of logistical and appropriate infrastructure for the empowerment of the local economy
Strong sectorisation of public policies
Difficulties in cooperation between the parties and the public and private sector
Opportunities
Existence of local varieties
Appearance of new farms with value-added production techniques
Cultivation of high-quality vegetables
Growing interest in healthy eating, impact on health
Awareness of the environmental and climatic implications of the traditional crop and its inputs
Existence of a recent associationism in the producing sector
Growing interest among the different agents in the chain to organise horizontally to improve the scalability of supply
Interest of restaurants in local food and gastronomic tradition
Existence of agromarkets of short food supply chains
Opportunity of collective catering for incorporation into short production channels
Initiatives to promote food excellence, local traditions and culture among consumers on a regular basis
Collaboration in research and technology transfer with universities and research centres
Consolidation of consumer associations and organisations for short food supply chains

5 Developing short food supply chains: the example of Spain

There are many different types of SFSC in regions such as Southern and Eastern Europe, but they share a common characteristic of reduced numbers of intermediaries between producer and consumer (Marsden et al., 2000; Renting et al., 2003). The JRC Scientific and policy report on SFSCs and local food systems in the EU (Kneafsey et al., 2013) identifies three main categories of SFSCs, all of which involve some form of direct 'connection' between the food consumer and producer:

- Face to face;
- Spatial proximity; and
- Spatially extended.

In a 'face-to-face' supply chain, the consumer purchases a product directly from the producer/processor. The best-known examples are farmers' markets, farm shops, roadside sales, community-supported agriculture and box schemes where consumers buy selections of produce from local producers.

Spatial proximity refers to a situation when the products are produced and retailed in a specific region of production, and consumers are made aware of the 'local nature' of production. Some examples are traditional, local markets, public procurement from local family farms, farmers' markets and initiatives that connect producers with restaurants, hotels and grocery stores within a locality. It may also include examples of cooperative supermarkets retailing locally sourced foods.

Spatially extended types of SFSC involve communicating information about the regional origin and process of production of a product to consumers who are outside the region of production and who may have no personal experience of the region or with a particular producer. The main product information is communicated by branding. The best-known examples are protection of designated origin (PDO) and protected geographical indication (PGI). In this context, the connection between consumer and product is the association of the product with local production.

It is important to highlight that not all these types of SFSCs have the same positive impact on objectives such as adding value to local products, re-connection of local food production with local markets, reducing waste or carbon emissions, preservation of the cultural landscape, improving food security, strengthening the local economy etc. (Yacamán Ochoa et al., 2020). Face-to-face and spatial proximity SFSCs show a greater capacity for improving the overall sustainability of local food systems. They play an important role in putting decision-making back in the hands of local producers and consumers in terms of quality, methods of production, prices etc. (Yacamán Ochoa et al., 2019; Yacamán Ochoa et al., 2020). Spatially extended food systems do not have to be retailed locally - they can be exported - though they can offer greater opportunities for local producers to benefit from bigger markets and potentially better returns (Kneafsey et al., 2013).

Examples of organisations promoting SFSCs in Spain include:

- production/consumption groups;
- producer-consumer groups;
- producer associations;
- local shops; and
- farmers' markets.

Production/consumption groups can be defined as self-managed collectives where there is a direct relationship between those who produce and those who consume who, in some cases, are also involved in production. These groups may be groups of local families who pool and share what each family produces. These groups tend to be very informal with little infrastructure (e.g. sometimes only with weighing scales to allow measuring and pricing or exchange of agreed quantities of food). These groups are widely scattered across Spain and other Southern European countries (Yacamán Ochoa et al., 2019). There are many examples in Spain, some of which have been operating for decades (Vicente-Almazán Castro et al., 2019).

Producer-consumer associations link producers and consumers, for example, through a local shop which may offer a discount to consumer members regularly buying local produce. There are examples in most southern European countries (Matarán Ruiz, 2013). There are many across the whole of Spain with some in operation for over 30 years (Vicente-Almazán Castro et al., 2019; Matarán Ruiz et al., 2021).

Associations of producers are groups of producers (e.g. farmer cooperatives) who have generally organised themselves to improve their capacity to market their products, although they may also use these organisations for other purposes such as the exchange of seeds, collective discounted purchase of inputs, development of training activities or lobbying of official bodies for funding or more supportive government policies. Producer cooperatives can be either local or national and are typically product specific (e.g. wine and olive oil in Spain). Though this is not always the case, a number of producer cooperatives have been specifically established by smaller, local producers to give them a stronger collective voice. Some may be involved in retail operations to help market and sell members' produce. Spain has a well-established network of both national organisations (Vicente-Almazán Castro et al., 2019) and regional groups (Matarán Ruiz et al., 2021).

Although, as this chapter has shown, national supermarket chains have become a major force in food retailing, Spain retains a continuing network of regional supermarket chains and independent local food retailers, particularly in rural areas. These have survived because of factors such as long-established relationships with local customers and a reputation for sourcing high-quality, sustainably produced local produce (Matarán Ruiz et al., 2021). The continuing strength of these local networks is reflected in the fact that Spain was seen to experience fewer stock shortages during the COVID-19 pandemic compared to countries with more centralised supply chains characterised by low inventories and just in time production and supply. During early 2020, out of stock levels in Spain were estimated to be 7% compared to 12% in countries such as Italy and Belgium and 18% in France (Hobbs, 2020). To quote one study: The fact that regional chains gained so much ground in Spain during the pandemic is a unique phenomenon in the European region. Most of these chains share strategic characteristics: proximity, fresh produce and a commitment to providing local products (USDA, 2020).

Whilst supermarkets have responded to the greater interest of consumers in knowing the source of the food they purchase and how it has produced by stocking more organic produce, this shift in consumer attitudes has helped to sustain the independent food retail sector (Vicente-Almazán Castro et al., 2019; Matarán Ruiz et al., 2021). There is an increasing recognition of the economic and socio-cultural value of local food markets to the communities they serve (e.g. in terms of supporting local producers, boosting local economies, promoting tourism as well as a broader sense of community identity and cohesion), which has led to greater efforts (e.g. by local government agencies) to promote and support them (Crespi-Vallbona and Dimistrovski, 2017).

In many ways, farmers' markets function like a traditional street market but are typically organised by or on behalf of local producers and craftspeople to promote and sell their produce to local consumers. The north and east of Spain have always had traditional markets for local products that could be considered as farmer's markets. In the south and in larger metropolitan areas where supermarkets are particularly well established, farmers' markets have had to be created. There is now a thriving and growing network of these markets (De la Cruz et al., 2017; Vicente-Almazán Castro et al., 2019; Matarán Ruiz et al., 2021). The COVID-19 pandemic has also been seen to strengthen Spanish consumer preferences for locally produced food ('food patriotism') (USDA, 2020). This can be compared with other Southern European countries such as Italy which has retained a strong local food culture in many regions with a continued preference for traditional (often regional) dishes cooked with local ingredients (USDA, 2022b).

6 Conclusion

As has been noted, the countries of Southern and Eastern Europe shared a late entry into the globalised food system. Characteristics of modern agri-food supply chains such as concentration and specialisation of production, and the development of production for export, have marked the evolution of production systems in these regions. They are also places where the large distribution and retail chains that dominate the European food system have developed rapidly. All this has brought with it sustainability challenges in terms of the environmental impacts of intensive, monocultural production systems and complex, globalised distribution chains that have dramatically increased the distances travelled by food to reach the places where it is finally consumed.

Regions such as the EU have begun to recognise the need for change, e.g., through its 'From farm to fork' strategy which aims to promote SFSCs on the continent. The recovery of local food systems and the development of new types of SFSCs (e.g. by developing urban agricultural networks in and around cities) is seen as a key way to avoid the social, economic and environmental impacts of a globalised food system (Fanfani and Matarán Ruiz, 2020; Blay-Palmer et al., 2021).

This chapter has shown both the degree to which more traditional local production and supply networks have persisted in countries such as Spain and Poland (as well as other countries in Southern and Eastern Europe), as well as ways in which they can be extended or new short supply chain models developed to better serve local producers and consumers. There has been moves to rebuild local systems for distribution, marketing and consumption, recovering knowledge and infrastructures that existed before the 'conventionalisation' of the agri-food system (Yacamán Ochoa et al., 2019; Yacamán Ochoa et al., 2020; Matarán Ruiz et al., 2021).

Food hubs point to the 'missing middle' of the local food infrastructure, specifically they put this concept into the context of the debate about the adequacy of the local food infrastructure to innovate and scale-up SFSCs (Palacios-Argüello et al., 2017). The review of existing cases shows that food hubs can generate benefits such as:

- increasing access to markets for local and regional producers and protecting more sustainable, small-scale local food production systems; and
- potentially developing more responsive local markets that favour food production of sufficient quality (e.g. fresh, locally sourced produce) adjusted to the needs of local communities with reduced waste.

However, it is important to note that SFSCs are not necessarily a panacea, partly since transport can account for a relatively small proportion of overall emissions resulting from agri-food chain operations. It has also been found that SFSCs are not always more environmentally friendly than long food chains which are often run with very high levels of efficiency (Loiseau et al., 2020; see also: https://ourworldindata.org/food-choice-vs-eating-local). However, considering the need to re-localise our food systems and reduce the use of fossil fuels (Riechmann, 2013), there is a widespread movement to scale up SFSCs and make them much more efficient.

Whilst digital technologies have played a major role in enabling the development of ever more complex, long agri-food supply chains, they have also begun to open up new opportunities for growers of fresh produce to (re) connect directly to consumers. A recent development in Spain is the emergence of so-called 'hybrid' supply chains in which fresh fruit and vegetable producers (e.g. farm cooperatives involved in the export market) have begun to create outlets to sell direct to Spanish consumers, e.g. via online sales of boxes of fresh fruit and vegetables such as tomatoes (Navarro-del Aguila and de Burgos-Jiménez, 2022). As yet, this remains a small sector, with direct-to-consumer channels accounting for only 10.9% of the total food market (Ministerio de Agricultura, Pesca y Alimentación, 2022). In the case of Spanish organic produce, one study suggests that only 14% of sales of organic products come from direct-to-consumer channels (EcoLogical, 2018). This sector is even less developed in Eastern Europe (Bilewicz and Śpiewak, 2019).

Another solution to improving food supply chains lies in developments in eCommerce. Although online shopping only represents 2% of overall food sales in Spain, it has been boosted by the pandemic which restricted consumer access to physical stores, with 2019 seeing a 26% growth in the number of online transactions for food products (Savills, 2020). The national chain Mercadona has invested significantly in a network of local warehouses (known as 'colmenas' (hives)) specifically to support online shopping in cities such as Barcelona and Madrid, increasing online sales in these metropolitan areas to 5% (in line with other major European cities) (USDA, 2020). This is part of an 'omnichannel' approach developed by the larger supermarkets in particular to provide a range of channels for consumers, including services such as home delivery and 'click and collect' services. From a modest base, online shopping and other innovation in eCommerce are growing steadily in countries such as Italy, Poland and Romania (USDA, 2018, 2022a,b), Such developments are interesting for the extension of SFSCs and regional chains related to local food. However, eCommerce also demonstrates the range of potential solutions to explore in improving existing agri-food supply chains.

7 Where to look for further information

This section provides some references to further explore trends and opportunities in Europe's food supply chains:

- https://agriculture.ec.europa.eu/farming/organic-farming/organic-action-plan_en.

Action plan for organic production in the European Union:

- Agroecology Europe: https://www.agroecology-europe.org.
- Common Agricultural Policy (CAP) 2023-2027: https://agriculture.ec.europa.eu/common-agricultural-policy/cap-overview/cap-2023-27_es.
- European Commission. Farm to Fork strategy: https://food.ec.europa.eu/horizontal-topics/farm-fork-strategy_en.
- European Coordination Via Campesina: https://www.eurovia.org/es/.
- Food and Agriculture Organisation of the United Nations (FAO) database on agri-food system: https://www.fao.org/statistics/data-dissemination/agrifood-systems/en.

8 References

Aubry, C. and Kebir, L. (2013). Shortening food supply chains: a means for maintaining agriculture close to urban areas? The case of the French metropolitan area of Paris. Food Policy 41, 85–93. https://doi.org/10.1016/j.foodpol.2013.04.006.

Bilewicz, A. and Śpiewak, R. (2019). Beyond the "northern" and "southern" divide: Food and space in Polish consumer cooperatives. *East European Politics and Societies: And Cultures* 33(3), 579–602.

Blay-Palmer, A., Santini, G., Halliday, J., Malec, R., Carey, J., Keller, L., Ni, J., Taguchi, M. and van Veenhuizen, R. (2021). City region food systems: Building resilience to COVID-19 and other shocks. *Sustainability* 13(3), 1325. https://doi.org/10.3390/su13031325.

Chang, B. P. I., Massri, C., Reipurth, M., Petropoulou, E., Hüttl-Maack, V., Gawlik, D., Kujáni, K., Szente, V., Hegyi, A., Fricz, Á. S., Santa Cruz, E., Benos, T., Aouinaït, C., Campos, D., Alfaro, B., Jansseni, F., Theodorakopoulou, I., Iliopoulos, C. and Hieke, S. (2022). Barriers and facilitators to purchasing from short food chains: Evidence from consumer focus groups in Germany, Spain, Greece and Hungary. *International Journal of Food Studies* 11(2), SI208–SI218.

Crespi-Vallbona, M. and Dimistrovski, D. (2017). Food markets from a local dimension – la Boqueria (Barcelona), Spain. *Cities* 70, 32–39.

De La Cruz, C., Matarán Ruiz, A., Hung, J. and Knudson, E. (2017). Relaciones entre productores y consumidores. Confianza integral en la agroecología a través de canales cortos de comercialización. *Leisa. Revista de Agroecología* 33(4), 8–11. Ed. Asociación Ecología, Tecnología y Cultura en los Andes.

EcoLogical. (2018). El Sector Ecológico en España 2018. Available online: www.ecological.bio/es/sectorbio2018.

Ellen MacArthur Foundation. (2015). Growth within: A circular economy vision for a competitive Europe. Available Online: http://www.ellenmacarthurfoundation.org/assets/downloads/publications/EllenMacArthurFoundation_Growth-Within_July15.pdf.

Ellen MacArthur Foundation. (2019). Cities and circular economy for food. Available online: Cities and circular economy for food | Shared by Business (thirdlight.com).

Escobar-Lopez, S. Y., Amaya-Corchuelo, S. and Espinoza-Ortega, A. (2021). Alternative food networks: Perceptions of short food supply chains in Spain. *Sustainability* 13(5), 2578.

Fanfani, D. and Matarán Ruiz, A. (2020). *Bioregional Planning and Design (Volume I)*. Gewerbestrasse, Springer Nature Switzerland AG.

Figurska, J., Peterson, A. and Galang, L. (2023). Food processing sector report: Poland. Report No. PL2022-0028. United States Department of Agriculture -. Foreign Agricultural Service (USDA-. FAS). Available online: https://www.fas.usda.gov/data/poland-food-processing-sector-report-0.

Firlej, K., Kowalska, A. and Piwowar, A. (2017). Competitiveness and innovation of the Polish food industry. *Agricultural Economy Czech* 63, 1–9.

Fonte, M. (2008). Knowledge, food and place. A way of producing, a way of knowing. *Sociologia Ruralis* 48(3), 200–222.

Giménez, R., Espejo, C., García, R. and Ruiz, V. (2021). The pig sector in Spain: Characterization, production, trade and environmental problems. *Terra* 8, 194–230.

González De Molina, M. and Lopez-Garcia, D. (2021). Principles for designing agroecology-based local (territorial) agri-food systems: A critical revision. *Agroecology and Sustainable Food Systems* 45(7), 1050–1082.

Gorlach, K., Starosta, P., Pilichowski, A., Adamski, T. and Dzwonkowska, K. (2006). Local food production and knowledge dynamics in rural sustainable development in Poland. In: Kucerova, E., et al. (Eds.), *Local Food Production, Non-agricultural Economies and Knowledge Dynamics in Rural Sustainable Development*. Political Science Institute of the Hungarian Academy of Sciences. Budapest, Országház.

Hermann, P., Moser, A. and Weber, S. (2009). Grocery retailing in Poland: Structural changes and foreign direct investment. Discussion Document 40. Centre for International Development and Environmental Research (ZEU). Germany: Justus Liebig University Giessen.

Hobbs, J. E. (2020). Food supply chains during the COVID-19 pandemic. *Canadian Journal of Agricultural Economics* 68(2), 171–176.

Horska, E., Ürgeová, J. and Prokeinova, R. (2011). Consumers' food choice and quality perception: Comparative analysis of selected Central European countries. *Agricultural Economics* 57(10), 493–499.

Hruška, V., Konečný, O., Smutná, Z. and Duží, B. (2020). Evolution of alternative food networks in an old industrial region of Czechia. *Erdkunde*, 143–159. https://doi.org/10.3112/erdkunde.2020.02.04.

Jurgilevich, A., Birge, T., Kentala-Lehtonen, J., Korhonen-Kurki, K., Pietikäinen, J., Saikku, L. and Schösler, H. (2016). Transition towards circular economy in the food system. *Sustainability* 8(1), 69.

Kneafsey, M., Venn, L., Schmutz, U., Balázs, B.,Trenchard, L., Eyden-Wood, T., Bos, E., Sutton, G. and Blackett, M. (2013). Short food supply chains and local food systems in the EU (EUR 25911 EN; JRC Scientific and Policy Reports). https://doi.org/10.2791/88784.

Klepacki, B. (2018). The role of logistics for Polish economic development. In: Proceedings of the 2018 International Scientific Conference on Economic Services for Agribusiness and Rural Economy, Warsaw, Poland.

Klepacki, B. (2019). Organization of supply chains in the fruit sector in Poland. *Annals of the PAAAE* XXI(3), 191-199.

Klepacki, B. and Perkowska, A. (2020). Organization of food supply chains in dispersed production: The example of the vegetable sector in Poland. *Annals PAAAE* XXII(1), 149-158.

Loiseau, E., Colin, M., Alaphilippe, A., Coste, G. and Roux, P. (2020). To what extent are short food supply chains (SFSCs) environmentally friendly? Application to French apple distribution using life cycle assessment. *Journal of Cleaner Production* 276, 124166.

MAGRAMA. (2008). La distribución agroalimentaria y transformaciones estratégicas en la cadena de valor. Ministerio de Agricultura, Pesca y Alimentación. Available online: https://www.mapa.gob.es/es/alimentacion/temas/consumotendencias/cadena_valor_tcm30-89259.pdf.

Marsden, T. and Symes, D. (1987). Survival strategies in post-productionist agriculture. European Coordination Centre for Research and Documentation in Social Sciences.

Marsden, T., Banks, J., and Bristow, G. (2000). Food supply chain approaches: exploring their role in rural development. Sociologia ruralis, 40(4), 424–438.

Martin, D. and del Fuente, R. (2022). Global and local food agendas: The Milan Urban Food Policy Pact and innovative sustainable food policies in Euro-Latin American cities. *Land* 11(2), 202.

Matarán Ruiz, A., López Medina, J. M., Fuentes-Guerra Soldevilla, R., de la Cruz, C. y Soler Montiel, M. (2021). Diagnóstico de las redes agroecológicas locales de Andalucía y diseño participativo de propuestas de consolidación. Agencia de Gestión Agraria y Pesquera de la Junta de Andalucía, Universidad de Granada, Hábitat Cuatro S.C.A.

Matarán Ruiz, A. (2013). Social participation in an active protection of periurban agrarian spaces: State of the question. *Boletín de la Asociación Española de Geografía* N.° 63, 57-79.

McGreevy, S. R., Rupprecht, C. D., Niles, D., Wiek, A., Carolan, M., Kallis, G. and Tachikawa, M. (2022). Sustainable agrifood systems for a post-growth world. *Nature Sustainability* 5 (12), 1-7.

Medina, O. and Oubiña, J. (1999). Análisis de las cuentas de resultados de las empresas minoristas de bienes de gran consumo. *Distribución y Consumo* 49, 27-33.

Mercasa. (2013). Alimentación en España 2013.Producción, industria, distribución, consumo. Available online: https://www.mercasa.es/wp content/uploads/2022/03/MERCASA2013_BAJA.pdf.

Mincyte, D. (2011). Subsistence and sustainability in post-industrial Europe: The politics of small-scale farming in Europeanising Lithuania. *Sociologia Ruralis* 51(2), 101-118.

Ministerio de Agricultura, Pesca y Alimentación. (2022). Informe interactivo del comercio exterior (Accessed August 10, 2022). Available online: https://www.mapa.gob.es/es/ministerio/servicios/analisis-y-prospectiva/powerbi-comex.aspx.

Ministry of Agriculture, Fisheries and Food. (2019). Informe del consumo alimentario en España 2019. Available online: https://servicio.mapa.gob.es/en/alimentacion/temas/c onsumotendencias/informe2019_v2_tcm38-540250.pdf.

Ministry of Agriculture, Fishiries and Food. (2022). Informe interactivo del comercio exterior. Available online: https://www.mapa.gob.es/es/ministerio/servicios/analisis-y-prospectiva/powerbi-comex.aspx.

Navarro-del Aguila, I. and de Burgos-Jiménez, J. (2022). Sustainable innovation through developing hybrid agri-food supply chains: A case in South-East Spain. *Toxics* 10(12), 752.

Palacios-Argüello, J., Morganti, E. and Gonzalez-Feliu, L. (2017). Food hub: Una alternativa para alimentar las ciudades de manera sostenible. *Revista Transporte y Territorio* 17, 10–33.

Pawlak, K. and Smutka, L. (2022). Does Poland's agri-food industry gain competitive advantage in trade with non-EU countries? *PLoS ONE* 17(9), e0274692.

Reina-Usuga, L., Parra-López, C., de Haro-Giménez, T. and Carmona-Torres, C. (2023). Sustainability of territorial short food supply chains versus large-scale food distribution. *Land Use Policy* 126, 106529.

Renting, H., Marsden, T. K., and Banks, J. (2003). Understanding alternative food networks: exploring the role of short food supply chains in rural development. Environment and planning A, 35(3), 393–411.

Riechmann, J. (2013). *El siglo de la Gran Prueba*. Ed. Baile del Sol. Madrid.

Rosset, P. and Martínez, M. E. (2014). Soberanía alimentaria: Reclamo mundial del movimiento campesino. *Ecofronteras* 18 (51), 8–11.

Rudawska, E. D. and Bilinska-Reformat, K. (2018). The development of food retail formats evidence from Poland. *British Food Journal* 120(2), 309–324.

Savills. (2020). Spanish Food Sector. Savill's Spotlight Research. Available online: www.savills-aguirrenewman.com.

Savills. (2022). Food Distribution in Spain. Savills Spotlight Research. Available online: www.savills-aguirrenewman.com.

Skaniska, E. and Zaluski, R. I. (2021). Activities of food retail companies in Poland during the COVID-19 Pandemic. *Sustainability* 13(13), 7323.

Solarz, K., Raftowicz, M., Kachniarz, M. and Dradrach, A. (2023). Back to locality. Demand potential analysis for short food supply chains. *International Journal of Environmental Research and Public Health* 20(4), 3641.

Spilková, J. and Perlín, R. (2013). Farmers' markets in Czechia: Risks and possibilities. *Journal of Rural Studies* 32, 220–229.

Spilková, J., Fendrychová, L. and Syrovátková, M. (2013). Farmers' markets in Prague: A new challenge within the urban shoppingscape. *Agriculture and Human Values* 30(2), 179–191.

USDA. (2018). Romania: Retail food sector. United States Department of Agriculture (USDA) – Foreign Agricultural Service. Global Agricultural Network (GAIN) Report RO1809.

USDA. (2020). Spanish retail and distribution sector update. United States Department of Agriculture (USDA) – Foreign Agricultural Service. Global Agricultural Network (GAIN) Report SP020-0020.

USDA. (2022a). Poland: Retail foods guide. United States Department of Agriculture (USDA) – Foreign Agricultural Service. Global Agricultural Network (GAIN) Report PL2020-0014.

USDA. (2022b). Italy: Retail foods. United States Department of Agriculture (USDA) – Foreign Agricultural Service. Global Agricultural Network (GAIN) Report IT2022-0012.

Vávra, J., Megyesi, B., Duží, B., Craig, T., Klufová, R., Lapka, M. and Cudlínová, E. (2018). Food self-provisioning in Europe: An exploration of sociodemographic factors in five regions. *Rural Sociology* 83(2), 431–461.

Vicente-Almazán Castro, L., Herrera Gil, M. and Escobar Cruz, M. (2019). Sistemas alimentarios territorializados en España 100 iniciativas locales para una alimentación responsable y sostenible. CERAI.

Vivas Esteve, E. (2015). El negocio de la comida: ¿Quién controla nuestra alimentación? Icaria.

Win, T. L. (2021). How clean are Europe's supply chains? The myths fuelling the massive growth of Spain's Pork Industry. Open Society European Policy Institute (www.opensocietyfoundation.org).

Woods, M. (2005). *Rural Geography: Processes, Responses and Experiences in Rural Restructuring*. London, SAGE.

World International Trade Solution. (2019). International trade indicators. Merchandise Trade Share (Accessed August 10, 2022). Available online: https://wits.worldbank.org/country-indicator.aspx?lang=en.

Yacamán Ochoa, C. (2022). Acaparamiento de tierras y globalización del sistema agroalimentario: Impactos, resistencias y alternativas desde los comunes. En: Ezquerra, S., Di Masso, M. y Rivera, M. (Eds.), *Comunes reproductivos Cercamientos y descercamientos contemporáneos en los cuidados y la agroecología*. Catarata. ISBN: 978-84-1352-460-3.

Yacamán Ochoa, C., Mataran, A., Mata, R., López, J. M. and Fuentes-Guerra, R. (2019). The potential role of short food supply chains in strengthening periurban agriculture in Spain: The cases of Madrid and Barcelona. *Sustainability* 11(7), 1–19.

Yacamán Ochoa, C., Matarán Ruiz, A., Mata Olmo, R., Macías Figueroa, Á. and Torres Rodríguez, A. (2020). Peri-urban organic agriculture and short food supply chains as drivers for strengthening city/region food systems—two case studies in Andalucía, Spain. *Land* 9(6), 177.

Chapter 5

Agri-food supply chains and food security in Asia

Jose Ma. Luis Montesclaros and Paul S. Teng, S. Rajaratnam School of International Studies, Nanyang Technological University, Singapore*

1 Introduction: Analysing Asia's agri-food supply chains and from a food security perspective

Food security has been at threat globally in recent years, as noted by the UN Food and Agriculture Organization's (UN FAO) report of an increasing trend in global undernourishment from 2014 onwards (Montesclaros, 2021). This is relevant to Asia, given that despite significant regional progress in economic growth (albeit to varying degrees across the region), food security continues to be a challenge today with 44.2% of Asia's population unable to afford a healthy diet, and 8-10% being either undernourished or severely food insecure (UN FAO, 2023).

This chapter explores how such food security outcomes of undernourishment and lack of food affordability, are driven by supply chain management (SCM) within the region's agri-food supply chains (AFSCs) which contribute to food production and marketing, among others. SCM is defined by the Council of Supply Chain Management Professionals (CSCMP) as 'the active management of supply chain activities to maximize customer value and achieve a sustainable competitive advantage', wherein supply chain activities include 'product development, sourcing and production to logistics, as well as the information systems needed to coordinate these activities' (CSCMP, 2023).

Mentzer et al. have also defined supply chains as 'the systemic, strategic coordination of the traditional business functions and the tactics across these

* corresponding author, ismontesclaros@ntu.edu.sg

http://dx.doi.org/10.19103/AS.2023.0122.08

business functions within a particular company and across businesses within the supply chain, for the purposes of improving the long-term performance of the individual companies and the supply chain as a whole' (Mentzer et al., 2001: 18). The focus of such definitions on meeting demands of buyers/consumers at each stage of the supply chain, in fact, lends SCM to the idea of being a process of 'demand chain' management (Woods, 2003: 20). Following this logic, AFSC management in the case of the agri-food sector can thus be seen as the management of production, distribution and marketing processes to provide consumers with their agri-food consumption requirements.

The approach taken in this chapter (Section 1.2) is to assess the relevance of Asia's AFSCs in shaping regional food security outcomes. Following this perspective, it discusses the key driving factors of these outcomes along the lines of food availability (Section 2.1), physical food access (Section 2.2) and economic food access (Section 2.3). These include challenges from climate change, as well as supply-chain disruptions from the coronavirus disease 2019 (COVID-19) pandemic and Russia's war on Ukraine.

In the second half of this chapter, it argues that moving forward, other regions globally can take a page from the various ways by which Asian countries (and its sub-regions) have been working to address food security challenges (Section 3.1), building on four examples cited such as China's state-led upstream interventions (Section 3.1); India's digitalisation of public distribution (Section 3.2); Singapore's state supported agri-food tech ecosystem (Section 3.3); and private sector-led digitalisation in Southeast Asia (Section 3.4). The final section (Section 4) concludes the chapter.

1.1 Diversity of demand and value-addition across Asia's agri-food supply chains

The diversity of consumption patterns in Asia can be seen in Table 1, which reflects that in total tonnage, rice is the primary agricultural commodity that is most consumed in the region, followed by wheat and vegetables. Across Asia's five sub-regions, however, it can be seen that rice is the top consumed commodity only in South Asia (14%) and Southeast Asia (27%). The top commodity consumed in West Asia is wheat (17%) and in Central Asia (22%) is milk. In the case of East Asia, vegetables (10%) and rice (10%) take equal shares in top commodities consumed.

AFSCs cover a broad range of activities from input provision to agricultural production, processing and delivery to consumers. Figure 1 shows that for the most populous countries in Southeast Asia, there is a significant degree of heterogeneity in the contributions of different activities across the supply chain. Whereas food production has the largest value addition in Indonesia, Thailand and Vietnam, food manufacturing takes up the largest share in the Philippines.

Table 1 Top 20 commodities consumed in % share in total tonnage of food supply quantity by region/sub-region (primary commodities–crops and livestock products)

	Top 20 items consumed in Asia by % share in food supply (tonnes)	Asia	Central Asia	East Asia	Southeast Asia	South Asia	West Asia
1	Rice, milled	**12%**	1%	**10%**	**27%**	**14%**	3%
2	Wheat and meslin flour	9%	**14%**	6%	4%	**12%**	17%
3	Other vegetables, fresh N.E.C.	7%	4%	**10%**	9%	5%	1%
4	Raw milk of cattle	6%	**22%**	3%	1%	**11%**	11%
5	Potatoes	5%	**10%**	5%	1%	6%	5%
6	Tomatoes	3%	6%	4%	0%	2%	7%
7	Cucumbers and gherkins	3%	3%	5%	0%	0%	2%
8	Refined sugar	2%	2%	1%	5%	3%	4%
9	Watermelons	2%	3%	4%	1%	1%	3%
10	Meat of pig with bone, fresh/chilled	2%	0%	4%	3%	0%	0%
11	Skim milk of buffalo	2%	0%	0%	0%	6%	0%
12	Onions and shallots, dry (excluding dehydrated)	2%	3%	2%	1%	3%	2%
13	Bananas	2%	0%	1%	4%	3%	1%
14	Eggplants (aubergines)	2%	0%	3%	0%	1%	1%
15	Cabbages	2%	2%	3%	1%	1%	1%
16	Hen eggs in shell, fresh	2%	1%	2%	2%	1%	1%
17	Apples	1%	2%	2%	0%	0%	2%
18	Meat of chickens, fresh/chilled	1%	1%	1%	3%	1%	3%
19	Sweet potatoes	1%	0%	2%	1%	0%	0%
20	Flour of maize	1%	1%	1%	3%	1%	2%
Subtotal for top 20 items		67%	75%	69%	66%	71%	66%

N.E.C. under 'vegetables' (item 3) refers to commodities that have been classified as vegetables but where no further sub-classifications have been included.Source: FAO (2023). UN FAOSTAT database, accessed 7 August 2023, https://www.fao.org/faostat/en/#data.

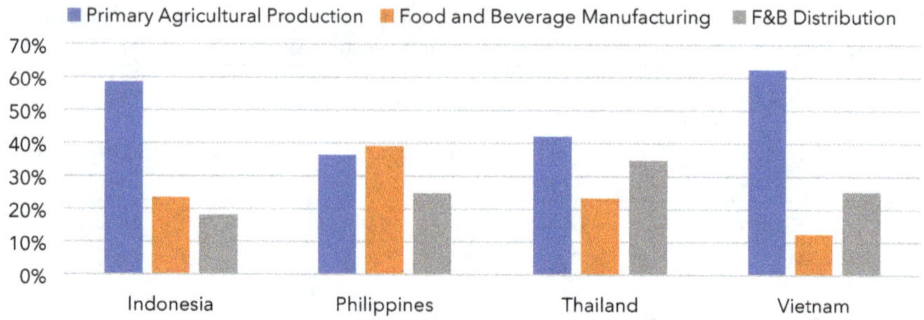

Figure 1 % Direct value addition in AFSCs across Southeast Asia's most populous countries. Only the direct contributions from primary agricultural production were included; further indicators in Oxford Economics publication include the indirect and induced economic contributions. Source: Compiled from Oxford Economics (2021). *The Economic Impact of the Agri-Food Sector in Southeast Asia*. Oxford Economics and Food Industry Asia.

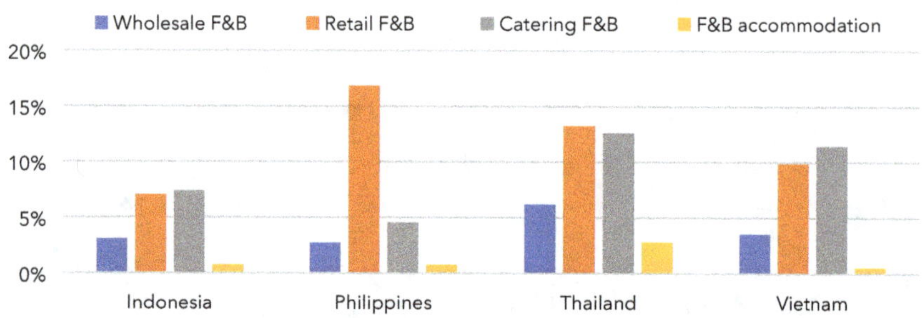

Figure 2 Breakdown of % direct value addition from food distribution across Southeast Asia's most populous countries. Source: Compiled from Oxford Economics (2021). 'The Economic Impact of the Agri-Food Sector in Southeast Asia'. Oxford Economics and Food Industry Asia.

The structure of food distribution likewise varies, with retailers having the largest contribution in the Philippines and Thailand, and with catering having the largest contributions in Indonesia and Vietnam (Fig. 2).

1.2 Prioritising 'needed' vs. 'desired' consumption in analysis

As far as the ends or goals of AFSC are concerned, an important distinction needs to be made, between meeting commercial demand for agri-food products based on what is desired by consumers as is the case in most commercial supply chains, and meeting the needed food security requirements in terms of nutritional needs of individuals.

- The 'desired' consumption of individuals results from market dynamics of changing tastes and prices. For instance, there have been increases in the shares of meat and proteins, as well as vegetables, out of the overall

diet of consumers, as countries reach higher levels of income per capita. In other cases, consumers may also desire greater intakes of sweets and unhealthy junk foods.
- In contrast, more pertinent to states is ensuring that the 'needed' food security requirements are met. These reflect the minimum amount of food that will allow individuals to have sufficient calories per day so that they can live active and healthy lives. These are not determined by the market but rather based on the physical caloric requirements determined by dieticians and nutritionists. For any population size, therefore, there is a corresponding quantity of food stocks which need to be available in local markets.

The perspective of 'needed' food security aligns with the measurement of food system outcomes within the United Nations Food and Agricultural Organisation's (UN FAO) approach to food security. Such outcomes are termed as 'food utilisation', i.e. whether food consumption is sufficient to allow individuals to achieve active and healthy lifestyle. Challenges in this regard include food safety or the lack of it, owing to production processes (e.g. debates surrounding the harmful effects of pesticide use and of genetically modified food). Equally important are the consumption choices among individuals, in favouring nutritious food as opposed to less nutritious junk food.

Figure 3 shows, based on UN FAO (2023), that the prevalence of undernourishment across Asia is 8.8% (based on macro-level data estimates by the UN FAO), which coincides to some extent with the prevalence of severe food insecurity at 10.4% (based on micro/individual/household-level surveys collated by the UN FAO). Although these measurements vary owing to differences in methodologies in data collection, they both reflect individuals missing meals throughout an entire day or more, or where food is significantly in

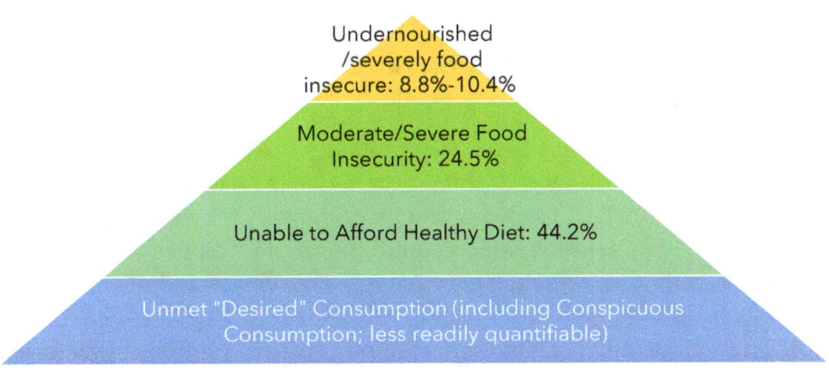

Figure 3 Measurements of unmet food needs in Asia. Source: Compiled by authors based on UN FAO (2023).

lower quantity relative to population requirements owing to their circumstances (including lack of income).

A further extension of the 'needed' consumption is reflected by the consumption of healthy food, beyond just caloric requirements, but venturing into healthy diets, to include a balance of fruits and vegetables, starchy staples, oils and fats, etc. Figure 3 shows that 44.2% or close to half of the population in Asia is unable to afford healthy diets. In between these two measurements, is the prevalence of moderate food insecurity, with 24.5% of individuals across Asia being moderately or severely food insecure based on surveys; such individuals do not consume sufficient quantities to meet caloric requirements to some extent.

While the 'needed' food consumption is more readily quantifiable as it is based on scientific assumptions, the 'desired' food consumption is less quantifiable, as this relates to changing tastes and market prices. Furthermore, focussing on the 'needed' consumption allows for a steady benchmark against which to measure the progress and development of AFSCs, in contrast to meeting 'desired' consumption which only provides limited scope for discussion, i.e. generic recommendations on maintaining competitive markets and ensuring efficient logistics infrastructure.

The next section of this chapter goes further into the AFSC-related drivers shaping these food utilisation outcomes in Asia, building on the UN FAO's food security framework that considers food availability, physical food access and economic food access, including supply chain vulnerabilities from climate change, COVID-19 and Russia's war in Ukraine. The third section discusses approaches being taken by states to address these challenges, including China's upstream state-led interventions in protecting food supply chains, Singapore's state-supported agri-foodtech system, India's digital public sector food distribution and Southeast Asia's private sector-driven approaches to the digitalisation of AFSCs, and the fourth section concludes.

2 Drivers of food utilisation outcomes in agri-food supply chains

Globally, food security has been under threat, since as early as 2017, the UN FAO saw that global undernourishment started to increase from 2014 onwards (Montesclaros, 2021). This raises the importance of functioning agri-food supply chains in Asia, as the region contributes more than half (52%) of global production of its key cereal staples, namely rice, wheat and maize/corn (UN FAO, 2022).

Food utilisation outcomes presented above on the prevalence of undernourishment, moderate/severe food insecurity and unaffordability of healthy diets, are shaped by a number of factors. Following the perspective of the 'needed' consumption, three key drivers shaping the food utilisation outcomes above are food availability, physical food accessibility and economic food accessibility:

- Food availability focusses on whether there is sufficient food produced by farmers to meet growing consumer demand. From this perspective, key challenges involve climate change and how it reduces food productivity or reduces the areas which can be tilled.
- Physical food access within the country/city concerns disruptions to the trade, transport of food or in storage, which prevent countries from accessing food which is available globally.
- Economic food access concerns food prices and how food price inflation, alongside employment disruptions, can contribute to reduced access to food that is already physically accessible within countries/areas.

2.1 Food availability

The first driver of undernourishment outcomes in Asia has been the changing configuration of productivity growth in meeting growing consumption requirements. The key disruptor in this regard has been climate change and its impacts on farming across the region's major crops.

2.1.1 Long-term production challenges from climate change

Asia contributes 89% of world rice production, 46% of world wheat production and 31% of world corn production, in total tonnage, according to the UN FAO database (UN FAO, 2022). Figure 4 shows that across the region, South Asian countries have the largest plots of land allocated for growing rice (62 million hectares) and wheat (51 million hectares). Southeast Asia has the second largest land allocation for rice (44 million hectares), followed by East Asia (33 million hectares). In the case of maize, East Asia has the largest land allocation (more than 41 million hectares), followed by South Asia (13 million hectares) and Southeast Asia (9.4 million hectares).

Teng (2020a) has earlier noted the impacts of disruptions to food production even before COVID-19 started, including the fall armyworm which impacted crops in Asia and Africa; the African swine fever which impacted on hogs in China and other Asian countries. The European Commission Directorate-General for European Civil Protection and Humanitarian Aid Operations (2020) has likewise noted droughts and floods in Pakistan and Thailand which are key grain exporters in Asia.

Longer-term trends show that the region has been facing harsher environmental conditions. The Fifth Assessment Report of the Intergovernmental Panel on Climate Change (IPCC AR5) has noted the impacts of climate

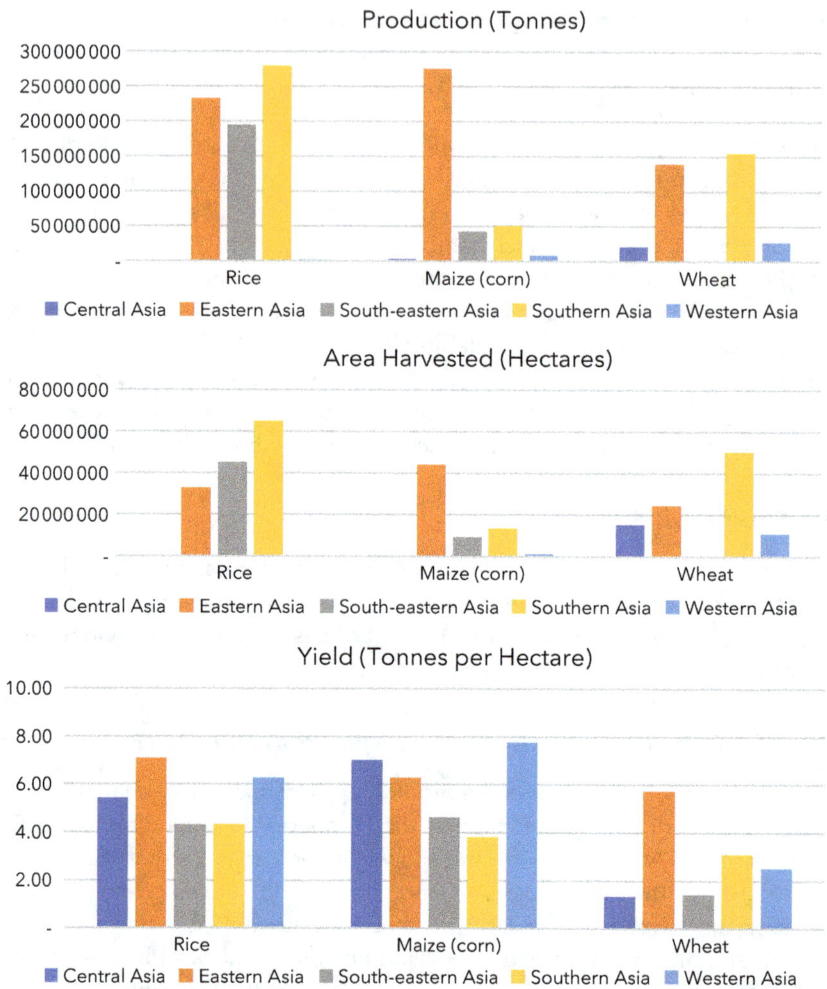

Figure 4 Asia's production, area harvested and productivity in rice, wheat and maize. Source: FAO (2023). UN FAOSTAT database, accessed 7 August 2023, https://www.fao.org/faostat/en/#data.

change on food production systems since temperature changes can result in 'geographics shifts through changes in seasonal extremes' (Porter et al., 2014: 500). Moreover, a later IPCC report has noted 'delays in crop harvesting, declining crop yields and quality of produce, increasing incidence of pests and diseases, stunted growth, livestock mortality and low farm income' (Shaw et al., 2022: 1490).

Major crops, in particular, rice, maize and wheat are sensitive to varying extents to changes in abiotic factors like temperatures, precipitation and flooding. These also disrupt fisheries, aquaculture and livestock sectors. The report likewise anticipated negative impacts on average productivity from the 2030s onwards, with negative effects set to be more likely and graver by 2050 (Porter et al., 2014: 505).

Further factors like weeds, pests and diseases can cause 'yield loss in major crop species due to animal pests and (non-virus) pathogens, in the absence of any physical, biological, or chemical crop protection, has been estimated at 18% and 16%, respectively... but weeds produce the highest potential loss' (Porter et al., 2014: 500). Greater carbon dioxide (CO_2) and ozone (O_3) can impact the distribution (whether increase or decrease) of plant diseases (Porter et al., 2014: 500). In a later report by the IPCC, it noted the potential for as much as '29% cereal price increase in 2050 due to climate change' (Mbow et al., 2019: 439).

Historical trends show that yield growth in many of Asia's crops has been slowing over the past three decades (1990–2020) compared to the previous three decades (1960–1990), based on FAO (2023) data (Fig. 5), a trend that can be observed across the three major grains consumed in the region (rice, maize and wheat).

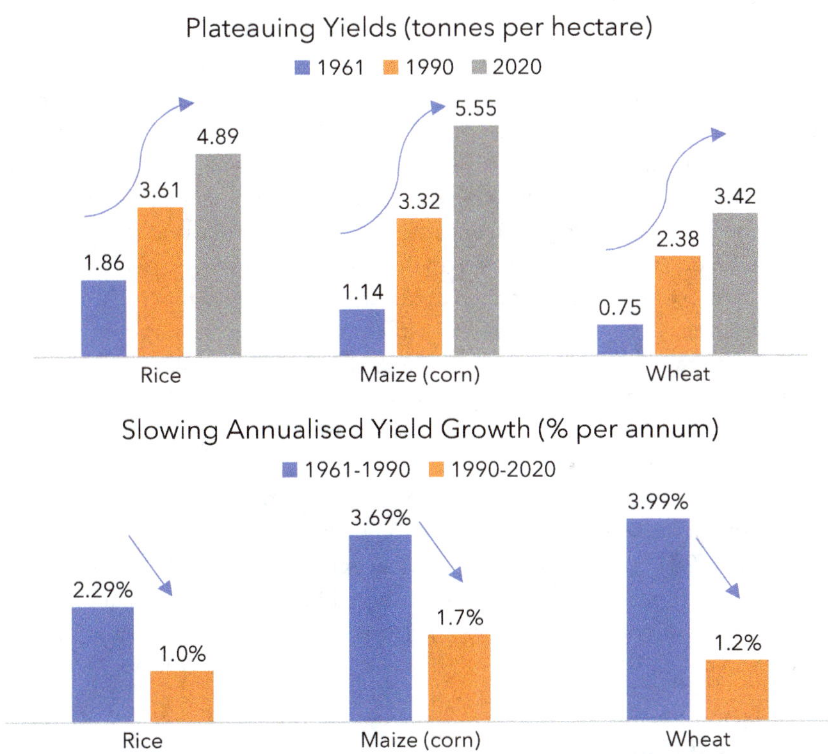

Figure 5 Slowing annualised yield growth across Asia's key staple crops (1960–1990 vs. 1990–2020). Source: FAO (2023). UN FAOSTAT database, accessed 7 August 2023, https://www.fao.org/faostat/en/#data.

- Rice: Asian rice yields on average were at 1.86 tonnes per hectare in 1961, which increased to 3.61 tonnes per hectare by 1990, representing an almost doubling in yields over the period, or an annualised growth of 2.29% per year. However, from 1990 to 2020, it grew to only 4.8 tonnes per hectare on average, or an annualised growth of 0.96% per year (UN FAO, 2022).
- Maize: For maize, yields have tripled from 1.14 tonnes per hectare in 1961 to 3.32 tonnes per hectare in 1990, an increase of 3.7% per annum. However, yield growth slowed down to 1.7% per annum in 1990–2020.
- Wheat: For wheat, yields more than tripled from 0.7 tonnes per hectare in 1961 to 2.3 tonnes per hectare in 1990, or an increase of 3.99% per annum. However, yield growth slowed down to an annualised growth of 1.2% per annum in 1990–2020.

As a result of climate change, countries have reached potential 'inflection points' in food security. For instance, Southeast Asian sub-region was seeing steady progress in reducing the number of undernourished from 101.7 million (18.1% sub-region's population) in 2005 to 60 million people (9.7% of sub-region's population) in 2014. Yet, it saw an increase of 3 million undernourished from 2014 to 2016 (Montesclaros, 2020). The rest of the countries in the region have all been heading towards this inflection point, though at various stages of plateauing reduction in undernourishment.

Moving forward, climate change is forecasted to alter the types of crops that are suitable for various environments across the region, as well as the farming approaches (including irrigation adoption), as have been found from studies in South Asia, such as in Nepal, India and Bangladesh. Similarly, in China, flooding patterns are expected to alter crop areas and land uses, while in South Korea, water availability is seen to be a key influencer of farming system changes. Within Southeast Asia, some areas in Cambodia, Northwest Vietnam, Northeast Thailand and Philippines, among others, are estimated to have significantly high yield reductions, although these are within longer timeframes of 2050, 2080 and 2100.

Poorer (low-income) consumers were found to be at greater risk and most likely to see an increase of as many as 183 million more people facing hunger (Mbow et al., 2019: 439). The study likewise echoed changes in the distribution of pests and diseases, as well as the potential for lower nutritional content of food products, including less protein, iron and zinc for every kilogram of crops produced.

2.2 Physical food access

The potential threats to the physical accessibility of food lie in the extent of food import dependence by countries in the region, given that food supplies may be disrupted in the process of transporting them from one country to another. Two key aspects in this regard are the production self-sufficiency of Asian countries for key items which contribute a significant share of domestic consumption, as well as the region's overall trade position as a net trade deficit region which imports more of its consumption than it exports.

2.2.1 Production self-sufficiency at sub-regional level

Asia's sub-regions are in general not highly import reliant when it comes to commodities that take up significant shares of its total consumption in meeting food consumption requirements. For instance, Table 2 shows that East Asia, Southeast Asia and South Asia depend on imports for at most, approximately 8% of rice consumption requirements. Vegetables, as East Asia's largest consumed commodity at par with rice, are mostly self-produced, with only 0.2% of food supplies imported. Similarly, for milk, which is the most consumed commodity in Central Asia, only 0.2% of food supplies are imported. Finally, for wheat, which is the primary commodity consumed in West Asia, trade dependence is at 9%.

Therefore, juxtaposing Table 2 on import dependence against Table 1 on key commodities consumed in Asia, a key vulnerability of the region would appear to be in the relatively high wheat dependence of the West Asia sub-region since wheat takes a significant share of domestic consumption in the said sub-region of West Asia, and yet the sub-region has a significant degree of dependence on food source beyond the sub-region.

2.2.2 Food trade: Asia as a 'net food-trade deficit' region

A further issue is that Asia is a net food-trade deficit region, importing more of its consumption requirements than it exports. This is shown in Table 3,

Table 2 Asia's import dependence in top consumed food items by Asian sub-region (imports as % of total food supplies)

	Asia	Central Asia	East Asia	Southeast Asia	South Asia	West Asia
Rice, milled	4.2%	26.2%	**2.1%**	**5.2%**	**1.8%**	108.9%
Wheat and meslin flour	2.7%	8.1%	0.9%	6.4%	1.6%	**9.3%**
Other vegetables, fresh N.E.C.	0.4%	2.0%	**0.2%**	0.2%	0.2%	27.1%
Raw milk of cattle	0.7%	**0.2%**	2.2%	4.4%	0.0%	1.2%

N.E.C. under 'vegetables' (item 3) refers to commodities which have been classified as vegetables but where no further sub-classifications have been included.Source: FAO (2023). UN FAOSTAT database, accessed 7 August 2023, https://www.fao.org/faostat/en/#data.

Table 3 Asia's agricultural exports and imports by sub-region (in billions US$), 2020

	Exports	Imports	Net exports
Asia	311.31	552.27	–240.96
Central Asia	5.53	7.94	–2.41
East Asia	80.81	279.35	–198.54
Southeast Asia	135.83	103.74	32.10
South Asia	43.69	57.56	–13.86
West Asia	45.45	103.69	–58.25

Source: FAO (2023). UN FAOSTAT database, accessed 7 August 2023, https://www.fao.org/faostat/en/#data.

Table 4 Asia's agricultural exports and imports by commodity group (in billions US$), 2020

	Exports	Imports	Net exports
Cereals and preparations	38.91	83.09	–44.18
Fats and oils (excluding butter)	40.64	41.17	–0.53
Meat and meatp preparations	12.67	72.06	–59.39
Sugar and honey	11.16	18.43	–7.27
Fruit and vegetables	78.48	80.55	–2.07
Dairy products and eggs	5.98	27.06	–21.08
Beverages	12.15	21.43	–9.27
Other food products	58.26	118.68	–60.42
Fodder and feeding stuff	10.98	30.69	–19.72
Non-edible crude materials	24.12	36.04	–11.92
Non-edible fats and oils	5.63	4.71	0.92
Tobacco	12.32	18.35	–6.03
Total: All agricultural products	311.31	552.27	–240.96

Source: FAO (2023). UN FAOSTAT database, accessed 7 August 2023, https://www.fao.org/faostat/en/#data.

which reflects net imports of more than US$240 billion. Across the region, only the Southeast Asian sub-region is a net exporter in agricultural commodity products, with net exports of US$32 billion, while the rest are net food importers.

Table 4 further breaks down Asia's trade in agricultural products according to major commodity groups. The largest contributor to Asia's total exports is fruits and vegetables (US$78 billion), followed by other food products (i.e. food preparations, tea leaves and coffee, at US$58 billion), and fats and oils (US$40 billion). Although, based on net exports, the greatest net imports of Asia lie in other food products (US$60 billion), followed by meat and meat preparations (US$59 billion), and cereals (US$44 billion).

To provide a clearer picture of food deficits faced by the Asian region, Table 5 provides a further breakdown of the net agricultural trade position of Asia by Asian sub-region. The figure indicates that East Asia is the largest driver of agricultural net imports mostly in meat and meat preparations (US$53 billion), other food products (US$50 billion) and cereals (US$27 billion). In contrast, Southeast Asia is the only net food exporting region, owing to its significant exports of fats and oils (US$30 billion). Further disaggregated data indicates that majority of the net exports of Southeast Asia come from palm oil (US$25 billion).

2.2.3 War in Ukraine: Implications for Asia's food and fertiliser supplies

A key example of disruptions to physical accessibility of food has been the impacts of Russia's war in Ukraine. In late February 2022, Russia invaded Ukraine as a result of Ukraine's bid to join the North Atlantic Treaty Organization (NATO), an intergovernmental military alliance between 30 member states

Table 5 Asia's net Agricultural exports and imports by sub-region and commodity group (in billions US$), 2020

	Central Asia	Eastern Asia	Southeast Asia	South Asia	Western Asia
Cereals and Preparations	0.51	-27.43	-3.82	3.92	-17.37
Fats and Oils (excluding Butter)	-0.57	-13.47	30.40	-14.12	-2.77
Meat and Meat Preparations	-0.51	-53.41	-1.39	2.72	-6.80
Sugar and Honey	-0.60	-2.77	-1.87	0.27	-2.30
Fruit and Vegetables	0.22	-7.95	10.02	-2.26	-2.09
Dairy Products and Eggs	-0.30	-10.68	-4.53	-0.57	-4.99
Beverages	-0.30	-8.35	0.18	-0.20	-0.60
Other food	-0.91	-50.22	5.18	-0.53	-13.95
Fodder and Feeding Stuff	-0.19	-6.78	-7.17	-0.48	-5.09
Non-edible Crude Materials	0.48	-11.60	3.20	-1.89	-2.11
Non-edible Fats and Oils	-0.01	-0.49	1.86	-0.37	-0.07
Tobacco	-0.22	-5.37	0.03	-0.34	-0.12

Source: FAO (2023). UN FAOSTAT database, accessed 7 August 2023, https://www.fao.org/faostat/en/#data.

from both European countries (28) and two North American countries. This war had a significant impact on the supplies of wheat since Russia and Ukraine are known as the 'breadbaskets' of the world which contribute the largest share of wheat exports, even relative to Canada and the United States combined (Montesclaros and Caballero-Anthony, 2022).

The prominent areas impacted by the war were the ports in Ukraine, where most grains were stored before being exported through the Black Sea, and transport routes through the Black Sea. Across Asia, the most impacted countries which imported the largest share of Ukraine's wheat exports were Egypt (18%), Indonesia (14%) and Turkey (8%), based on data from the previous year in 2021. Other countries impacted were Pakistan (8%), Bangladesh (4%) Yemen (4%) and Lebanon (3%), according to data from the International Trade Centre (2022). Ukraine takes up 10% of global wheat exports and is also the cheapest wheat producer among the top 5 wheat exporters, with average wheat export prices of US$244 per tonne of wheat. As a result, a continued war in Ukraine will contribute to higher import prices for the Asian countries which are dependent on Ukraine for their imports.

Similarly, in the case of maize, for which Ukraine takes up 12% of global shares and is the third cheapest provider at US$239 per tonne. The cheapest maize exporter is Brazil at US$205 per tonne followed by Argentina at US$227 per tonne. As such, the countries which source from Ukraine likely saw higher maize import prices if they were unable to procure sufficient quantities of the grain from Brazil and Argentina, and if they were forced to import from the United States which exports maize at US$278 per tonne. The key countries impacted within Asia are the top importers from Ukraine, in particular, China which takes up 32% of Ukraine's exports. Other countries impacted include Egypt (9%), Iran (7%), Turkey (4%) and Israel (3%).

The potential of Russia's war on Ukraine to spark a crisis similar to the 2007–2008 global food price crisis is questionable, though. During the 2007–2008 GFPC, rice prices more than tripled, owing to largescale panic imports by Philippines, as well as price speculation and hoarding by traders in Vietnam and Thailand (Dawe and Slayton, 2012). At this time, three traders contributed 58% of global rice exports, namely, Thailand, Vietnam and India, making it more likely for such trader speculation to impact prices.

Unlike the 2007–2008 GFC, the key commodities impacted by Russia's war on Ukraine do not include rice but rather wheat and maize. On the one hand, there is a similar degree of concentration in the exports of maize and wheat, similar to rice in 2007–2008. It is slightly less concentrated in the case of wheat where five exporters compose 58% of global wheat exports (Russia, Australia, the United States, Canada and Ukraine). In maize, four exporters compose 77% of global exports (United States, Argentina, Ukraine and Brazil).

A key difference though lies in the size of trade relative to total production, based on data from the UN FAO. In the case of rice, less than 1% of rice produced globally was traded based on data over the previous decade. In contrast, an average of 14% of maize production is traded, and an average of 24% of wheat production is traded. As such, disruptions such as the panic imports by Philippines which occurred during the 2007-2008 crisis are less likely to have as large an impact today should this occur in the case of wheat and maize.

In the case of fertilisers, Fig. 6 shows that out of the total fertiliser exports of 105 million tonnes in 2020, the biggest share is taken by nitrogen (N) fertilisers at 47.7 million tonnes, followed by potash (K_2O) at 34.8 million tonnes and phosphate (P_2O_5) at 22.3 million tonnes. In the case of nitrogen fertilisers, the largest share of global exports come from Russia (15%) and China (14%). Ukraine contributes 2% to global exports. Although, as much as 39% of total fertiliser production is traded, which makes for a less shaky market should any single country make sudden larger purchases.

For potash, the second and third largest exporters are Russia and Belarus, contributing 19% and 15%, respectively of global exports. This is more resilient, though, in the sense that over 77% of potash production is traded. In the case of phosphate, Russia is the third largest exporter, contributing 14% of global exports. It is also resilient, in the sense that 50% of global phosphate production is traded. Therefore, based on this analysis, it is unlikely that a significant disruption in the form of a sudden purchase of fertilisers from a single country will spark a similar crisis as in the case of grains in the 2007-2008 crisis.

2.3 Economic food access

The potential threats to the economic accessibility of food lie in factors which may prevent food from becoming affordable to consumers in Asia. These are in part impacted by trade disruptions, which can lead to reduced imports and in turn local food shortages. Further disruptions lie in those which prevent consumers from earning a sufficient income to afford food, as will be shown in the case of the COVID-19 pandemic's impacts.

2.3.1 COVID-19 impacts on economic access

Apart from major events like Russia's war in Ukraine, a further disruption of COVID-19 was an important disruptor to Southeast Asian (ASEAN) supply chains. Figure 7 illustrates the movement restrictions imposed across ASEAN countries from February 2020 to the first half of 2022. Movement restrictions prevented individuals from earning incomes, especially those working in

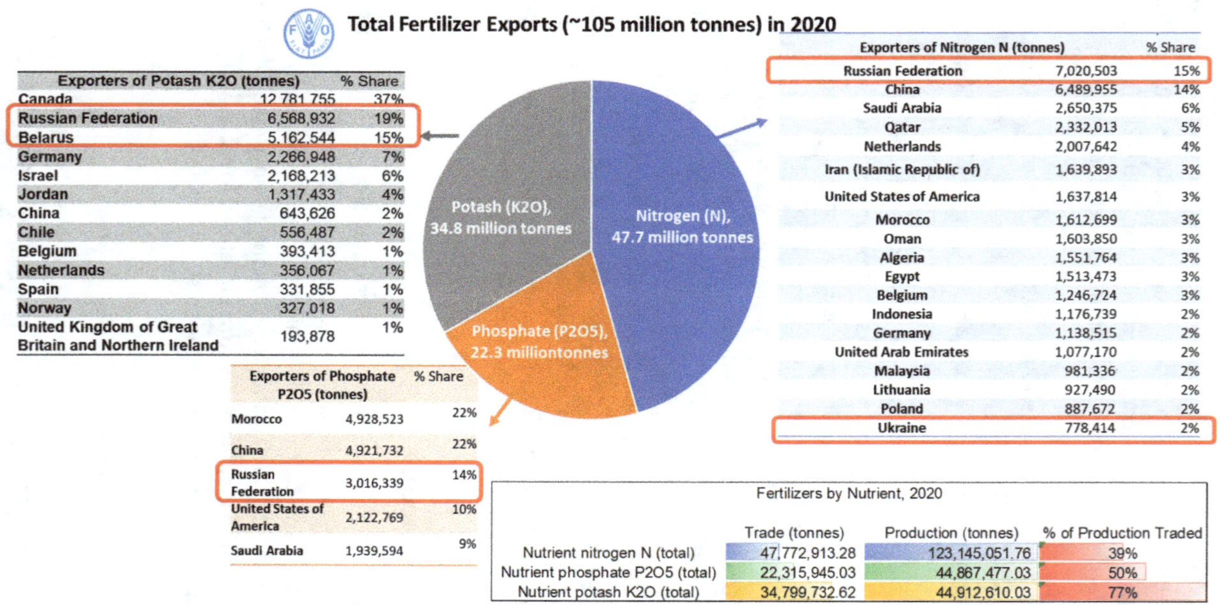

Figure 6 Risk exposure of the fertiliser sector to Ukraine crisis. Source: Montesclaros, Jose Ma. Luis, 2022a. 'Impact of Persistent Supply Chain Constraints on Food Prices and Inflation: Food Security Implications'. Presentation at the 37th Environmental Scanning Exercise with the Bangko Sentral ng Pilipinas (English translation: Central Bank of Philippines). With permission from the author. Data were sourced from FAO (2022). UN FAOSTAT database, accessed 27 November 2022, https://www.fao.org/faostat/en/#data.

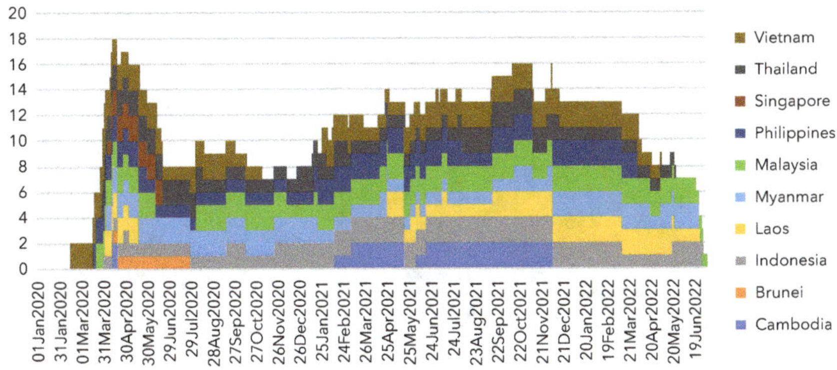

Figure 7 Movement restrictions from COVID-19 in Southeast Asia (ASEAN). Source: Thomas Hale et al., 2021, 'A global panel database of pandemic policies (Oxford COVID-19 Government Response Tracker)'. *Nature Human Behaviour*. https://doi.org/10.1038/s41562-021-01079-8.

high-contact industries like restaurants, conferences, entertainment and other services.

The agricultural sector was impacted by lockdowns, which disrupted the flow of labour and production inputs like fertilisers and seeds, as well as pesticides. Additionally, as farming is a time-sensitive activity, such delays led to lower harvests (Caballero-Anthony et al., 2020; Aday and Aday, 2020). For instance, planting and harvesting schedules of the top three rice producers globally (India, Thailand and Vietnam) were affected by the initial lockdowns in March to May 2020. Transport capacities for ships were likewise impacted, as social distancing measures reduced the number of workers travelling (Teng, 2020b). In sum, lockdowns in 2020 were estimated to have led to up to 167 million falling into unemployment in Asia, and up to 140 million falling into extreme poverty globally (ADB, 2020).

Moreover, COVID-19 contributed to reduced incomes of individuals, and increased needs for government support (Montesclaros and Caballero-Anthony, 2020). Figure 8 illustrates the income support provided by countries to their citizens amid COVID-19. It is noteworthy however that even as the region saw another uptick in COVID-19 infections from the delta variant in 2021, countries were not able to provide the same extent of income support as they did in 2020. Singapore was able to provide consistent high levels of income support to its constituents. In contrast, countries like Malaysia, Philippines, and Indonesia could provide only moderate levels of support. Moreover, in Brunei, Thailand and Cambodia, higher levels of income support were provided in the first wave of COVID-19 in the year 2020, but the support fell in the later stages in 2021.

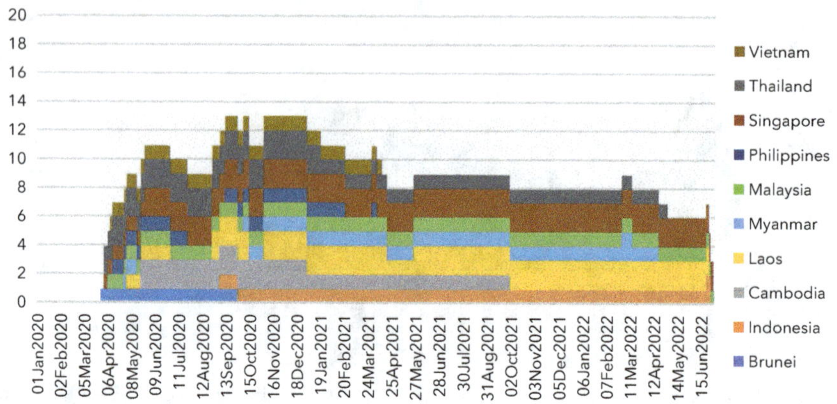

Figure 8 Income support amid COVID-19 in Southeast Asia (ASEAN). Source: Thomas Hale et al., 2021, 'A global panel database of pandemic policies (Oxford COVID-19 Government Response Tracker)'. *Nature Human Behaviour*. https://doi.org/10.1038/s41562-021-01079-8.

Given the limited potential for lower income countries to provide economic support amidst disruptions to agricultural input supplies and input prices, some countries in Southeast Asia imposed bans on the exports of their products. Within Southeast Asia, rice exports were subject to licensing-related restrictions in Myanmar from mid-March 2020 to end-September 2020, while exports were banned altogether by Cambodia (early April to mid-May 2020). Vietnam halted new rice export contracts (throughout April 2020), while Thailand banned its exports of eggs for 7 days initially , but extended this to the end of the following month (end-March to end-April 2020) (Laborde and Mamun, 2022). The purpose of these bans was to allow for meeting domestic consumption requirements and avoid food shortages in specific items. However, these also had the effect of raising food prices for imports, thus impacting countries which were net food importers.

2.3.2 Further drivers of food price increases: Fertiliser cost and supply issues due to COVID-19

As most of Asia's sub-regions are dependent on domestic production for commodities that are important for local consumption, the focus on food production brings to fore two other aspects of food security for Asia: food availability and economic accessibility. Both factors are impacted by changing costs of farming production which may hinder farmers from achieving their productivity potential.

In September 2021, in fact, China banned its exports of phosphate rock fertilisers (Qian et al., 2021), and introduced export licensing requirements as well for these fertilisers (Development and Reform Commission, 2021). Fertiliser prices were also on an uptrend amid the rebounding of demand for energy in 2021 (Montesclaros and Sembiring, 2022). When COVID-19 lockdowns started, economic activity was reduced, which translated to a reduction in demand for oil by about of 18 million barrels per day (mb/d) (International Energy Agency, 2022).

Fertiliser production is an energy-intensive process, with the production of nitrogen fertilisers contributing more than 1% of global energy consumption according to an earlier study by the International Fertilizer Industry Association (1998). Oil prices fell amid COVID-induced demand reductions, and oil producers responded by cutting their oil production targets, and these in turn led to a reduction in global oil supplies (Trading Economics, 2022). As a result, when economies reopened in 2021, the world saw a significant increase in oil demand for electricity generation in coal plants, of approximately 9% (Ambrose, 2021). Thus, the increased demand for oil, alongside reduced oil supplies, led to an oil supply shortage with high energy prices in 2021, also known as the 'coal crunch'. This in turn fed to higher fertiliser prices.

Table 6 shows that while some sub-regions like Central, East and West Asia were net exporters of fertilisers in 2020, Southeast and South Asia are net importers. As a whole, in fact, the Asian region is a net importer of fertilisers, especially phosphorous pentoxide. To contextualise these import dependence statistics, Table 7 indicates that on average Asia relies on imports for 34% of total agriculture use. Potassium oxide (K_2O) was the most imported fertiliser nutrient in Asia amounting to 80% of total agricultural use, whereas the region's total Phosphorous pentoxide (P_2O_5) is at 29%, and nitrogen (N) at 23%. Across Asia's sub-regions however, there are sub-regions with high rates of import reliance. In particular, Southeast Asia and West Asia depend on imports for 63% and 71% of total fertiliser consumption, respectively.

Table 6 Fertiliser trade deficits in Asia by sub-region and fertiliser type

	Asia	Central Asia	East Asia	Southeast Asia	South Asia	West Asia
Net exports all Fertilisers (million tonnes)	–4.14	0.27	4.24	–6.97	–14.13	12.45
Net exports N	5.48	0.21	5.72	–1.35	–6.00	6.91
Net exports P_2O_5	0.83	–0.03	4.41	–1.53	–4.37	2.35
N Net exports K_2O	–10.45	0.10	–5.89	–4.09	–3.76	3.20

Source: FAO (2023). UN FAOSTAT database, accessed 7 August 2023, https://www.fao.org/faostat/en/#data.

Table 7 Fertiliser import dependence in Asia by sub-region and fertiliser type

	Asia	Central Asia	East Asia	Southeast Asia	South Asia	West Asia
Import % of consumption of all fertilisers	34%	31%	18%	63%	38%	71%
Import % of consumption of N	23%	31%	4%	47%	29%	66%
Import % of consumption of P_2O_5	29%	31%	6%	62%	39%	59%
Import % of consumption of K_2O	80%	24%	63%	97%	98%	200%

Source: FAO (2023). UN FAOSTAT database, accessed 7 August 2023, https://www.fao.org/faostat/en/#data.

2.4 Summary of food security vulnerabilities across agri-food supply chains

This section has highlighted the multiple factors driving the food security outcomes in Asia which shape undernourishment levels across countries. The use of the food security perspective, which considers the 'needed' consumption, has provided an objective basis for measuring the progress in AFSCs over a long-term period. In particular, this section has shown that food availability is under threat from climate change, which reduces the productivity of farming and in turn leads to greater food scarcity amidst growing populations.

Further disruptions have been outlined, including physical access vulnerabilities the Asia's position as a net food trade deficit region, and West Asia's high import dependence for wheat which is a significant contributor to domestic consumption. The impacts of these were revealed partially by the war in Ukraine, which disrupted maize and wheat supplies. Economic access disruptions were also discussed, in how COVID-19 led to a reduction in state budgets, which limits the potential to give sustained economic support amidst greater unemployment owing to the pandemic.

3 Improving agri-food supply chain resilience for food security

Given the challenges faced above, states across Asia have taken varying measures in order to ensure food security, by making their AFSCs more resilient to disruption. Four cases are presented in the following sections, including China as the world's top consumer of food owing to its large population;

India as a key food consumer and exporter; Singapore as a net food importer; and Southeast Asia as an example for sub-regional cooperation in promoting technology adoption across the AFSC.

3.1 Upstream state-led approaches: The case of China

The role of the state in supporting agri-food supply chains today is perhaps best exemplified by a key development in China, as the largest food consumer globally. China has effected increasing vertical integration of supply chains, in other words, the merger of different parts of the supply chain into one company or group of companies.

In 2017, the Swiss-owned multinational Syngenta, which deals in agri-chemical inputs like seeds, pesticides and fertilisers, was bought over by the China National Chemical Corporation (Chinachem), one of China's biggest state-owned conglomerates, for US$43 billion (Shields, 2017). This move would allow China to overhaul its agricultural systems, which had become outdated over time. An anecdote shared in *South China Morning Post* (Wang, 2017) shows how a farmer in China in 2018 was losing money from a land purchase owing to his inability to increase productivity. Syngenta then provided the farmer with tailored recommendations, including seeds, pesticides, fertilisers and support services.

These solutions were identified using satellite mapping services provided by the Syngenta Group China's subsidiary Mod Agriculture Platform (MAP). This support allowed the farmer to slowly break-even and eventually turn a profit, with a 25% increase in productivity compared to five years ago (Wang, 2017). Several further elements of supply chain integration can be seen in China's recommendations, which leveraged fungicides from Israel-based Adama which was earlier bought over by Chinachem.

Beyond these, China also leveraged farm chemicals from German agri-chemical multinational Bayer, and corn seeds from American agri-chemical multinational Monsanto. As of July 21, this has resulted in '384 solution centres... over 900 MAP demonstration farms across 28 provinces... directly covering 11.62 million mu of arable land and over 1.45 million farmers' (Wang, 2021).

As the example above shows, the effective management of agri-food supply chains from the upstream perspective allows for getting farming inputs to farmers. The solutions provided to farmers were already available, whether in China, the United States, Germany or even Israel. However, it was state intervention through the state-owned company Chinachem of purchasing Syngenta, which allowed for information on these inputs to reach farmers, as well as for providing them with important information on the types of these products which would be most beneficial to them.

3.2 Public sector food distribution: The case of India

A further case of developments in agro-food supply chains focusses on India's public sector driven agricultural distribution. The country has a lower-middle income country/developing country status, and yet, has been lauded by the International Monetary Fund (IMF) (Bhalla et al., 2022) for its ability to 'keep extreme poverty to the pre-pandemic low of 0.8% amid the COVID-19 pandemic, owing to the country's food subsidy policies' (Montesclaros, 2022b: 2). India is likewise among the key exporters of food and agricultural products within Asia.

A notable aspect of India's ability to do so lies in its transparent reporting of food stocks at multiple levels and its leveraging of digitalisation in food distribution. Figure 9 illustrates the comprehensive reporting structure of India, whereby food stocks are tracked not only at the sub-national regional level,

	FOOD CORPORATION INDIA	STATE AGENCIES	TOTAL
Country Total	4 189 000.00	2 604 000.00	6 793 000.00
Eastern Region Total	227 000.00	309 000.00	536 000.00
Bihar	111 500.00	-	111 500.00
Jharkhand	30 000.00	-	30 000.00
Odisha	42 000.00	209 000.00	251 000.00
West Bengal	43 500.00	100 000.00	143 500.00
Northeastern Region Total	219 000.00	-	219 000.00
Assam	150 500.00	-	150 500.00
Arunachal Pr.	13 500.00	-	13 500.00
Tripura	6 500.00	-	6 500.00
Mizoram	6 000.00	-	6 000.00
Meghalaya	4 000.00	-	4 000.00
Manipur	16 000.00	-	16 000.00
Nagaland	22 500.00	-	22 500.00
Central Region Total	2 280 000.00	3 500.00	2 283 500.00
Delhi	23 000.00	-	23 000.00
Haryana	199 000.00	-	199 000.00
Himachal Pr.	10 000.00	-	10 000.00
J&K	53 500.00	-	53 500.00
Punjab	1 412 000.00	-	1 412 000.00
Rajasthan	12 500.00	-	12 500.00
Uttar Pr.	521 000.00	-	521 000.00
Uttrakhand	49 000.00	3 500.00	52 500.00
Southern Region Total	**1 044 000.00**	**1 572 000.00**	**2 616 000.00**
Andhra Pr.	173 000.00	527 500.00	700 500.00
Karnataka	285 000.00	-	285 000.00
Kerala	103 500.00	8 500.00	112 000.00
Tamil Nadu	200 000.00	877 500.00	1 077 500.00
Telangana	282 500.00	158 500.00	441 000.00
Western Region Total	**419 000.00**	**719 500.00**	**1 138 500.00**
Gujarat	141 000.00	12 000.00	153 000.00
Maharashtra	161 500.00	6 000.00	167 500.00
Madhya	28 000.00	377 000.00	405 000.00
Chhattisgarh	88 500.00	324 500.00	413 000.00

Figure 9 Region-wide rice stocks reporting in India. Source: Lead author's analysis of data from Food Corporation India, 2022, Region wise stock position as of 16 November 2022, https://fci.gov.in/stocks.php?view=82.

but also at the provincial level. Food is first procured by the state agency, Food Corporation India (FCI), and subsequently re-distributed into district-level warehouses under the auspices of state agencies.

Another unique element of India's system lies in its digitally enabled public distribution system (PDS) for food. The PDS' goal is to allow for efficient targeting of individual recipients and minimising leakages in food distribution. To this end, it leverages the 'Aadhaar' national digital identity system whereby each citizen/individual is given a unique 12-digit identification number that is tied to his or her biometrics; as of March 2022, over 1.3 billion individuals were covered by Aadhaar (Montesclaros, 2022b). Moreover, when food is distributed in 'Fairprice shops' at subsidised prices, Automated e-Point of Sale (e-POS) machines are leveraged to allow for electronic weighing of grains in grain distribution. In fact, from January 2017 to February 2022, over 166 million metric tonnes of grains distributed using e-POS machines (Montesclaros, 2022b).

3.3 Ecosystem approaches in Singapore's agri-foodtech system

Singapore is a small island city-state which imports more than 90% of its consumption requirements. In 2013, it launched its Food Security Roadmap, which envisioned reducing the import dependence for key crops that could be produced locally, including vegetables (target of 10% self-sufficiency), fish (target of 15% self-sufficiency) and hen's shell eggs (target of 30% self-sufficiency) (AVA, 2013). By 2018, however, only the first target on leafy vegetables was met and in fact exceeded, achieving 13% leafy vegetable self-sufficiency. In March 2019, Singapore made a bolder statement, envisioning to achieve 30% self-sufficiency by 2030 (Ministry of Sustainability and the Environment, 2019).

Singapore's approach to food security involves budgetary funding support to grow the country's agri-foodtech ecosystem, presented in Fig. 10. This can be broken down into core and supporting components. Core components include On-Farm agricultural technologies (Agricultural tech on-farm), Distribution technologies, Post Farmgate/Pre-Consumer processing technologies, Foodtech (ingredients/processing) and Industrial use technologies (e.g. biofuels, biomaterials) (Box 1), while supporting components include Research Institutions, Education and Extension, Regulation, Industry Representation, Quality Assurance, International linkages, Financing intermediaries/support and Incubators and Accelerators.

The 'Singapore Food Story' Programme in 2019 for instance allocated SG$144 million to support the said ecosystem (Singapore Food Agency, 2021). Prior to these, other forms of assistance as compiled in Montesclaros and Teng (2023) include the 'Food Fund' (SG$1 million allocated in 2009, SG$19.37

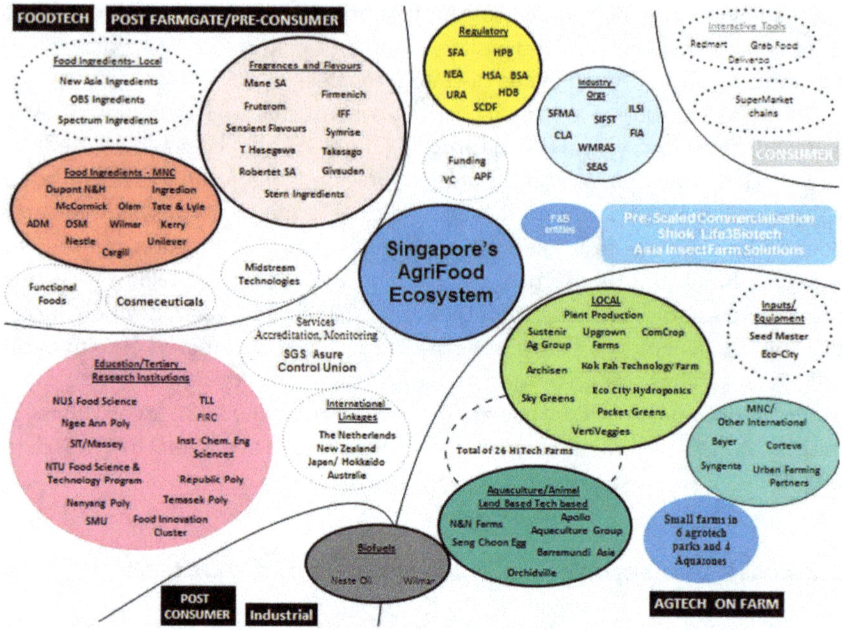

Figure 10 Asian Agri-Food Ecosystems – example of Singapore. Source: Copied, with authors' permission, from: Teng, Montesclaros, Hulme and Powell, 'The Evolving Singapore Agrifood Ecosystem', NTS Insight, No. IN19-03, RSIS Centre for Non-Traditional Security Studies (NTS Centre), Nanyang Technological University Singapore, Singapore, 4.

million allocated in 2011, SG$10 million in 2013), which supported import diversification and the Agricultural Productivity Fund (SG$63 million in 2014). Further discussion on the mechanisms adopted by Singapore to achieve food security despite its limited land area, can be found in Teng and Montesclaros (2023).

Box 1: Singapore's agri-foodtech system

Core components

1. ***Agtech on-farm***: *high-tech farms using advanced agricultural technologies in order to produce more food with less inputs, crop waste and land;*
2. ***Distribution***: *optimised supply chain platforms that allow the products to transition through distribution and retail networks to reach consumers in a timely manner with consistent forecasting, quality, transparency and provenance. This includes wholesalers,*

retail supermarkets, wet markets, ecommerce vendors and direct sales from the farms;

3. ***Post farmgate/pre-consumer sector**: efficient processing of primary produce into ingredients that can be readily used by the market;*
4. ***Foodtech**: development of food ingredients, flavours and processing technology;*
5. ***Industrial**: utilisation of crop and food waste to generate biofuels, repurposed food products and biomaterials.*

Supporting components

1. ***Research institutions**: industry and government-backed research and development programmes to innovate and optimise food production, processing and distribution;*
2. ***Education and extension**: polytechnic institutes and national universities, in collaboration with strategic country partner education programmes, to deliver a new generation of graduates aimed at building capacity and capability in domestic food production and processing.*
3. ***Regulation**: regulators to harmonise policies and interests to allow the scaling of the firms involved in the direct production, processing and distribution of food products without compromising safety;*
4. ***Industry representation**: both industry and NGO interest groups to serve as effective linkages to connect, communicate and collaborate in order to address challenges and realise opportunities faced by companies;*
5. ***Quality assurance**: accreditation and certification bodies to ensure that products are produced to standard and safe for consumption, and to develop and implement standards for healthy competition across companies;*
6. ***International linkages**: to provide catalysts for best-practice extension and technology transfer, learnings from other countries that have been able to successfully develop agri-food ecosystems;*
7. ***Financing**: intermediaries to link producers, processers, distributors and technology developers with funding sources, so that the former can be scaled.*
8. ***Incubators and accelerators**: funded programs aimed at supporting startups (idea generation, business planning, mentoring, pitch deck presentations, capital raise, market access and scaling).*

Source: Adapted, with authors' permission, from: Teng, Montesclaros, Hulme and Powell (2019), 'The Evolving Singapore Agrifood Ecosystem', NTS Insight, No. IN19-03, RSIS Centre for Non-Traditional Security Studies (NTS Centre), Nanyang Technological University Singapore, Singapore, 4.

3.4 Private sector-driven digitalisation of agricultural supply chains: The case of Southeast Asia

Beyond the state-led and state-supported approaches to enhancing food security through agro-food ecosystems, there has also been growing private sector activity in supporting the digitalisation of such ecosystems. An earlier report by PwC, Rabobank and Temasek (2019) has noted that there is an US$800 billion investment opportunity presented by Asian agriculture, in meeting future consumption requirements. These include approximately US$550 billion 'to satisfy demand for better quality food (including safer, healthier food, and more sustainable food)' and approximately US$250 billion to 'drive increased quantities of food to feed Asia's growing population' (2019: 8).

Within the production process, digitalisation involves the use of sensors that allow for measuring and recording input use, environmental conditions, output and productivity; analytics software to identify ways of optimising practices and in turn maximising productivity (including digital advisory); and automation technologies for accurate implementation of the recommendations from analytics (Montesclaros et al., 2019). This has also been termed as 'smart farming' (GSMA, 2020) and as 'knowledge-intensive agriculture' (Teng, 2017).

Beyond the production process, the use of e-commerce likewise allows for linking consumers to food producers directly, whereby consumers make online payments while food producers leverage platforms for accepting and delivering orders (Demirgüç-Kunt et al., 2018). Such technologies likewise include digital applications such as automation and information sharing in food trading, processing/manufacturing and distribution, as well as for ordering agricultural inputs (digital procurement).

According to a report by the GSM Association (GSMA) in 2020, the number of active digital agriculture services globally has grown from approximately 53 in 2009, to as much as 713 as of 2019, inclusive of digital advisory, digital procurement, smart farming, digital financial services in agriculture and agriculture-related e-commerce. The report notes that within Southeast Asia, there are over 31 providers of digital agriculture services (e.g. Farmerlink in Philippines; Agrimedia in Vietnam; MyCrop in Indonesia; GoldenPaddy in Myanmar and CropBase in Malaysia).

In relation to digital financial services in agriculture, there are approximately 16 companies (e.g. FarmTrek in Myanmar and Farmcloud in Indonesia, Philippines and Cambodia; Crowde in Indonesia; and the Lenddo Score in Thailand, Philippines, Vietnam, Indonesia and Malaysia, among others). There are likewise 32 companies involved in digital procurement, 12 in agriculture e-commerce and 9 in smart farming.

ASEAN presents a potential role model in this regard, in its ability to engender cooperation among its 10 member states despite the significant degree of heterogeneity across the countries in terms of economic development and food security approaches. Unlike the European Union's Commonwealth Agricultural Policy (CAP), however, ASEAN does not have significant pooled regional funding to support digital agriculture that parallels the CAP. A potential way forward in this regard, lies in 'developing a regional platform for sharing of geospatial data at the farm level across companies and governments to allow for rationalising and creating synergies in approaches amid resource constraints' as well as 'standards for interoperability of geospatial data' (Montesclaros, 2023: 4).

4 Conclusion

Asia is a dynamic economic region with a fast-growing population, and vast cases of successes and failures in sustaining food utilisation outcomes. This chapter has discussed the food security challenges in Asia, and the role of Asia's AFSC in addressing these challenges.

The food security perspective on the 'needed' food quantities to meet minimum daily consumption requirements has been adopted, given that the alternative 'desired demand' perspective is less objectively measurable. Following this perspective, the second section showed that food availability in Asia, as the first element of food security based on the UN FAO's framework, is under threat from long-term impacts of climate change on food production, as observed from declining levels of productivity growth in rice as a key staple in Asia. This has also partly resulted from increased frequencies of variability in rainfall and warming temperatures.

Increasingly, food access from both the physical and economic perspectives (second and third elements of food security based on the UN FAO's framework) have also been under threat amidst the COVID-19 pandemic which contributed to disruptions to food exports and the supplies of agricultural inputs, and also the disruptions in the form of greater unemployment which have made food less affordable, especially for poorer populations. Russia's war on Ukraine further disrupted food shipments passing through the Black Sea, and further led to rising food prices that have resulted from rising fertiliser costs, since Russia and Ukraine are key sources of fertilisers. These altogether contribute to worse food utilisation (the fourth element) outcomes as evidenced by the increase.

Asia's vulnerability to these threats is partly a result of the structure of its AFSCs, given its status as a net-trade deficit region, and to some extent from its import dependence for some food its key food items. The approaches raised in the third section of this chapter present ways by which countries/sub-regions have sought to make AFSCs in the region more resilient to such challenges. Strong state-led approaches can be seen in China to overhaul outdated food systems. China's model can be applicable to other countries where much coordination is needed in optimising the use of scarce financial resources in adapting agriculture to climate change and ensuring food availability to meet growing food demand.

Similarly, in the case of countries where food is heavily subsidised, India's digitalised public sector food distribution to reduce leakages in the supply chain and in government subsidies which are becoming scarcer owing to India's fast-growing population and trajectory of becoming the most populous country within the decade. For higher-income countries who have capacities to support technology upgrading, a potential alternative can be seen in Singapore's state-supported approaches, where the state provides financial support to its private sector in developing and adopting new productivity-improving technologies, alongside ecosystem level approaches including education, incubators/accelerators for supporting burgeoning innovative enterprises, and international linkages in diversifying import sources. Finally, ASEAN's private sector-driven approaches provide a potential model for regional cooperation in promoting AFSCs, since the capability and expertise to develop novel digital technologies in farming are increasingly within the ambit of the private sector.

The bottom line is to ensure that AFSCs are resilient in overcoming long-term climate challenges as well as potential future supply chain disruptions. Moving forward, other regions globally can take a page from the various ways by which Asian countries (and its sub-regions) have been working to address food security challenges, whether through strong state-led approaches, digital approaches to minimise leakages across the supply chain or state-supported financing approaches. Southeast Asia's experience likewise provides the insight that states need not always take the lead but rather, may partner with the private sector in promoting technologies for improving AFSC resilience.

5 References

Aday, S. and Aday, M. S. (2020). Impact of COVID-19 on the food supply chain. *Food Quality and Safety* 4(4), 167–180.

Agri-Food & Veterinary Authority of Singapore. (2013). AVA's food security roadmap for Singapore. *AVA Website*, 4 March 2013 (updated 27 February 2020). Available at: www.sfa.gov.sg/food-for-thought/article/detail/avapercent27s-food-security-roadmap-for-singapore, accessed 30 March 2022.

Agri-Food & Veterinary Authority of Singapore. (2013). *AVA Annual Report 2018/19*. Agri-Food & Veterinary Authority of Singapore, Singapore.

Ambrose, J. (2021). Global demand for coal could hit all-time high in 2022. *The Guardian*, 17 December. Available at: https://www.theguardian.com/business/2021/dec/17/global-demand-coal-high-electricity-plants-covid-economic-recovery, accessed 7 August 2023.

Asian Development Bank (ADB). (2020). Updated assessment of the potential economic impact of COVID-19, *ADB Brief*, No. 133, May 2020, Asian Development Bank,

Manila. Available at: https://www.adb.org/publications/updated-assessment-economic-impact-covid-19, accessed 2 December 2022.

Bhalla, S. S., Bhasin, K. and Virmani, A. (2022). Pandemic, poverty, and inequality: evidence from India. *IMF Working Paper 22/69*.

Caballero-Anthony, M., Teng, P. and Montesclaros, J. M. L. (2020). COVID-19 and Food Security in Asia: How Prepared are We? *NTS Insight, No. IN20-03*. RSIS Centre for Non-Traditional Security Studies (NTS Centre), Nanyang Technological University Singapore, Singapore.

Council of Supply Chain Management Professionals (CSCMP). What is Supply Chain Management? *CSCMP Website*. Available at: https://cscmp.org/CSCMP/Certify/Fundamentals/What_is_Supply_Chain_ Management.aspx?WebsiteKey=0b3f453d-bd90-4121-83cf-172a90b226a9, accessed 31 October 2023.

Dawe, D. and Slayton, T. (2012). The world rice market crisis of 2007–2008. In: *The Rice Crisis*, pp. 39–52. Routledge, London, UK.

Demirgüç-Kunt, A., Klapper, L., Singer, D., Saniya, A. and Hess, J. (2018). *The Global Findex Database 2017: Measuring Financial Inclusion and the Fintech Revolution*. World Bank, Washington, DC.

Development and Reform Commission, China. (2021). Notice of the National Development and Reform Commission and other departments on doing a good job in ensuring the supply and price of domestic fertilizers in the coming period (English translation). *Development and Reform Economic and Trade* No. 1351, issued 22 September. Available at: http://www.gov.cn/zhengce/zhengceku/2021-09/24/content_5639088.htm, accessed 2 December 2022.

European Commission Directorate-General for European Civil Protection and Humanitarian Aid Operations. (2020). Thailand –drought, DG ECHO daily flash of 10 January 2020. *Reliefweb*. Available at: https://reliefweb.int/report/thailand/thailand-drought-dg-echo-government-media-echo-daily-flash-10-january-2020.

Food and Agriculture Organization of the United Nations (UNFAO). (2002). *The State of Food Insecurity in the World 2001*. UN FAO, Rome.

Food and Agriculture Organization of the United Nations (UNFAO). (2023). UN FAOStat database. Available at: https://www.fao.org/faostat/en/#data, accessed 7 August 2023.

Food Corporation India (FCI). (2022). Region wise stock position as of 16.11.2022. *FCI Website*. Available at: https://fci.gov.in/stocks.php?view=82, accessed 1 December 2022.

GSM Association (GSMA). (2020). Digital agriculture maps 2020 state of the sector in low and middle-income countries. Available at: gsma.com/r/wp-content/uploads/2020/09/GSMA-Agritech-Digital-Agriculture-Maps.pdf, accessed 7 August 2023.

Hale, T., Angrist, N., Goldszmidt, R., Kira, B., Petherick, A., Phillips, T., Webster, S., Cameron-Blake, E., Hallas, L., Majumdar, S. and Tatlow, H. (2021). A global panel database of pandemic policies (Oxford COVID-19 Government Response Tracker). *Nature Human Behaviour* 5(4), 529–538.

International Energy Agency (IEA). (2022). Market report: oil 2021. Available at: https://iea.blob.core.windows.net/assets/1fa45234-bac5-4d89-a532-768960f99d07/Oil_2021-PDF.pdf, accessed 7 August 2023.

International Fertilizer Industry Association. (1998). The fertilizer industry, world food supplies and the environment. Available at: large.stanford.edu/courses/2014/ph240/yuan2/docs/1998_IFA_UNEP _worldfoodsupplies.pdf, accessed 7 August 2023.

International Trade Centre (ITC). (2022). Trade map - trade statistics for international business development. *ITC Website*. Available at: https://www.trademap.org/, accessed 1 December 2022.

Laborde, D. D. and Mamun, A. (2022). Documentation for food and fertilizers export restriction tracker: tracking export policy responses affecting global food markets during crisis. In: *Food and Fertilizer Trade Policy Tracker Working Paper 2*. International Food Policy Research Institute (IFPRI), Washington, DC. https://doi.org/10.2499/p15738coll2.135857.

Mbow, C., Rosenzweig, C., Barioni, L. G., Benton, T. G., Herrero, M., Krishnapillai, M., Liwenga, E., Pradhan, P., Rivera-Ferre, M. G., Sapkota, T., Tubiello, F. N. and Xu, Y. (2019). Food security. In: Shukla, J. P. R., Skea, E., Buendia, C., Masson-Delmotte, V., Pörtner, H. O., Roberts, D. C., Zhai, P., Slade, R., Connors, S., van Diemen, R., Ferrat, M., Haughey, E., Luz, S., Neogi, S., Pathak, M., Petzold, J., Portugal-Pereira, J., Vyas, P., Huntley, E., Kissick, K., Belkacemi, M., and Malley, J. (Eds). *Climate Change and Land IPCC Special Report on Climate Change, Desertification, Land Degradation, Sustainable Land Management, Food Security, and Greenhouse Gas Fluxes in Terrestrial Ecosystems*. In press. Available at: https://www.ipcc.ch/site/assets/uploads/sites/4/2021/02/08_Chapter-5_3.pdf, accessed 7 August 2023.

Mentzer, J. T., DeWitt, W., Keebler, J. S., Min, S., Nix, N. W., Smith, C. D. and Zacharia, Z. G. (2001). Defining supply chain management. *Journal of Business Logistics* 22(2), 1–25.

Ministry of Sustainability and the Environment. (2019). *Speech by Mr Masagos Zulkifli, Minister for the Environment and Water Resources, at the Committee of Supply Debate 2019, on 7 March 2019* (line 64).

Montesclaros, J. M. L. (2020). Southeast Asia's food security: inflection point? In: Caballero-Anthony, M. and Gong, L. (Eds), *Non-Traditional Security Issues in ASEAN: Agendas for Action*. ISEAS-Yusof Ishak Institute, Singapore. https://doi.org/10.1355/9789814881098-004.

Montesclaros, J. M. L. (2021). Changing the narrative of ASEAN progress in addressing hunger: 'Snoozing' the alarm for SDG# 2? *Food Security* 13(5), 1283–1284.

Montesclaros, J. M. L. (2022a). Impact of persistent supply chain constraints on food prices and inflation: food security implications. Presentation at the 37th Environmental Scanning Exercise with the Bangko Sentral Ng Pilipinas (English translation: Central Bank of the Philippines).

Montesclaros, J. M. L. (2022b). India's approach to food security resilience amid COVID-19: relevance to ASEAN? *NTS Insight, No. IN22-01*. RSIS Centre for Non-Traditional Security Studies (NTS Centre), Nanyang Technological University Singapore, Singapore.

Montesclaros, J. M. L. (2023). Empowering digitalisation in ASEAN agriculture: lessons from EU regional platforms for geospatial technologies. *RSIS Policy Report* July 2023 S. Rajaratnam School of International Studies, pp. 1-21.

Montesclaros, J. M. L., Babu, S. C. and Teng, P. S. (2019). IoT-enabled farms and climate-adaptive agriculture technologies: investment lessons from Singapore. *IFPRI Discussion Paper No. 1805*. International Food Policy Research Institute, Washington DC.

Montesclaros, J. M. L. and Caballero-Anthony, M. (2020). The burden of COVID-19: urgent need for social safety nets. *RSIS Commentaries*, 043-20.

Montesclaros, J. M. L. and Caballero-Anthony, M. (2022). Ukraine war and food security: how should ASEAN respond? *RSIS Commentaries*, 053-22. Republished in *The Straits Times* as "ASEAN should deal with food security concerns", 8 May 2022. Available at: https://www.straitstimes.com/opinion/asean-should-deal-with-food-security-concerns.

Montesclaros, J. M. L. and Sembiring, M. (2022). Food insecurity beyond borders: untangling the complex impacts of Ukraine war on global food security. *NTS Insight, No. IN22-03*. RSIS Centre for Non-Traditional Security Studies (NTS Centre), Nanyang Technological University Singapore, Singapore.

Montesclaros, J. M. L. and Teng, P. (2023). Impact of COVID-19 on Singapore's rice supplies. In: Davidson, J. S. (Ed.), *Just Another Crisis? The Impact of the COVID-19 Pandemic on Southeast Asia's Rice Sector*. ISEAS Yusof Ishak Institute, Singapore, pp. 161–189.

Oxford Economics. (2021). *The Economic Impact of the Agri-Food Sector in Southeast Asia*. Oxford Economics and Food Industry Asia. Oxford, UK.

Porter, J. R., Xie, L., Challinor, A. J., Cochrane, K., Howden, S. M., Iqbal, M. M., Lobell, D. B. and Travasso, M. I. (2014). Food security and food production systems. In: Field, C. B., Barros, V. R., Dokken, D. J., Mach, K. J., Mastrandrea, M. D., Bilir, T. E., Chatterjee, M., Ebi, K. L., Estrada, Y. O., Genova, R. C., Girma, B., Kissel, E. S., Levy, A. N., MacCracken, S., Mastrandrea, P. R. and White, L. L. (Eds), *Climate Change 2014: Impacts, Adaptation, and Vulnerability. Part A: Global and Sectoral Aspects*. Contribution of Working Group II to the Fifth Assessment Report of the Intergovernmental Panel on Climate Change. Cambridge University Press, Cambridge and New York, pp. 485–533. Available at: https://www.ipcc.ch/site/assets/uploads/2018/02/WGIIAR5-Chap7_FINAL.pdf, accessed 7 August 2023.

PwC, Rabobank and Temasek. (2019). The Asia Food Challenge: harvesting the future. PwC, Rabobank and Temasek, Singapore. Available at: https://www.theasiafoodchallenge.com/2019-report.html, accessed 7 August 2023.

Qian, C., Crossley, G., Patton, D. and Cho, E. (2021). China's major fertiliser makers to suspend exports amid tight supplies. *Reuters*, 30 July. Available at: https://www.reuters.com/article/us-china-exports-fertilisers-idUSKBN2F007W, accessed 7 August 2023.

Shaw, R., Luo, Y., Cheong, T. S., Abdul-Halim, S., Chaturvedi, S., Hashizume, M., Insarov, G. E., Ishikawa, Y., Jafari, M., Kitoh, A., Pulhin, J., Singh, C., Vasant, K. and Zhang, Z. (2022). Asia. In: Pörtner, H.-O., Roberts, D. C., Tignor, M., Poloczanska, E. S., Mintenbeck, K., Alegría, A., Craig, M., Langsdorf, S., Löschke, S., Möller, V., Okem, A. and Rama, B. (Eds), *Climate Change 2022: Impacts, Adaptation and Vulnerability. Contribution of Working Group II to the Sixth Assessment Report of the Intergovernmental Panel on Climate Change*. Cambridge University Press, Cambridge and New York, pp. 1457–1579, https://doi.org/10.1017/9781009325844.012, accessed 30 October 2022. Available at: https://www.ipcc.ch/report/ar6/wg2/downloads/report/IPCC_AR6_WGII_Chapter10.pdf accessed 7 August 2023.

Shields, M. (2017). ChemChina clinches landmark $43 billion takeover of Syngenta. *Reuters*, 4 May.

Singapore Food Agency. (2021). *Annual Report 2020/21*. Singapore Food Agency, Singapore.

Teng, P. (2017). Knowledge intensive agriculture: the new disruptor in world food? *RSIS Commentary*, 23 June.

Teng, P. (2020a). Swine fever, climate change, armyworm: a perfect storm for Asia's food prices. *South China Morning Post*, 10 January.

Teng, P. (2020b). COVID-19: maintaining food security in Asia Pacific. *Inter Press Service News*, 16 April.

Teng, P. and Montesclaros, J. M. L. (2023). Agricultural transformation for small (island and developing) states. *Asian Journal of Agriculture and Development* 20(1), 13–30. https://doi.org/10.37801/ajad2023.20.1.2.

Teng, P., Montesclaros, J. M. L., Hulme, R. and Powell, A. (2019). The evolving Singapore agrifood ecosystem. *NTS Insight, No. IN19-03*. RSIS Centre for Non-Traditional Security Studies (NTS Centre), Nanyang Technological University Singapore, Singapore.

Trading Economics. (2022). Brent crude oil. *Trading Economics*. Available at: https://tradingeconomics.com/commodity/brent-crude-oil?embed.

Wang, O. (2021). China food security: Syngenta buyout shows Beijing's haste to plug technology gap with foreign expertise. *South China Morning Post*, 1 July.

Woods, E. J. (2003). Supply-chain management: understanding the concept and its implications in developing countries. In: Johnson, G. I. and Hofman, P. J. (Eds). *Agriproduct Supply-Chain Management in Developing Countries*. Workshop Proceedings, 19–22 August 2003, Bali, Indonesia.

Part 2

Assessing agri-food supply chain performance: Tools and techniques

Chapter 6

End-to-end performance measurement systems for agri-food supply chains

Lusine Aramyan, Wageningen University and Research, The Netherlands; and Jos van Iwaarden, Kearney, The Netherlands

1 Introduction

Measuring supply chain performance across various industry sectors has gained increasing attention recently (see, for instance, the works of Neely et al., 1995, 2005; Beamon, 1998, 1999; Christopher, 1998; Li and O'Brien, 1999; Gunasekaran et al., 2001, 2004; Lambert and Pohlen, 2001; Lohman et al., 2004; Govindan et al., 2017; Maestrini et al., 2018; Simão et al., 2022). While there have been many studies on measuring supply chain performance in manufacturing industry, research in the agricultural sector has been lacking. Studies have begun to develop from 2000s (see, for instance, Van der Vorst, 2000, 2005; Aramyan et al., 2006, 2007, 2009; Gellynck et al., 2008; Widyaningrum and Masruroh, 2012; Fattahi et al., 2013; Piotrowicz and Cuthbertson, 2015; Beske-Janssen et al., 2015; Govindan et al., 2017; Chopra et al., 2017; Moazzam et al., 2018; Trivellas et al., 2020). The reason for the growing attention on this topic in the agricultural sector is increasing competition as a result of globalization, which is paired with quality standard requirements, ever higher customer expectations for fresh, healthy, and safe food produced in an environmentally and socially responsible way and available 24 h per day.

Measuring the performance of the entire supply chain is important because it provides the key data for benchmarking and improving performance (Aramyan et al., 2007). Performance measurement of the supply chain brings

http://dx.doi.org/10.19103/AS.2023.0122.09

together business information and market intelligence data about market conditions external to the individual firm. The general idea behind performance measurement of the supply chain is to optimize activities to reach the highest overall performance of the chain by adding value for the least possible cost. In the following, we summarize the arguments for measuring (chain) performance.

Performance measurement:

- Allows decision-makers to align their strategic activities to a strategic plan.
- Provides an overview of how well the targets are achieved.
- Provides an overview of strategic improvements and adjustments needed.
- Allows decision-makers to identify best practices and expand their usage elsewhere in the supply chain.
- Allows benchmarking against competitors.

Performance measurement systems (PMS) for agri-food supply chains face particular challenges due to specific supply chain characteristics such as seasonality, perishability and short shelf life for fresh produce, variability in output quality and quantity, the involvement of many different actors, long production times, and specialized transportation requirements (Van der Vorst, 2000; Aramyan et al., 2007).

In addition, agri-food supply chains currently face challenges from climate change, partly because the global food system is an important contributor to climate change. The Intergovernmental Panel on Climate Change estimates that '23% of total anthropogenic greenhouse gas emissions (2007–2016) derive from Agriculture, Forestry and Other Land Use (AFOLU)' (2020, p.8). In turn, climate change also threatens the future of the agri-food sector due to extreme weather events such as droughts, floods, and other effects such as changing patterns in the incidence of pests and diseases (De Vries, 2019; De Haan et al., 2019). These challenges have resulted in the rise of the concept of circular agriculture. This originates from industrial ecology (Jurgilevich et al., 2016) and aims to reduce resource consumption and emissions to the environment by closing loops in the use and loss of materials and other system inputs. This implies that, wherever possible, inputs are recovered for reuse, remanufacturing, and recycling (de Boer and van Ittersum, 2018).

More generally, in 2015, all United Nations Member States adopted the 2030 Agenda for Sustainable Development, which provides a shared blueprint for peace and prosperity for people and the planet, now and in the future. 'At its heart are the 17 Sustainable Development Goals, which are an urgent call for action by all countries - developed and developing - in a global partnership. They recognize that ending poverty and other deprivations must go hand-in-hand with strategies that improve health and education, reduce inequality, and spur economic growth - all while tackling climate change and working to

Published by Burleigh Dodds Science Publishing Limited, 2024.

preserve our oceans and forests' (https://sdgs.un.org/goals). This implies that, in the agri-food sector, farmers, agribusinesses, governments, and society must cooperate to promote inclusive, fair, and efficient food systems, including better integration of smallholder farmers into supply chains and agribusiness and improving their access to markets which are characterized by ever-changing consumption patterns (Naseer et al., 2019).

In addition, over the last few years, several countries have implemented mandatory Environmental, Social, and Governance (ESG) reporting for large companies. ESG reporting involves disclosure of performance of business operations related to ESG aspects, i.e. performance related to a business impact on the environment and society (e.g. how fairly it treats its staff and suppliers such as farmers) as well as how transparent and accountable its reporting on such issues are. Since 2022, ESG reporting, e.g. has become mandatory in UK and Canada, and in 2023, it will become mandatory in New Zealand. In the EU, currently some large companies must already disclose ESG information under the Non-Financial Reporting Directive (NFRD). However, from 2023, the NFRD will be replaced by a new ESG reporting directive – the Corporate Sustainability Reporting Directive, which will extend reporting requirements to approximately 50 000 companies across the EU.

While the performance of conventional agri-food supply chains has traditionally been measured in terms of productivity and profit, requirements for a more sustainable agriculture and circular economy require consideration of other dimensions such as measuring environmental and social performance of agri-food supply chains. Despite its importance, a recent literature review on PMS shows that studies including the concept of circular agriculture and/or circular supply chain management are limited (Vegter et al., 2021). The authors also concluded that most of the available performance measures for circular supply chain management have not yet been tested in practice and are therefore still in an early phase of development.

In this chapter, we discuss the main components of PMS, what methods of measuring supply chain performance exist, how they have evolved over time, what the challenges related to each method are, and how environmental and social indicators can be incorporated in PMS.

2 Key performance indicators for agri-food supply chains

An essential part of developing a PMS is the identification of key performance indicators (KPIs). While there are many indicators of performance that can be deployed in a supply chain, there is a small set of critical factors that disproportionally affect success or failure in the market that can be defined as KPIs.

KPIs provide quantitative information to decision-makers about their products, services, and production processes and inform decision-makers whether they are meeting their goals, whether customers are satisfied, and where improvements are necessary. KPIs are thus used to help measure progress toward achieving the strategic goals set for the supply chain.

The choice of appropriate supply chain performance indicators in agri-food supply chains is complicated due to the presence of multiple inputs and outputs in the system. While there is an extensive literature on supply chain KPIs and PMS, the literature on KPIs and effective PMS in agri-food sectors is still limited compared to other industries such as the automotive sector. Moreover, actual implementation of supply chain PMS in agri-food chains in practice is also still limited (Aramyan, 2007). It has, however, gained more importance in recent years, particularly KPIs related to green and sustainable supply chains. In 2015, Ahi and Searcy studied 445 journal articles and found 2555 performance indicators for green and sustainable supply chain management. Given the large number of performance indicators in practice, selection of the right KPIs is crucial. Supply chain actors need to develop strategic goals and set KPI targets and objectives accordingly to ensure the long-term success of the supply chain.

Creating a circular agriculture and developing sustainability strategies require quantifiable and measurable environmental and social goals. However, this has proved to be challenging (Gold et al., 2010; Beske-Janssen et al., 2015). While circularity is supposed to lead to more sustainability, it does not necessarily contribute to it. For example, 'reuse and recycle' strategies promote the return of products after end-of-use in order to reuse products or their components. This leads to less use of raw materials and hence less resource depletion and less costs. However, these returns may also require more transportation and an energy-intensive process for reusing the product or for the recycling of materials which, in turn, may generate additional CO_2 emissions and increased costs (Batista et al., 2018; Sehnem et al., 2019; Vegter et al., 2021). As a consequence, trade-offs are often required between traditional economic goals and other KPIs (e.g. relating to sustainability outcomes).

2.1 Trade-offs between KPIs

Supply chain actors typically seek to simultaneously optimize multiple objectives in their supply chain management practices, for example, improving product quality while lowering costs. However, such optimization often requires a compromise in some objectives in order to obtain benefits in others which, in its turn, requires decision-makers to make trade-offs. Examples of such trade-offs are customer service levels versus inventory holding costs, product quality versus production costs and, of course, trade-offs, between environmental, social, and economic objectives. In order to be successful, companies have to make decisions that enhance their margins, allow them to fulfill customer demands on quality and price in a fast and flexible manner and, at the same time, be sustainable. They must therefore evaluate trade-offs between efficient and sustainable solutions (Darvish et al., 2019). This makes it particularly important to focus on minimizing and balancing trade-offs among different objectives.

Well-defined PMS often help to guide supply chain actors in making trade-offs between performance components in order to improve the performance of the entire supply chain (Aramyan, 2007). For instance, the implementation of a specific management system (or plan or strategy) in the chain may increase the costs of the product but, at the same time, it may increase flexibility or improve food quality. If supply chain members consider food quality or flexibility as the most important aspect of performance, a trade-off can be made between increased costs versus increased flexibility or improved food quality.

Trade-offs are usually made based on an overall supply chain strategy because they influence overall chain performance and profits. In order to develop a supply chain PMS and define relevant performance indicators, supply chain operators need first to develop strategic goals and establish key objectives with KPIs to measure how well targets are being achieved. Different objectives are usually developed at strategic, tactical, and operational levels in order to allow a supply chain to measure progress toward the overall goal. The strategic level is the top level of supply chain management responsible for long-term decisions by the company (e.g. 5 years or more). The tactical level is the second level of supply chain management and involves medium-term decisions for the supply chain (e.g. over a year). The operational level involves day-to-day decision-making and planning to keep the supply chain running.

Four key supply chain pillars can be distinguished: Plan, Source, Make, and Deliver (Supply Chain Operations Reference (SCOR®), 2004). Each of these pillars has a clearly defined function in the end-to-end supply chain (see Fig. 1). Each pillar thus also affects supply chain KPIs. This has traditionally been a challenge because it requires a trade-off between economic indicators (such as inventory costs, production costs, and distribution costs) while meeting customer/market requirements for service levels. Modern day supply chain management is increasing this trade-off challenge, because responsive and efficient supply chains are no longer sufficient. In a world of disruption, resilience has emerged rapidly as a supply chain KPI together with ESG requirements (see Fig. 1). A balance should be found between efficient production with lower costs, responsive service, chain flexibility and agility, meeting sustainability requirements (e.g. ESG), and being resilient to disruption (i.e. the ability to

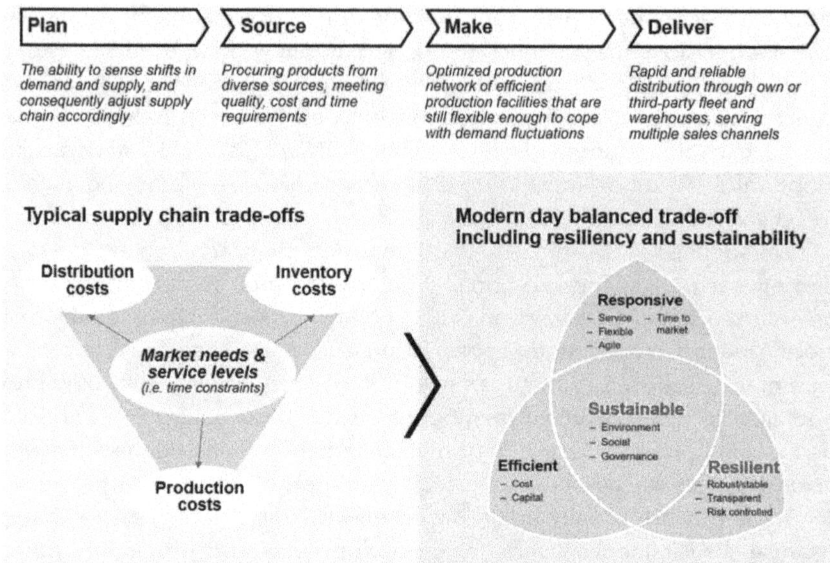

Figure 1 Supply chain trade-offs: typical vs modern. Source: Adapted from Kearney (www.kearney.com/operations-performance-transformation).

quickly and efficiently adjust operations to manage disruptions and even avoid them by minimizing the impact of events before they occur). The winning supply chains of tomorrow will be the ones that are able to break free from today's trade-off thinking and instead create win-win situations where e.g. efficiency increases while at the same time being sustainable.

3 Performance indicators and frameworks used in agri-food supply chains

As mentioned, measuring performance of agri-food supply chains is complex given their specific characteristics (Van der Vorst, 2000; Van der Spiegel, 2004; Aramyan et al., 2007). This implies that PMS developed for general supply chains will need to be adapted for agri-food supply chains, as measurement indicators such as food quality, safety, and seasonal factors will have to be considered. Important distinctions are made between daily fresh products (vegetables and fruits), chilled products (salads, dairy products, etc.), frozen products (fish, ice cream, etc.), and non-perishables such as sugar and coffee (Van der Vorst, 2000). Qualitative performance indicators that reflect consumer acceptance of the product (qualitative aspects such as taste, texture, and color) need to be considered along with other non-qualitative performance indicators (Apaiah, 2006; Aramyan et al., 2009).

In 2000, in one of the few studies on the topic, Van der Vorst highlighted several performance indicators for food supply chains on three levels:

- Supply chain;
- Organization; and
- Process.

At supply chain level, there are five indicators, which are as follows:

- Product availability;
- Quality;
- Responsiveness;
- Delivery reliability; and
- Total supply chain costs.

At organization level, there are another five indicators, which are as follows:

- Inventory level;
- Throughput time;
- Responsiveness;
- Delivery reliability; and
- Total organizational costs.

Finally, at process level, there are four indicators, which are as follows:

- Responsiveness;
- Throughput time;
- Process yield; and
- Process costs.

In 2007, Aramyan et al. suggested a performance measurement framework with a set of performance indicators for agri-food supply chains, consisting of four components with a comprehensive set of KPIs per component (see Table 2 in Annex). The four components are as follows:

1. Efficiency, i.e. indicators dealing with how well resources are utilized, including such KPIs as company turnover, profit, and quantity of products sold.
2. Flexibility, i.e. indicators dealing with how well supply chains cope with a changing environment and with extraordinary customer service requests.

3. Responsiveness, i.e. indicators dealing with the lead time between products requests and their delivery.
4. Food quality, i.e. dealing with product safety and health/nutritional properties, sensory properties and shelf life, and product quality aspects.

Gellynck et al. (2008) added additional performance indicators to those of Aramyan et al. (2007), such as growth in terms of market share, stability, and chain balance in agri-food chains. Later on, Fattahi et al. (2013) argued that financial, quality and safety, flexibility, customer service, efficiency, and chain coordination are useful indicators to define performance measurement of agri-food value chains. Chopra et al. (2017) applied the framework developed by Aramyan et al. (2007) to a rice supply chain and found that the rice supply chain stakeholders did not measure the importance of performance indicators equally but that all four performance indicators were relevant to rice millers and other actors in the supply chain.

4 Environmental and social performance indicators

The challenge of feeding 9 billion people by 2050 puts significant pressure on land and water resource use. Due to intensification of agri-food production, natural resources are being overexploited and negative impacts on human health are increasing (e.g. in terms of the impact of ultra-processed foods in promoting obesity). This means the environmental and socioeconomic costs associated with the externalities of the intensification of agri-food production are also increasing dramatically (Jackson et al., 2009). These costs are typically not considered in prices of food products and thus are not reflected in the performance of agri-food supply chains. There has been a more recent focus on developing performance indicators used in agri-food supply chains to reflect the cost of these externalities[1]. Gaitan Cremaschi et al. (2017) e.g. have developed a framework for measuring agri-food supply chain sustainability performance with two unique metrics, based on a total factor productivity indexing approach, to compare products in terms of their sustainability performance. Both metrics are adjusted to internalize food production's social and environmental externalities and consider the sustainability effects of different stages along agri-food supply chains. Similarly, Walkiewicz et al. (2021) have incorporated externalities related to sustainable performance activities (e.g. the time and costs of activities that create externalities or value-added) into a PMS for the German food industry. The resulting KPIs and identification

[1] Internalization of externalities refers to all measures (public or private) that guarantee that unpaid benefits or costs are considered in the composition of goods and services prices (Ding et al., 2014).

Published by Burleigh Dodds Science Publishing Limited, 2024.

of performance and cost indicators reflected four interrelated dimensions: ecological, social, financial, and knowledge.

Dutch companies in agri-food supply chains have recently started developing schemes related to the performance of their farmers. In these schemes, farmers are rewarded for good performance and penalized for bad ('bonus-malus'). The focus of these arrangements is on sustainable agriculture in a broader sense: improving soil quality, reduced use of pesticides, enhancing water quality, lower energy use, and promoting biodiversity (Runhaar, 2016). Beyond these indicators, the implementation of a Dutch circular agricultural policy requires KPIs focused on integrated management which accounts for climate, biodiversity, and nature (Louis Bolkinstituut, 2020). An integral set of KPIs (later renamed Critical Performance Indicators (CPIs)) has been proposed to measure farmer performance/contribution to societal goals such as restoration of biodiversity, soil quality, landscape, climate and the wider environment. By adopting best management practices farmers (land users) can improve their score against CPIs. The development and implementation of CPIs are still underway. Baayen et al. (2022) suggest that it is crucial to define what issues CPIs can resolve and how a CPI system can fit into relevant governance and legislative frameworks. One of the conclusions of this study is that the prime contribution of a CPI system to enhancing sustainability in agriculture is the definition of objectives and desirable behavioral outcomes. This requires a common language, measuring methodology, and targets that all stakeholders can understand and accept.

In addition to capturing the effects of internalization of externalities on performance, there has also been a focus in recent years on issues concerning just and fair distribution of costs and profits between agri-food supply chain partners. For example, when adopting a new biomass sustainability certification system in agri-food supply chains, it was found that actors at the beginning of the supply chain (farmers/plantations) bear most of the annual certification costs, while actors at the end of the supply chain often receive most of the external benefits. There is thus a discrepancy in balancing costs and benefits in the chain. Certification systems do not have defined rules to ensure that a share of the external benefits return to the farmer (van Dam et al., 2012). Farmers in the EU often complain about the 'unfairness' of the benefits that retailers/processors receive from farmers' efforts to improve environmental or social sustainability performance. These imbalances require agri-food supply chain actors to find the ways to deal with these perceptions of unfairness. Addressing this issue is important as alignment of the perceptions of fairness among supply chain actors can improve chain performance (Bouazzaoui et al., 2020). See the example on how chocolate producer Tony Chocolonely addresses the issues of Fair Price in Box 1.

Box 1 The Living Income Reference Price of Tony Chocolonely

In 2019, Dutch chocolate brand Tony's Chocolonely whose mission is to 'make all chocolate 100% slave free' has introduced the Living Income Reference Price (LIRP). That's the price a farmer should receive for one kilo of cocoa beans to enable him or her to earn a living income. The price is based on complicated calculations, taking into account productivity, country, and family size and diversification of income. 'We think the cocoa industry has a responsibility to pay this reference price for a livable income. You too ...?' The brand is known for its 'unequally divided' chocolate bars, designed to underline the inequality in the cocoa and chocolate industries. It claims cocoa farmers in West Africa have been hard hit by increasing inflation which has forced them 'further into poverty'.

Source: https://tonyschocolonely.com/uk/en

5 Methods and techniques to measure performance

Performance measurement has evolved over time. Initially, it was a measurand solely at a firm level. Globalization and changing customer requirements have forced industries to adapt new and efficient manufacturing practices and change traditional ways of business to remain competitive. Examples of methods applied to firm-level performance measurement are Activity-Based Costing (ABC) and Balanced Scorecard and Economic Value-added (EVA).

As agri-food supply chains have become more complex, and with increasing performance requirements in terms of cost, speed, flexibility, resilience, and sustainability, there is a growing need to measure overall supply chain performance. Some existing methodologies, like Balanced Scorecard, have been extended to measure supply chain performance. New methodologies have also been developed. One of the first methods to measure supply chain performance was the SCOR model developed by the Supply Chain Council in 2004. While previous methods focused largely on economic performance and aimed to enhance supply chain returns and profitability, more recent performance measurement methodology has also focused on sustainability aspects and accounting for environmental and social impacts. Examples of such methods are Life-Cycle Assessment (LCA) and True Price and/or Fair Price assessments. Table 1 provides a short overview of different performance measurement methods, including a summary of the advantages and disadvantages of these methods.

Table 1 Performance measurement methods

Method	Advantages	Disadvantages
Activity-Based Costing (ABC)	• Gives more than just financial information • Recognizes the changing cost behavior of different activities	• Costly data collection • Difficulties to collect initial required data • Heavily relies on financial indicators • Difficulties to determine appropriate and acceptable costs drivers
Balanced Scorecard	• Balanced view about the performance • Financial and non-financial factors • Top-level strategy and middle management-level actions are clearly connected and appropriately focused	• Not a quick fix • Complete implementation should be staged
Economic Value-Added (EVA)	• Considers the cost of capital • Allows projects to be viewed separately	• Computation's difficulties • Difficult to allocate EVA among divisions
Life-Cycle Assessment (LCA)	• Allows to establish comprehensive baselines of information on a product's or processor's resource requirement • Allows to identify areas where the greatest reduction of environmental burdens can be achieved • Possibility to assess the cost and environmental effects associated with life cycle of a product or process	• Data-intensive methodology • Product-level analysis • Lack of confidence in the LCA methodology
Supply Chain Council's SCOR Model	• Takes into account the performance of overall supply chain • Balanced approach • Performance of supply chain in multiple dimensions	• Does not attempt to describe every business process or activity • Does not explicitly address training, quality, information technology, and administration
True price	• Takes into account the performance of overall supply chain • Considers costs of externalities which can be reflected in the performance of agri-food supply chains	• Data-intensive methodology • Calculating it requires defining what can be considered sustainable (and unsustainable) in the economy • Difficult to monetize externalities • Still under development
Fair price	• Takes into account a fair distribution of costs and benefits for all chain actors • Analyses the aspects that affect the distribution of costs and the prices in agricultural value chains in relation to fairness	• Data-intensive methodology • Difficult to monetize fairness perceptions • Still under development

Methods for measuring supply chain performance include the following:

- *ABC:* The ABC method is based on accounting methods and involves breaking down activities into individual tasks or cost drivers, while estimating the resources (i.e. time and costs) needed for each one. Costs are then allocated to these cost drivers, including allocating overhead costs either equally or based on ranking cost drivers. This approach allows for better assessing the productivity and costs of a supply chain process. Using the ABC method means that companies can more accurately assess, for instance, the costs of services for a specific customer or the costs of marketing a specific product. Businesses can then understand the factors that drive each major activity, the costs of activities, and the relationship between activities and products. ABC analysis does not replace traditional financial accounting but provides a better understanding of performance by looking at the same numbers in a different way (Lapide, 2000).
- The *Balanced Scorecard* is a popular performance measurement scheme initially developed by Kaplan and Norton (1992). This method employs performance metrics from different perspectives:
  - Financial (e.g. cost of manufacturing and cost of warehousing);
  - Customer (e.g. on-time delivery and order fill rate);
  - Business process (e.g. manufacturing adherence-to-plan); and
  - Innovation and technology perspective (e.g. new product development cycle time).

By combining these different perspectives, the Balanced Scorecard helps a manager to understand the interrelationships and trade-offs between alternative performance metrics and leads to improved decision-making. This method is not specifically designed for supply chains but could be adapted to focus on supply chain performance. The Balanced Scorecard is more tactical and strategically oriented compared with the SCOR® model which is an operation-oriented method.

- The *Supply Chain Council's* SCOR® Model is a standard supply chain process reference model designed to fit all industries (Supply-Chain Council, 2004). This model provides guidance on the types of metrics decision-makers can use to develop a balanced approach toward measuring the performance of an overall supply chain. The SCOR® Model advocates a set of supply chain performance indicators as a combination of the following:

1. Reliability measures (e.g. fill rate, perfect order fulfillment);
2. Cost measures (e.g. cost of goods sold);
3. Responsiveness measures (e.g. order fulfillment lead time); and
4. Asset measures (e.g. inventories).

The SCOR® Model directly addresses the needs of supply chain management at the operational level. One of the tenets of the SCOR® Model is that a supply chain must be measured and described in multiple dimensions. These dimensions include reliability, responsiveness, flexibility, cost, and efficiency of asset utilization. The SCOR model is a cross-industry model that breaks down the processes within a supply chain into five primary processes: plan, sources, make, deliver, and return. Supply chain success depends on the success of each process. The model provides a best practice view of supply chain processes.

- *EVA*: This method is based on the assumption that shareholder value is increased when a company earns more than its cost of capital. Unlike balanced scorecards, which offer a functional focus toward performance, the EVA offers a project focus. EVA attempts to quantify value created by an enterprise, based on the operating profits in excess of capital employed (through debt and equity financing). EVA metrics are less useful for measuring detailed supply chain performance. They can be used, however, as supply chain metrics within an executive-level performance scorecard and can be included in other measurement systems such as The Logistics Scoreboard approach (Lapide, 2000).
- *LCA* involves making detailed measurements of input use and environmental waste during the production of a product, from the mining of the raw materials used in its production and distribution through to its use, possible reuse or recycling, and its eventual disposal. LCA has thus far focused on the environmental burden a product poses throughout its life. It offers possibilities for extension to include economic performance, when combined with life-cycle cost assessment methods (Azapagic and Clift, 1999; Hagelaar and Van der Vorst et al., 2009; Carlsson-Kanyama et al., 2003). Using the life-cycle cost assessment method, it is possible to integrate economic and environmental cost information into the LCA framework and assess the cost and environmental effects associated with the life cycle of a product or process.
- *True price* is a more recent method which has grown in popularity. It is a methodology that accounts for the external costs of societal impacts (be it environmental or social) in the value chain of an agricultural

product. True pricing entails the quantification of externalities and their expression in monetary terms in a way that is comparable to a product's market price (Galgani et al., 2021). The true price is a way of accounting for negative consequences of production and consumption at the product level, expressed as costs in monetary terms. The sum of environmental and social costs is called the true price gap. The true price of a product is defined as the sum of the market price and the true price gap, and this is the price a buyer would have to pay for a product if the cost of remediating its unsustainable externalities were added on top of its price. The lower the true price gap, the less social and environmental costs a product has (Galgani et al., 2021). See the example of the shop in Amsterdam that sells products with a 'true price' in Box 2.

- *Fair price* is very often used in combination with True Price and refers to True and Fair Price methodology. Despite this, there is a substantial difference between these two methods. While True price considers the costs of externalities in the price of the product, in Fair Price, a key issue is whether the (additional) costs of sustainability and risks that are linked to producing a product in a more sustainable manner are fairly distributed between the actors in the value chain. Furthermore, if a market needs to be established for more sustainable products, the question arises whether (enough) consumers are willing to pay. What prices, or any other market outcomes, are (un)fair in agricultural value chains follows from an economic notion of (in)efficient markets and economic behavior, as well as from personal and societal ethics and perceptions of market actors involved. The discussion about fair or just prices appeals to moral notions such as a fair distribution of remuneration for all parties and their performance in the food chain. This methodology is currently under development. Analysis of factors that affect the distribution of costs and the prices in agricultural value chains in relation to fairness, including practical guidance for value chain actors, is being developed in research programs such as follows: True and Fair Price for Sustainable Products (see more at https://www.wur.nl/en/project/True-and-fair-price-for-sustainable-products.htm). The framework links economic models (e.g. neo-classical, industrial, transaction costs) to a construct of subjectively perceived fairness (see e.g. Diller, 2008). The framework is built on the links between fairness in value chains and chain performance and includes goals (justice), norms and rules, fairness perceptions, behavior, relationship quality, and performance.

Box 2 New Amsterdam store sells goods at the price required to avoid pollution, exploitation and climate change

Paying the true price for a better world

Now open every Saturday in the heart of the famed Amsterdam shopping district Haarlemmerdijk, True Price Store is offering products 'free from poverty, exploitation, pollution, destruction and climate change', according to Michel Scholte, director of external affairs for the social enterprise True Price.

The store will begin by selling coffee, chocolate, and bread, before expanding its repertoire and opening hours. Meanwhile, the store hopes to already stand as an example to other retailers, restaurants, and supermarkets, encouraging them to follow suit ...

The true - and fair - cost ...

So, what is the true price of goods? 'It differs per product', explains Scholte. 'We show what a product actually costs society in addition to the market price. For example, jeans here cost €40 plus €33 more, a t-shirt cost €15 plus €8, a chocolate bar €2.79 plus 90 cents, a café latte €3.50 plus 25 cents, and a loaf of bread €3.25 plus 18 cents'.

'At this moment we, only voluntarily charge a small part of that: the costs for climate change and underearning of farmers. And this money is then redirected to reforestation and supporting the extreme poor. However, this is just a part of the total true price gap', says Scholte. 'We still need to not only improve these remediations, but also address 29 other social and environmental impacts, including water, soil, air and the treatment of workers', says Scholte. 'Only then will we have a truly sustainable product'.

Source: Iamsterdam, 2020. https://www.iamsterdam.com/en/business/news-and-insights/news/2020/true-price-store

6 Conclusion and future trends

Methods for performance measurement of businesses have existed for centuries. However, the concept of measuring performance of entire supply chains is relatively young. Traditionally, most organizations have viewed themselves as entities that exist independently from others and need to compete with each other in order to survive. Increasingly, organizations are now focusing on their core business (i.e. the activities that they do really well and where they have competitive advantage over other organizations), while any other activities and products are procured from other organizations. This trend toward outsourcing and globalization has been a major development in global business. To effectively manage and control a chain of organizations, involvement of all supply chain members is necessary. Organizations are thus forced to look beyond the performance of their own organization, focusing also on the performance of the entire supply chain. Performance measurement of agri-food supply chains has, until the last decade, received little research attention. Recently, due to increasing globalization and competition in the world market, rising consumers' demands, governmental regulations of food quality, and the greater prominence of sustainability issues, measuring performance of agri-food supply chains has become a topic of interest for many researchers. Starting from the beginning of the twenty-first century, the number of papers devoted to this issue has increased and is expected to increase even further.

Organizations in the agri-food sector are generally interested in the implications and benefits that a supply chain approach to performance measurement could bring. However, the actual practice of implementing a supply chain PMS in agri-food supply chains is still limited. One of the reasons is the many difficulties in measuring the performance of organizations that have no legal power or other form of authority over each other. Moreover, most of the available performance measures related to circular supply chain management have not yet been tested in practice and are therefore still in an early phase of development.

Future research on supply chain PMS has to pay explicit attention to these challenging issues of enabling chain members to come to mutual agreement to implement such PMS. No system, however, skillfully designed, will make a difference unless supply chain members come to an agreement to support and use it in their daily practice. It is difficult to develop and adopt PMS that satisfy the needs of all stakeholders and ensures maximum value to end-users (consumers) (Mishra et al., 2018). For a PMS to be functional, it needs to fit the environment in which it operates (Ramos et al., 2022). Measurement criteria can be specific for each type of supply chain (Najmi and Makui, 2012). Different industries will require different metrics based on their supply chain performance characteristics and specific business environments (Bulsara et al., 2016; Govindan et al., 2017). A key issue in performance measurement is to determine which goal it is designed to support. Increasing social and environmental concerns are forcing supply chain actors to balance the economic, environmental, and social dimensions of their business. To be efficient, companies must now evaluate the trade-offs between efficient versus sustainable solutions. It is important therefore to focus on finding win-win solutions able to generate efficient and sustainable solutions that bring benefits to all supply chain actors.

Measuring performance of agri-food supply chains is complex due to specific characteristics such as seasonality, food quality, and safety issues. Sustainability challenges require inclusion of environmental and social performance indicators in PMS alongside more traditional economic performance indicators. While previous supply chain PMS aimed mainly at enhancing supply chain returns and profitability, focusing mainly on financial and economic performance indicators, current performance measurement urgently requires inclusion of more sustainability indicators and accounting for environmental and social impacts. As a result, new performance measurement methodologies such as True and Fair Price are currently being pioneered. It is expected that this trend will continue in years to come, given such urgent challenges as climate change and increasing concern about societal impacts of business activities.

As discussed, it is expected that future research will focus on further development of performance measurement methodologies such as True and Fair Price, since calculation of the costs of externalities will most likely become mandatory for all supply chain actors in future (including agri-food supply chains), requiring a reliable methodology to achieve this goal. This trend is already notable with growing mandatory ESG reporting, which requires reliable KPIs to objectively demonstrate a company's progress in ESG aspects of its operations against generally agreed benchmarks and targets. As the global food system is an important contributor to climate change, agri-food supply chain actors will be increasingly forced to comply with legislation related to reducing negative environmental impacts, thus introducing new business models with a robust set of KPIs where costs are balanced against performance, resilience, sustainability, and social equity.

7 Where to look for further information

To delve deeper into the complex world of agri-food supply chain performance measurement, various sources can be explored. The academic literature in the fields of supply chain management, agriculture, and sustainability can provide in-depth insights into the subject. For this, explore academic journals such as the *Journal of Supply Chain Management*, *Agricultural Systems*, *Food Policy*, and *Journal of Agricultural and Food Industrial Organization*. These journals often publish research articles, case studies, and reviews related to agri-food supply chain performance measurement. The following authors are the frontrunners on the topic of supply chain performance: Neely et al., 1995, 2005; Beamon, 1998, 1999; Christopher, 1998; Gunasekaran et al., 2001, 2004; Lambert and Pohlen, 2001; Lohman et al., 2004; Govindan et al., 2017;

Maestrini et al., 2018; and Simão et al., 2022. For performance measurement specifically in the agricultural sector, see Van der Vorst, 2000, 2005; Aramyan et al., 2006, 2007, 2009; Gellynck et al., 2008; Widyaningrum and Masruroh, 2012; Fattahi et al., 2013; Piotrowicz and Cuthbertson, 2015; Beske-Janssen et al., 2015; Govindan et al., 2017; Chopra et al., 2017; Moazzam et al., 2018; and Trivellas et al., 2020.

Additionally, industry reports, white papers and reports from organizations such as the Food and Agriculture Organization of the United Nations (FAO) (see for instance some developed performance tools at https://www.fao.org/policy-support/tools-and-publications/resources-details/en/c/1257355/), and global consulting firms can provide practical applications and case studies. Explore for instance performance measurement topics related to a supply chain resilience and strategic insights at the website of Kearney https://www.kearney.com/service/operations-performance/kearney-supply-chain-institute and https://www.kearney.com/service/operations-performance/kearney-supply-chain-institute. Visit websites of SCOR models such as https://scor.ascm.org or https://scmedu.org/scor/.

For studies related to true and fair prices visit websites:

- https://www.wur.nl/en/project/true-and-fair-price-for-sustainable-products.htm.Logatcheva, et al., 2023. True Cost Accounting (TCA): a methodology for making the global food system more sustainable (wur.nl).
- Taufik, D., Reinders, M. J. and van Haaster-de Winter, M. A. (2023). Creating trust and consumer value for true price food products, *Journal of Cleaner Production*, 390, 136145, ISSN 0959-6526. doi:10.1016/j.jclepro.2023.136145.
- Galgani, P., Woltjer, G., de Adelhart Toorop, R. and Ruiz, A. G. (2021). Fossil fuel and other non-renewable material depletion: true pricing method for agri-food products. https://edepot.wur.nl/558072.
- https://trueprice.org/.

8 References

Apaiah, R. K. (2006). Designing food supply chains-a structured methodology: A case on novel protein foods, PhD-Thesis, Wageningen University, Wageningen.

Aramyan, L. (2007). Measuring supply chain performance in the agri-food sector. PhD Thesis, Wageningen university, Available at: https://edepot.wur.nl/121904.

Aramyan, L., Ondersteijn, C., Kooten, O. and Oude Lansink, A. (2006). Performance indicators in agri-food production chains. In: Ondersteijn (Ed.). *Quantifying the Agri-Food Supply Chain*, Springer, Dordrecht 49–66.

Aramyan, L. H., Oude Lansink, A. G. J. M., van der Vorst, J. G. A. J. and van Kooten, O. (2007). Performance measurement in agri-food supply chains: a case study, *Supply Chain Management* 12(4), 304–315.

Azapagic, A. and Clift, R. (1999). The application of life cycle assessment to process optimisation, *Computers and Chemical Engineering* 23(10), 1509–1526.

Batista, L., Bourlakis, M., Smart, P. and Maull, R. (2018). In search of a circular supply chain archetype–a content-analysis-based literature review, *Production Planning and Control* 29(6), 438–451.

Beamon, B. M. (1998). Supply chain design and analysis: models and methods, *International Journal of Production Economics* 55(3), 281–294.

Beamon, B. M. (1999). Measuring supply chain performance, *International Journal of Operations and Production Management* 19(3), 275–292.

Beamon, B. M. (1999b). Designing the green supply chain, *Logistics Information Management* 12(4), 332–342.

Berry, L. (2006). *On Great Service: A Framework for Action*, Free Press, New York.

Beske-Janssen, P., Johnson, M. P. and Schaltegger, S. (2015). 20 years of performance measurement in sustainable supply chain management – what has been achieved?, *Supply Chain Management* 20(6), 664–680. https://doi.org/10.1108/SCM-06-2015-0216.

Bouazzaoui, M., Wu, H. J., Roehrich, J. K., Squire, B. and Roath, A. S. (2020). Justice in inter-organizational relationships: A literature review and future research agenda, *Industrial Marketing Management* 87, 128–137.

Bowersox, D. J. and Closs, D. J. (1996). *Logistical Management: the Integrated Supply Chain Process*, McGraw-Hill, New York.

Bulsara, H. P., Qureshi, M. N. and Patel, H. (2016). Green supply chain performance measurement: an exploratory study, *International Journal of Logistics Systems and Management* 23(4), 476–498.

Carlsson-Kanyama, A., Pipping Ekström, M. and Shanahan, H. (2003). Food and life cycle energy inputs: consequences of diet and ways to increase efficiency, *Ecological Economics* 4, 293–307.

Chopra, S., Laux, C., Schmidt, E. and Rajan, P. (2017). Perception of performance indicators in an agri-food supply chain: A case study of India's public distribution system, *International Journal on Food System Dynamics* 8, 130–145.

Christopher, M. G. (1998). *Logistics and Supply Chain Management; Strategies for Reducing Costs and Improving Services*, Pitman Publishing, London.

Darvish, M., Archetti, C. and Coelho, L. C. (2019). Trade-offs between environmental and economic performance and inventory-routing problems, *International Journal of Production Economics* 217(C), 269–280.

de Boer, I. J. M. and van Ittersum, M. K. (2018). *Circularity in Agricultural production*. Wageningen University & Research. Available at: https://edepot.wur.nl/470625.

de Haan, F. J., Ferri, D. Q., Walsh, S. J. and Bettencourt, L. M. A. (2019). Scales 1668 and transformative change: transitions in the Galapagos. In: Kvan, T. and Karakiewicz, J. (Eds.). *Urban Galapagos: Transition to Sustainability in Complex Adaptive Systems*, Springer, Cham, 43–58. https://doi.org/10.1007/978-3-319-99534-2_4.

de Vries, B. J. M. (2019). Engaging with the sustainable development goals by going beyond modernity: an ethical evaluation within a worldview framework, *Global Sustainability* 2, e18. https://doi.org/10.1017/sus.2019.15.

Diller, H. (2008). Price fairness, *Journal of Product and Brand Management* 17(5), 353–355.

Ding, H., He, M. and Deng, C. (2014). Lifecycle approach to assessing environmental friendly product project with internalizing environmental externality, *Journal of Cleaner Production* 66, 128–138.

Fattahi, F., Nookabadi, A. S. and Kadivar, M. (2013). A model for measuring the performance of the meat supply chain, *British Food Journal* 115(8), 1090–1111.

Gaitan Cremaschi, D., Meuwissen, M. P. M. and Oude Lansink, A. G. J. M. (2017). Total factor productivity: a framework for measuring agri-food supply chain performance towards sustainability, *Applied Economic Perspectives and Policy* 39(2), 259–285.

Galgani, P., Woltjer, G., de Adelhart Toorop, R. and Groot Ruiz, A. (2021). Valuation framework for true Price assessment of agri-food products, *Report of the Project w PPS AF18051/ TU18104 'True and Fair Price for Sustainable Products'*, Available at: https://trueprice.org/valuation-framework-for-true-price-assessment-of-agrifood-products/.

Gellynck, X., Molnar, A. and Aramyan, L. H. (2008). Supply chain performance measurement: the case of the traditional food sector in the EU, *Journal on Chain and Network Science* 8(1), 47–58. https://doi.org/10.3920/JCNS2008.x088.

Gold, S., Seuring, S. and Beske, P. (2010). 'Sustainable supply chain management and interorganizational resources: A literature review', Corp. Soc. Responsib, *Environmental Management* 17, 230–245.

Govindan, K., Mangla, S. K. and Luthra, S. (2017). Prioritising indicators in improving supply chain performance using fuzzy AHP: insights from the case example of four Indian manufacturing companies, *Production Planning and Control* 28(6–8), 552–573.

Gunasekaran, A., Patel, C. and Tirtiroglu, E. (2001). Performance measures and metrics in a supply chain environment, *International Journal of Operations and Production Management* 21(1/2), 71–87.

Gunasekaran, A., Patel, C. and McGaughey, R. E. (2004). A framework for supply chain performance measurement, *International Journal of Production Economics* 87(3), 333–347.

Hagelaar, G. J. L. F. and Van der Vorst, J. G. A. J. (2002). Environmental supply chain management: using life cycle assessment to structure supply chains, *International Food and Agribusiness Management Review* 4, 399–412.

Hobs, J. E. and Young, L. M. (2000). Closer vertical co-ordination in agri-food supply chains: a conceptual framework and some preliminary evidence, *Supply Chain Management: an International Journal* 5(3), 131–143.

Jackson, P., Ward, N. and Russell, P. (2009). Moral economies of food and geographies of responsibility, *Transactions of the Institute of British Geographers* 34(1), 12–24.

Jurgilevich, A., Birge, T., Kentala-Lehtonen, J., Korhonen-Kurki, K., Pietikäinen, J., Saikku, L. and Schösler, H. (2016). Transition towards circular economy in the food system, *Sustainability* 8(1), 69. https://doi.org/10.3390/su8010069.

Kaplan, R. S. and Norton, D. P. (1992). The balances scorecard- measures that drive performance, *Harvard Business Review* 70(1), January–February, 71–79.

Lai, K. H., Ngai, E. W. T. and Cheng, T. C. E. (2002). Measures for evaluating supply chain performance in transport logistics, *Transportation Research Part E: Logistics and Transportation Review* 38(6), 439–456.

Lambert, D. M. and Pohlen, T. L. (2001). Supply chain metrics, *The International Journal of Logistics Management* 12(1), 1–19.

Lohman, C., Fortuin, L. and Wouters, M. (2004). Designing a performance measurement system: a case study, *European Journal of Operational Research* 156(2), 267–286.

Lapide, L. (2000), *'What about measuring supply chain performance?', White Paper, (Ascet, 2, April 15)*, Available at: http://lapide.ASCET.com.

Li, D. and O'Brien, C. (1999). Integrated decision modelling of supply chain efficiency, *International Journal of Production Economics* 59(1-3), 147–157.

Luning, P. A., Marcelis, W. J. and Jongen, W. M. F. (2002). *Food Quality Management: A Techno Managerial Approach*, Wageningen Academic Publishers, Wageningen.

Maestrini, V., Luzzini, D., Caniato, F., Maccarrone, P. and Ronchi, S. (2018). Measuring supply chain performance: a lifecycle framework and a case study, *International Journal of Operations and Production Management* 38(4), 934–956.

Mishra, D., Gunasekaran, A., Rameshwar, D. and Papadopoulos, T. (2018). Supply chain performance measures and metrics: a bibliometric study: a bibliometric study', *Benchmarking: an International Journal* 25(3), 932–967.

Moazzam, M., Akhtar, P., Garnevska, E. and Marr, N. E. (2018). Measuring agri-food supply chain performance and risk through a new analytical framework: a case study of New Zealand dairy, *Production Planning and Control* 29(15), 1258–1274.

Najmi, A. and Makui, A. (2012). A conceptual model for measuring supply chain's performance, *Production Planning and Control* 23(9), 694–706.

Naseer, M. A. R., Ashfaq, M., Hassan, S., Abbas, A., Razzaq, A., Mehdi, M., Ariyawardana, A. and Anwar, M. (2019). Critical issues at the upstream level. In sustainable supply chain management of agri-food industries: evidence from Pakistan's citrus industry, *Sustainability* 11(5), 1326.

Neely, A., Gregory, M. and Platts, K. (1995). Performance measurement system design: a literature review and research Agenda, *International Journal of Operations and Production Management* 15(4), 80–116.

Neely, A., Gregory, M. and Platts, K. (2005). Performance measurement system design, *International Journal of Operations and Production Management* 25(12), 1228–1263.

Persson, F. and Olhager, J. (2002). Performance simulation of supply chain designs, *International Journal of Production Economics* 77(3), 231–245.

Piotrowicz, W. and Cuthbertson, R. (2015). Performance measurement and metrics in supply chains: an exploratory study, International Journal of Productivity and Performance Management 64(8), 1068–1091. https://doi.org/10.1108/IJPPM-04-2014-0064.

Ramos, E., Coles, P. S., Chavez, M. and Hazen, B. (2022). Measuring agri-food supply chain performance: insights from the Peruvian kiwicha industry, *Benchmarking: An International Journal* 29(5), 1484–1512.

Runhaar, H. (2016). Towards 'nature-inclusive' Agriculture, *Inaugural Lecture upon Taking up the Position of Special Professor of Management of Biodiversity and Agricultural Landscapes at Wageningen University on 22 September 2016*, Available at: https://edepot.wur.nl/389369.

Sehnem, S., Vazquez-Brust, D., Pereira, S. C. F. and Campos, L. M. S. (2019). Circular economy: benefits, impacts and overlapping, *Supply Chain Management* 24(6), 784–804.

Simão, L. E., Somensi, K., Dávalos, R. V. and Rodriguez, C. M. T. (2022). Measuring supply chain performance: the triple E model, *International Journal of Productivity and Performance Management* 71(7), 2951–2969.

Supply Chain Council (2004). Available at: http://www.supply-chain.org.

Trivellas, P., Malindretos, G. and Reklitis, P. (2020). Implications of green logistics management on sustainable business and supply chain performance: evidence from a survey in the Greek Agri-food sector, *Sustainability* 12(24), 10515.

Valeeva, N. I. (2005). *Cost-Effectiveness of Improving Food Safety in the Dairy Production Chain*, PhD thesis, Wageningen University, Wageningen.

van Dam, J., Ugarte, S. and van Iersel, S. (2012). Selecting a biomass certification system - a benchmark on level of assurance, costs and benefits, *NL Agency, Ministry of Economic Affairs, Agriculture and Innovation*, Available at: https://edepot.wur.nl /200883.

Van der Spiegel, M. (2004). *Measuring Effectiveness of Food Quality Management*, PhD thesis, Wageningen University, Wageningen SCOR, Supply Chain Council (2004). Available at: http://www.supply-chain.org.

Van der Vorst, J. G. A. J. (2000). Effective food supply chains. Generating, modelling and evaluation supply chain scenarios, PhD thesis, Wageningen University, Wageningen.

Van der Vorst, J. G. A. J. (2005). Performance measurement in agri-food supply chain networks. An overview. In: Ondersteijn, C. J., Wijnands, J. H., Huirne, R. B. and van Kooten, O. (Eds), Springer, Dordrecht. *Quantifying Supply Chains*.

Van Der Vorst, J. G. A. J., Tromp, S. and Van Der Zee, D. (2009). Simulation modelling for food supply chain redesign; integrated decision making on product quality, sustainability and logistics, *International Journal of Production Research* 47(23), 6611–6631.

Vegter, D., Hillegersberg, J. and Olthaar, M. (2021). Performance measurement systems for circular supply chain management: current state of development, *Sustainability* 13(21), 12082, Available at: https://doi.org.

Walkiewicz, J., Lay-Kumar, J. and Herzig, C. (2021). The integration of sustainability and externalities into the "corporate DNA": a practice-oriented approach, *Corporate Governance* ahead-of(ahead-of) 21(3), 479–496. https://doi.org/10.1108/CG-06-2020-0244.

Widyaningrum, D. and Masruroh, N. A. (2012). Development of the sea fishery supply chain performance measurement system: A case study, *International Journal of Supply Chain Management (IJSCM)* 1(3), 20–32.

Womack, J. and Jones, D. (2002). *Seeing the Whole: Mapping the Extended Value Stream*, Lean Enterprise Institute, Cambridge.

Published by Burleigh Dodds Science Publishing Limited, 2024.

Annex

Table 2 Definitions of performance indicators used in the framework of Aramyan et al. (2007)

Categories *Indicators*	Definitions	Measures
Efficiency[a]		
Production costs/ Distribution costs	Combined costs of raw materials and labor in producing goods/ Combined costs of distribution, including transportation and handling cost	The sum of the total costs of inputs used to produce output/services (fixed and variable costs)
Transaction costs	The costs other than the money price that are incurred in trading goods or services(e.g. searching cost, negotiation costs, and enforcement costs)	The sum of searching costs (the costs of locating information about opportunities for exchange), negotiation costs (costs of negotiating the terms of the exchange), and enforcement costs (costs of enforcing the contract)
Profit	The positive gain from an investment or business operation after subtracting all expenses	Total revenue less expenses
Return on investments	A measure of a firm's profitability and measures how effectively the firm uses its capital to generate profit	Ratio of net profit to total assets
Inventory	A firm's merchandise, raw materials, and finished and unfinished products which have not yet been sold.	The sum of the costs of warehousing of products, capital, and storage costs associated with stock management and insurance
Flexibility[a]		
Customer satisfaction	The degree to which the customers are satisfied with the products or services	The percentage of satisfied customers to unsatisfied customers
Volume flexibility	The ability to change the output levels of the products produced	Calculated by demand variance and maximum and minimum profitable output volume during any period of the time
Delivery flexibility	The ability to change planned delivery dates	The ratio of the difference between the latest time period during which the delivery can be made and the earliest time period during which the delivery can be made and the difference between the latest time period during which the delivery can be made and the current time period

(*Continued*)

Table 2 (*Continued*)

Categories *Indicators*	Definitions	Measures
Backorders	An order that is currently not in stock, but is being re-ordered (the customer is willing to wait until re-supply arrives) and will be available at a later time	The proportion of the number of backorders to a total number of orders
Lost sales	An order that is lost due to the stock out, because the customer is not willing to permit the backorder	The proportion of the number of lost sales to a total number of sales
Responsiveness[a]		
Fill rate	Percentage of units ordered that are shipped on a given order	Actual fill rate is compared to the target fill rate
Product lateness	The amount of time between the promised product delivery date and the actual product delivery date	Delivery date minus due date
Customer response time	The amount of time between an order has been done and its corresponding delivery	The difference between the amount of the time an order has been done and its corresponding delivery
Lead time	Total amount of time required to produce a particular item or service	Total amount of time required to complete one unit of product or service
Customer complaints	The registered complaints from customers about product or service	The total number of complaints registered
Shipping errors	Wrong products shipments	The percentage of wrong shipments
Product quality[b]		
Sensory properties and shelf life		
Appearance	First sight of the tomato, combination of different attributes (color, size and form, firmness, lack of blemishes and damages)	Number of damages, color scale, size, and form scale
Taste	Determined by the sweetness, mealiness, and aroma of a vegetable/fruit	Brix value, which is the measurement of a soluble dry substance in a liquid (providing an approximate measure of sugar content)

(*Continued*)

Table 2 (*Continued*)

Categories *Indicators*	Definitions	Measures
Shelf life	The length of time a packaged food will last without deteriorating	The difference in time between harvesting or processing and packaging of the product and the point in time at which it becomes unacceptable for consumption
Product safety and health		
Salubrity	The quality of the products being healthful and nutritious	Nutritional value and lycopene content
Product safety	Product does not exceed an acceptable level of risk associated with pathogenic organisms or chemical and physical hazards such as microbiological, chemical contaminant in products, and micro-organisms	Lab checks and monitoring processes according to certification schemes
Product reliability and convenience		
Product reliability	Refers to the compliance of the actual product composition with the product description	Number of registered complaints
Convenience	The information provided on the packaging is useful, complete, and easily understandable	Number of registered complaints
Process quality[b]		
Production system characteristics		
Traceability	Traceability is the ability to trace the history, application, or location of an product using recorded identifications	Information availability, use of barcodes, and standardization of quality systems
Storage and transport conditions	Standard conditions required for transportation and storage of the products that is optimal for good quality	Measure of relative humidity and temperature, complying with standard regulations
Working conditions	Standard condition that ensure a hygienic, safe working environment, with correct handlings and good conditions	Compliance with standard regulations
Environmental aspects		
Energy use	The amount of energy used during production process	The ratio of cubic meter gas used per square meter glasshouse

(*Continued*)

Published by Burleigh Dodds Science Publishing Limited, 2024.

Table 2 (*Continued*)

Categories *Indicators*	Definitions	Measures
Water use	The amount of water used during production process	The ratio of a liter water used per square meter land under the vegetables
Pesticide use	A permitted amount of pesticides used in production process	The amount and the frequency of the pesticide use complying with standard regulations
Recycling/reuse	Collected used product from crop, packaging etc. that is disassembled, separated, and processed into recycled products, components, and/or materials or re-used, distributed, or sold as used, without additional processing	Percentage of materials recycled/re-used
Marketing		
Promotion	Activities intended to increase market share for product (e.g. branding, pricing, and labeling)	Increase in number of customers and sales
Customer service	The provision of labor and other resources, for the purpose of increasing the value that buyers receive from their purchases and from the processes leading up to the purchase	Ratio of provision of resources used to increase customer service to increased sales
Display in stores	Demonstration of the product in the store	Increase in number of customers and sales

[a]Sources: Beamon, 1998, 1999a; Bowersox and Closs, 1996; Hobbs and young, 2000; Persson and Olhager, 2002; Lai et al., 2002; Womack and Jones, 2002; Gunasekaran et al., 2001; SCOR model, 2004; Berry, 2006.
[b]Sources: Luning et al., 2002; Van der Spiegel, 2004; Valeeva, 2005; Beamon, 1999b; Berry, 2006.

Chapter 7

State of the art in agri-food supply chain infrastructure design

Müge Çerçi, Kocaeli University, Turkey; Zehra B. Kanık, Cankaya University, Turkey; Sibel Erişkan, Burdur Mehmet Akif Ersoy University, Turkey; and Mehmet Soysal[1], Hacettepe University, Turkey

1 Introduction

Supply chain management (SCM) has become an important part of management in the manufacturing and retail sectors since the 1990s (Chopra and Meindl, 2012). The growing interest in SCM has expanded beyond the manufacturing and retail industries to become an indispensable element of food and agriculture management (Bloemhof and Soysal, 2017). In today's world, where it is getting harder to meet unlimited needs with scarce resources, the ever-increasing global population and changes in current consumption habits have significantly increased food consumption. Increased consumption leads to increased demand for food production and distribution, which requires the successful coordination and integration of key business processes in SCM (Kamble et al., 2020).

The global population is projected to increase by more than a third, or 2.3 billion individuals, between 2009 and 2050. During the same time period, urbanization is projected to increase rapidly while rural populations decline. These trends indicate that food demand will increase. According to the Food and Agriculture Organization of the United Nations (FAO), cereal demand for both human consumption and animal feed is expected to reach 3 billion tonnes by 2050, up from approximately 2.1 billion tonnes currently. Demand for other

1 Corresponding author, mehmetsoysal@hacettepe.edu.tr.

http://dx.doi.org/10.19103/AS.2023.0122.20

food crops (meat and dairy products, vegetable oils, etc.), which are more income-sensitive in developing countries, will grow much faster than cereals. To feed a world population of 9.1 billion in 2050, total food production will have to increase by about 70% between 2005 and 2050. In developing countries, production will need to almost double.[2] All of these developments point to the growing importance of food supply chain management (FSCM).

The FSCM encompasses the entire process of producing plant- and animal-based food products, as well as their subsequent distribution to the ultimate consumer. Concerns about food supply have grown as a result of consumer awareness and global population expansion, and FSCM has gained importance for both public and commercial organizations. The primary focal points of FSCM encompass traceability, freshness, perishability, and product quality (Van Der Vorst et al., 2009). Moreover, the growing requirements and complicated supply chain networks, combined with product variety, are accelerating the development of FSCM with economic, social, and environmental sustainability (Soysal et al., 2014).

As supply chains become more complex and logistics operations more challenging, the transition from traditional supply chains to FSCM becomes imperative. FSCM differs from traditional SCM. It involves planning production, inventory, and distribution policies based on the food's characteristics, taking into consideration perishability, cost, and sustainability parameters throughout the food supply chain. Maintaining the quality of food at retail, shelf life, availability, and sale at an acceptable price are important factors to be considered in FSCM (Bloemhof and Soysal, 2017). Globalization, urbanization, and agricultural industry advancements have placed agri-food supply chain management (AFSCM), including agri-food chains and associated organizations, on the agenda today.

Although there are significant similarities between FSCM and AFSCM, there are also fundamental differences to consider. The principal distinction is that FSCM encompasses all phases of food processing, packaging, storage, transportation, and delivery from production to consumption, while AFSCM covers all SCM processes from seed to fork. This means that FSCM addresses both processed and unprocessed food products, whereas AFSCM addresses the growing and harvesting of the crop. FSCM concentrates on essential factors, including food safety, traceability, freshness, perishability, and the suitable storage and transportation of food. In contrast, AFSCM must take into account agricultural factors, such as harvesting, storing, and processing products at the appropriate time and under the right conditions (Ahumada and Villalobos, 2009; Taskiner and Bilgen, 2021). In this context, various documents and certifications are required to maintain food safety and quality standards in

2 Food and Agriculture Organization of the United Nations (FAO), https://www.fao.org.

AFSCM. Some of these are the Euro-Retailer Produce Working Group (EUREP GAP), ISO 22000, British Retail Consortium (BRC), and International Food Standard (IFS) (Aramyan et al., 2009; Hu et al., 2023).

The evolution of the traditional structure from SCM to AFSCM has caused the chain to become multi-dimensional. In the traditional supply chain, infrastructure includes physical, digital, and human resources that enable the flow of information, materials, and fund. However, in a supply chain infrastructure with multiple stakeholders, inefficiencies, quality, cost, and waste issues are inevitable and environmental, social, and economic sustainability must be considered. Lack of real-time data and communication, inadequate use of technology, and lack of integration of strategic goals and objectives in organizations are common sources of inefficiency in a supply chain (Derakhti et al., 2003).

Infrastructure investments play a crucial role in the context of supply chains. Proper road, rail, and port infrastructure ensures timely and efficient deliveries to consumers and provides significant cost savings by reducing waste and spoilage. In terms of product quality and safety, adequate storage and refrigeration facilities also offer substantial benefits. Communication and information technology infrastructure facilitate real-time monitoring and information sharing throughout the supply chain. When analyzing the matter of cooperation in AFSCM, the focal points that emerge are stakeholder collaboration, supply chain transparency, and data exchange. Collaboration plays a crucial role in facilitating supply chain integration, offering notable advantages in terms of enhanced planning, risk mitigation, and operational optimization through the prevention of market fluctuations (Bosona and Gebresenbet, 2023; Demestichas et al., 2020, Soysal et al., 2022). Finally, there is the issue of digital technologies of Industry 4.0, which inevitably finds a place in the field of AFSCM as in every field. The supply chain benefits significantly from data analytics, blockchain technology, and Internet of Things-based technologies in the areas of food safety, product quality, transportation, storage, transparency, and traceability (Feng et al., 2020; Ünal et al., 2020; Senturk et al., 2023). In addition to analyzing market trends and consumer expectations, it also facilitates the functioning of operations by reducing the role of intermediaries through online markets.

In light of this information, it can be concluded that it is crucial to achieve sustainability gains by resolving inefficiencies and other obstacles in the transportation of agricultural products from farms to processing facilities and the final consumer. The relationship between digital technologies and physical infrastructure is inevitable within AFSCM. This relationship offers significant benefits such as increased efficiency, real-time monitoring and visibility, data analytics and decision support systems, collaboration and communication, customer satisfaction, and risk management. As a result, digital technologies and physical infrastructure make AFSC more efficient, flexible, cost-effective, and customer focused. This helps companies in the chain to increase their competitive advantage and manage AFSC more effectively (Derakhti et al., 2023).

According to World Food Program (WFP) data, stakeholders received food, goods, and services worth $6 billion in 2022.[3] Investing in the infrastructure needed to ensure that this supply chain is uninterrupted not only connects people to opportunities but also stimulates economic growth and improves living standards. Agricultural products are produced on 570 million farms, the majority of which are family farms in developing countries.[4] However, to achieve the desired sustainable development goals, developing countries need to spend $1.5 trillion per year or about 4.5% of GDP. The World Bank provides information, advice, and financial support through a range of short- and medium-term investments and platforms.[5]

The main purpose of this chapter is to examine the field of AFSCM and to provide an overview of the relevant literature on the subject. To fulfil this objective, an extensive examination of review papers was undertaken within the Web of Science database. The reason for focusing on literature review articles within the scope of the review is to address the need for in-depth knowledge on critical supply chain issues such as network design, physical infrastructure, and digitalization through investigating the field of AFSCM comprehensively. This chapter contributes to the literature in three main ways: (1) by conducting a review of literature review articles from the AFSCM academic literature; (2) by analyzing the current state of the AFSCM literature and presenting future study suggestions of the articles scanned in the literature for the development of AFSCM; and (3) by examining the use of network design, physical infrastructure, and digital technologies that play a key role in AFSCM and their relationship with SCM.

The chapter is divided into the following sections: The second section presents descriptive information on the literature review and detailed reviews of the articles included in the study. The third section examines the network design in AFSCM in light of physical and digital infrastructures. Finally, section four presents the research findings and conclusions.

2 Analyses of the reviewed papers

This section presents implications to decision-makers and researchers regarding current developments in the agricultural food supply chain and the findings of the analysis performed within the scope of the study. The methodology

3 World Food Programme (WFP), https://www.wfp.org.
4 World Bank, https://openknowledge.worldbank.org.
5 World Bank, https://www.worldbank.org.

employed for conducting the literature analysis, as well as the specific keywords utilized in the search, are concisely outlined in Fig. 1. The flowchart depicted in Fig. 1 illustrates the execution of a search in the Web of Science database within the research domains of engineering, business economics, and operations Research. The search is performed using the specified keywords 'supply chain' and 'agri-food'. The investigation focused on literature review articles in order to present a comprehensive summary of the current body of knowledge while excluding research articles. The analysis focused solely on studies written in the English language, with a total of 40 literature review articles on the topic being examined.

2.1 Descriptive analyses

This subsection presents a keyword analysis of literature review articles on the agricultural supply chain and a descriptive analysis of the distribution of related studies by year. The aim of the section is to provide researchers with an overview of the existing literature and to discuss the topic's relevance. Figure 2 presents

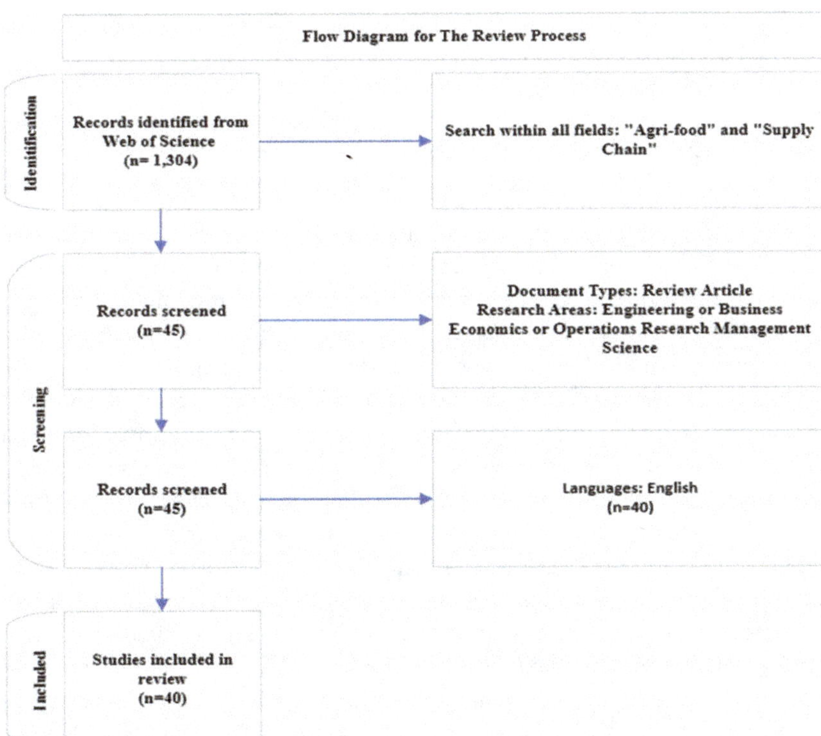

Figure 1 Summary of the systematic review articles selection process. Source: Adapted from: Ahmed et al. (2022).

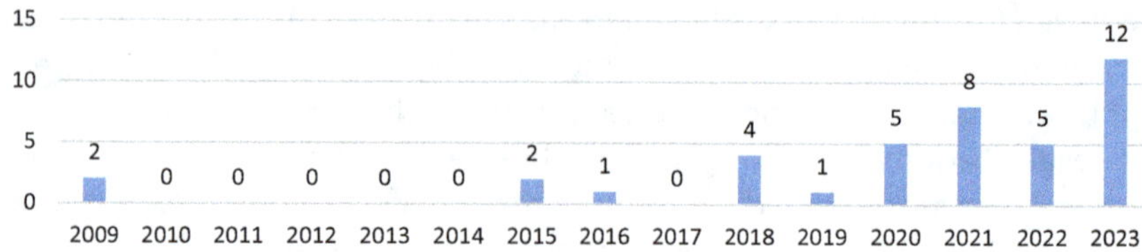

Figure 2 The review articles of agri-food supply chain on Web of Science.

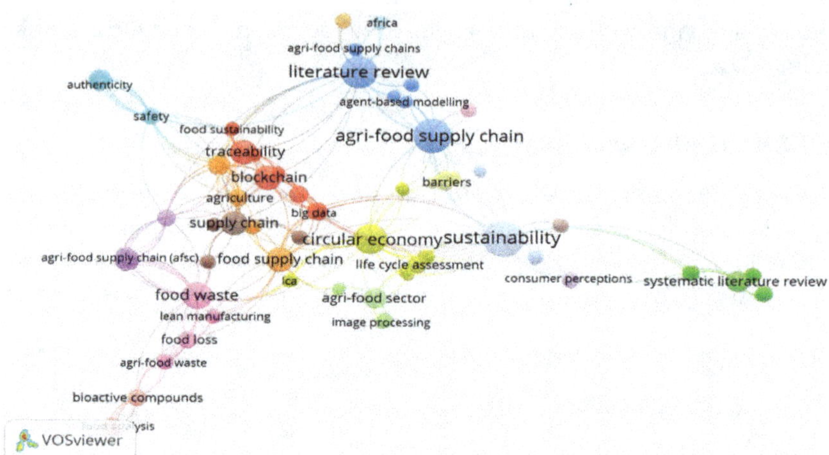

Figure 3 A network visualization illustration of co-occurrence of keywords.

the distribution of the articles analyzed within the scope of the study over the years. The statistics show that most of the 40 articles date from the last few years. The number shows that interest in the topic has recently emerged and has continued to expand over time. The increased interest in the topic reflects the importance of SCM in the agricultural sector and the field's importance.

Figure 3 is a visual created with VOSviewer that groups together the prevalent keywords in the reviewed studies. This graphic demonstrates that the reviewed literature can be categorized into 18 distinct color-coded clusters. Sustainability, literature review, and agriculture supply chain are prominent keywords. Since the review focuses on literature review articles and is conducted on the agricultural supply chain, it is unsurprising that these two keywords come to the forefront. Nonetheless, it is evident that the concept of sustainability is addressed as a crucial issue in the studies. Keywords such as 'circular economy,' 'blockchain,' 'big data,' 'traceability,' and 'life cycle assessment' are frequently used in studies addressing sustainability. The 'food waste' is another keyword deemed essential. In this cluster, 'bioactive compounds,' 'food SC,' 'agrifood SC,' 'lean manufacturing,' 'food loss,' and 'agri-food waste' are also discussed in conjunction with 'food waste'-focused studies. Therefore, the shown map illustrates the interconnections among many subjects within the field of research.

2.2 Detailed analyses

In this section of the study, we present detailed analyses of the literature review studies in the agricultural supply chain. The contributions/scopes, major findings, and potential research directions mentioned in the studies are presented in Table 1.

Several conclusions can be derived from the presented table. As indicated in the introduction section, the unique characteristics of the agricultural supply chain play a prominent role in the studies. The difficult management and planning process, the high dependence on environmental conditions, and the intensity of external intervention are recurrent themes among the AFSC's distinctive characteristics. The risk of product perishability, high product differentiation, and additional security requirements are considered important AFSC features (Lugo et al., 2023; Stone and Rahimifard, 2018).

The literature shows that there is great interest in the international community to introduce internationally recognized standards and frameworks for AFSC. There are already widely used certifications in AFSC and the most widely used certification is EUROPE GAP. Since 1997, EURO GAP has been offered as a certification that provides a set of principles to be applied to on-farm production and post-production processes that result in safe and healthy food and non-food agricultural products, taking into account economic, social, and environmental sustainability. Similarly, there are other widely used certifications. BRC was introduced in 1998 and aims to assure product quality. IFS is another certification that has been used since 2003 to reduce costs and bring transparency to the whole chain.

The uncertain future of the linear economy in the agri-food system is another issue that draws attention in the studies. In contrast, the proposed circular economy approach is important to ensure the sustainability needed in the agri-food system. The necessity of a circular economy is emphasized for CE applications that will be used globally to eliminate waste, especially in the agricultural sector. It appears from the relevant literature that government intervention to promote CE initiatives plays a critical role in the transition process. The literature takes an external pressure approach to understanding the motivators for the adoption of CE practices. A gap emerges in the literature linking CE practices in agri-food eco-industrial parks with social value as well as economic and environmental performance (Adelodun et al., 2021; Atanasovska et al., 2022; Mehmood et al., 2021).

Another point that draws attention in the literature is that scrap and nutrient loss, especially at the processing level, are seen as the main forms of loss/

Table 1 Detailed analysis of literature review articles in agri-food supply chain

Research papers	Main contribution - scope	Findings	Future directions
Ahumada and Villalobos, 2009	Scanning production and distribution planning for AFSC	Gaps in the literature for AFSC planning models are as follows: integrated planning models, realistic stochastic models for tracking shelf-life at different stages of the perishable supply chain, operational planning models.	The development of a model for planning operational decisions for the production/harvest and distribution of perishable products is recommended.
Aramyan et al., 2009	Analyzing the impact of different quality assurance systems (QASs) requirements on supply chain performance	Based on a review of QAS, the study developed seven critical requirements for a four-tier supply chain consisting of growers, producers, wholesalers, and retailers (variety and rootstock records, product recall and control of non-conforming products, management commitment to food safety, residue records, analysis and chemical use, Worker health, safety and welfare, hazard analysis, traceability).	The proposed methodology and performance measurement model can reveal the perceived impact of QAS on performance and the perceived impact of other management systems on the performance of the entire supply chain.
Matopoulos et al., 2015	Investigating the interaction between supply chains and natural resources and identify the critical characteristics of resource-efficient supply chains	The literature tends to focus primarily on the environmental impact of using resources. In an era of resource scarcity and depletion, it is essential to explore how logistics/supply chain decisions will have an impact on the overall structure of future food supply chains and what the trade-offs might be.	Application of the proposed framework and key characteristics of resource efficient supply chains to other sectors.
Siddh et al., 2015	Reviewing studies addressing quality performance in the perishable food supply chain	There are fewer articles addressing the problems of developing countries compared to developed countries. Many articles deal with 'agri-food,' 'dairy,' and 'pork' as perishable products. Information sharing, logistics management, strategic management, demand forecasting, and integration among the various stakeholders of the perishable food supply chain are some of the critical issues.	Carrying out studies on the quality of the beef, fish, frozen food, and poultry supply chains. Conducting empirical studies on risk management, supply chain efficiency, supply chain security, industrial supply chain quality, product safety, supply chain quality performance, strategic alignment, conflict management, relationship quality, visibility, each node in the FSC, final delivery quality, sustainability, etc.

De Steur et al., 2016	Demonstrating the potential of Value Stream Mapping (VSM) to identify and reduce food loss and waste, preserving the nutrients of SC	Primary production, processing, storage, food service, or consumption are identified as vulnerable points for losses and waste.	Prioritization of research that focuses on how simulation models can be reliably integrated into the VSM method.
Dania et al., 2018	Examining the critical factors in shaping the characteristics and effectiveness of collaboration in sustainable supply chains	Ten key behavioral factors have been identified to enable an effective collaboration system for sustainable AFSC management: Joint efforts, sharing activities, collaborative value, cohesion, trust, commitment, strength, continuous improvement, coordination, and stability.	Using the factors presented as a guide for future studies to help model and measure AFSC collaboration performance more effectively and efficiently.
Pearce et al., 2018	Identifying the determinants driving lean sustainable performance in primary production of horticultural crops (apple, pear)	There is a positive relationship between the size of the farm, the capacity of the labor force and the ability of the farm to achieve better performance. Horizontal cooperation between primary producers helps small producers to overcome resource constraints. Data-driven decision control is seen as a key determinant of sustainable performance, both at the producer level and centrally at the cooperative level.	Development of lean management research and operationalized lean methods in the horticultural sub-sector and the agri-food context.
Stone and Rahimifard, 2018	Reviewing resilience in AFSC and resilience features suitable for AFSC	The AFSC is a complex process that is inherently subject to external interference, so disruptions are constant. The AFSC must be flexible, not only in maintaining its core function, but also in its ability to adapt to changing circumstances.	Adaptation of the AFSC framework of resilience characteristics and strategies to economies in transition.

(*Continued*)

Table 1 (*Continued*)

Research papers	Main contribution - scope	Findings	Future directions
Utomo et al., 2018	Reviewing existing AFSC research using agent-based simulation (ABS), evaluation of modeling approaches reviewed	In terms of data, most papers using ABS for policy development use empirical data, while those using ABS for theory development use hypothetical data. Food processors, supermarkets, and retailers are not yet widely covered in AFSC research using ABS.	Examining cooperation and competition in supply chains, buyer–seller relationships, and services, incorporating general theoretical frameworks such as transaction cost economics and the resource-based view into model design taking into account local and seasonal changes in agricultural production, developing reliable and transparent data relationships through alternative methods such as machine learning, developing water management tools, and methodologies for supplier engagement.
Jia et al., 2019	Identifying research gaps in the existing literature and propose a more integrated water management framework in AFSC	Drivers, implications, water management practices, and barriers are the overarching themes in the water management in AFSC. The study aimed to better understand integrated water management by classifying the main findings in the AFSC water management literature under four main themes.	Focusing more on operational decisions in water resources management at CLSC and increasing case studies, investigating the impact of corporations on water management in China, taking into account local and seasonal changes in agricultural production, developing reliable and transparent data relationships through alternative methods such as machine learning, developing water management tools, and methodologies for supplier engagement.
Kamble et al., 2020	Screening sustainable data driven AFSC	The review provides an overview of the level of analysis (descriptive, predictive, and prescriptive), the objectives achieved (social, environmental, and economic), the supply chain processes in which data is collected and the supply chain resources used. The integration of corporate and supply chain resources provides enhanced visibility and advanced data analysis.	Addressing key data analytics implementation issues for sustainable AFSC.

Feng et al., 2020	Reviewing the benefits, challenges, and development methods of blockchain-based food traceability systems	Blockchain has far-reaching implications for research and applications in agri-food traceability. When integrated with IoT devices, it can address concerns about information security and transparency. There are five main areas of challenges for sustainable traceability management applications of blockchain: technical challenges, blockchain infrastructure, interoperability and standardization, social and organizational challenges, and system performance.	Evaluating operational decisions in blockchain-based traceability applications with pilot applications.
Pomarici and Sardone 2020	Providing a comprehensive overview of the EU's current situation in wine supply	Viticultural activities can play a role in maintaining or restoring the socioeconomic balance. A combination of rural development and sectoral measures can be used to develop new supply chains in problem areas, effectively linked to profitable markets.	Analyzing the impact of sectoral regulation, including a comprehensive set of rules on geographical indications, on the performance of the EU wine sector.
Demestichas et al., 2020	Reviewing the applicability of blockchain technology in traceability systems of AFSC products	AFSC is reluctant to adopt blockchain for the following reasons: it is a complex process, with different levels of technological knowledge among stakeholders, products go through many transformations along the chain, the large number of stakeholders and the heterogeneity of their roles, the dispersion of FSC over large geographical areas (even across continents), and the emergence of interoperability and distribution barriers.	Focus on how blockchain can be made more secure and therefore more efficient.
Ravi et al., 2020	Analyzing insect processing and black soldier fly processing methods and the functional properties of the foods obtained from them	Temperature is an important abiotic factor that directly and indirectly affects various aspects of insect life history traits. Larvae fed the diet with the lowest crude protein content have higher larval protein.	Investigating the genetic modification of black soldier fly larvae.

(Continued)

Table 1 (*Continued*)

Research papers	Main contribution - scope	Findings	Future directions
Santagata et al., 2021	Analyzing food waste recovery and recycling pathways, environmental assessment, and combined assessment of opportunities and constraints	Food waste offers a carbon-rich waste that can be used to recover a wide range of energy and materials, from fuel/energy to chemical compounds and bioplastics, among others, but it needs to be competitively priced in the global market. Tools and methods such as LCA (life cycle assessment) and EMA (EMergy Accounting) should be actively incorporated into the decision-making process, and guidelines proposed by different categories of stakeholders should be carefully considered and comprehensively coordinated.	Increasing interest in local characteristics, other technological options and time series analysis.
Sid et al., 2021	Examining the short- and long-term impacts of the COVID-19 pandemic on AFSCs. Assessing the regulatory mechanisms taken by different government agencies, NGOs, and the food industry to manage demand and supply disruptions	This paper addresses the demand and supply side of AFSCs in India, possible short- and long-term impacts and resolution strategies. Strategies and policies are needed across all AFSCs to manage the post-pandemic period.	Focusing on innovations, policies, machinery, etc. will make AFSC more productive and attractive for future generations and provide farmers with an efficient AFSC to deal with epidemic-like SC risk situations.
Aramyan et al., 2021	Examining the factors and barriers that influence supply chain operators' decision to adopt innovations (technological, organizational, and marketing) to reduce food waste	Innovative technologies are functional in preventing waste, but they need to be economically viable. Reducing food waste requires a combination of different types of innovation.	Focus on evaluating the relationship between innovations that reduce food waste and a company's performance. To evaluate the relationship between reducing food waste through marketing innovation and a company's image (as a performance indicator).

Mehmood et al., 2021	Identifying drivers and barriers to implementing the circular economy in AFSC	The popularity of circular economy (CE) drivers and barriers for AFSC, ranked according to how often they appear in research studies, are as follows. Environmental, political, economic, and financial benefits are considered the top three drivers. Institutional, financial, and technological risks are the top three barriers to implementing CE in AFSC.	Focusing on the final stages of agri-food supply chains from a circular economy perspective, which are less studied in the existing literature.
Adelodun et al., 2021	Researching emission mitigation and decarbonization, food waste and circular economy practices in the agro-ecosystem and food supply chain during the pandemic	Agriculture 4.0 tools, research activities and innovative ideas can support public efforts in the transition phase to the CE model by guiding transition methods and facilitating the implementation of CE for a sustainable food system.	Focusing on assessing people's food consumption habits in order to establish recycling programs to reduce the carbon footprint on the environment. Focus on the impact of the circular economy on the agri-food system.
Tharatipyakul and Pongnumkul, 2021	Building on blockchain-based traceability applications to improve food quality and safety; user interfaces for application development and reporting	The underlying technology in blockchain-based agri-food traceability applications is relatively complex compared to other applications. More emphasis should be placed on error prevention so that farmers with limited knowledge of the technology do not enter incorrect information. Most of the literature reviewed provided no or limited evaluation of user interfaces.	Users should be involved in interface development and evaluation of interface and design decisions, with a focus on user involvement.
Hassan and Abbasi, 2021	Examining the impact of SC integration on SC performance	Careful selection of the right degree of supply chain integration across network factors (actors, activities, and resources) will help firms realize cost savings and significant profits.	Consider the different levels of SC integration (strategic, tactical, and operational) in different sectors and the contextual differences in relation to the environment.

(*Continued*)

Table 1 (*Continued*)

Research papers	Main contribution - scope	Findings	Future directions
Taskiner and Bilgen, 2021	Reviewing on optimization models, harvest, and production planning in AFSCs	There is relatively little interest in perishability for fresh SC in the literature. Deterministic models are often preferred. Exact solution methodologies are used to solve optimization models. Farmer and producer collaborations are recommended against real-life problems. AFSCs should be extended to more specific objectives, such as maximizing product quality.	Study integrated decision-making in AFSC. Establish a close relationship between academia and AFSC and produce more applied research. Incorporate new methods in agriculture into AFSC's optimization models. Monitor crop waste.
Van Bussel et al., 2022	Reviewing the consumer perceptions of food-related environmental sustainability	Consumers consider price, taste, and personal health to be more important than sustainability. For consumers to make healthy and sustainable food choices, they need nutritional guidance, including better access to and availability of sustainable alternatives.	Dissemination of sustainability knowledge to consumers in a peer-to-peer, evidence-based and controlled manner.
Atanasovska et al., 2022	Investigating the social value arising from circular economy (CE) practices in agri-food eco-industrial parks	The literature shows that CE practices have traditionally been analyzed at the corporate/company level and not really explored at the eco-industrial park level. Social value has received a lot of attention when interpreted as a value-added element (e.g., CSR, human rights, social justice), but the social value resulting from CE practices and actual business activities has not been sufficiently explored.	Investigate the relationship between economic, environmental, and social value of CE practices (in eco-industrial parks) under external pressure. Investigate whether social and environmental value together can improve financial performance.
Echegaray et al., 2022	Reviewing Meat 4.0 and evaluating the Industry 4.0 innovations in digital transformation and process automation of the meat industry	Industry 4.0 has significant potential to improve the way meat is processed, preserved, and analyzed, reduce food waste and loss, develop high-quality, safe meat products, and prevent meat fraud.	Conducting research to improve existing technologies and reduce implementation costs.

Moreno-Miranda and Dries, 2022	Presenting a baseline analysis of sustainability assessment for AFSC	Important considerations for the implementation of sustainability assessment in AFSC are comprehensiveness, user-friendliness, transdisciplinarity and robustness; coordination mechanisms; the common dimension approach (provides an understanding of the relationships between economic, environmental, and social aspects of sustainability).	In-depth exploration of the relationship between coordination and sustainability.
Silvestri et al., 2022	Reviewing sustainability and circular economy (CE) perspective at AFSC	An integrated approach of indicators (environmental, social, and economic) offers the best solution to ensure an easier transition to sustainability.	Providing an overview of indicators derived from the analysis of scientific research. Conducting a field study for AFSC assessing the level of implementation of indicators by stakeholders.
Yadav et al., 2022	Investigating issues in agriculture/food supply chain network design and reviewing AFSC performance indicators	The literature on AFSC and Network Design suggests that the primary challenges are sustainability, food waste, food safety and security, etc. Efficiency, flexibility, food quality and responsiveness and labor productivity, net profit, grain yield, nitrogen use efficiency, water use efficiency are offered as performance measures in the connected field.	Taking into account the interests of all actors in AFSC network design, from farmers to financial institutions, in a broad viewpoint is encouraged. To investigate sustainable AFSC solutions for modeling methodologies in network design, food waste, Industry 4.0, technology integration, and performance assessment of AFSC is recommended.
Hu et al., 2023	Proposing a conceptual framework for the private food certification market, synthesizing research findings on seven global personal food safety standards	While the adoption by producers of the seven specific food safety certifications identified in the paper and the effects of specific certification on producers have received considerable attention, other determinant interactions identified in the conceptual framework of the paper have generally been ignored.	Exploring the legitimacy (or limits) of private food safety certifications relative to public or semi-public certificates and possible market failure in the private food safety certification market, working more on other standards in addition to the existing literature.

(Continued)

Table 1 (*Continued*)

Research papers	Main contribution – scope	Findings	Future directions
Berenguer et al., 2023	Discussing and identifying some of the key sustainable challenges affecting the valorization of agri-food loss/waste and by-products, as well as addressing the many opportunities, trends, and innovations	Fruit and vegetable wastes, which contain edible and non-edible parts rich in valuable compounds, can be marketed by adding value. To make this strategy economically viable, innovative techniques based on the principles of green analytical chemistry are effective in extracting and recovering valuable compounds.	Conducting continuous research and improvements in innovative extraction techniques that enable the development of appropriate strategies for the real-life sustainable development and utilization of valuable compounds from agri-food waste and their application in various industries.
Ciampa et al., 2023	Demonstrating the potential of NMR (nuclear magnetic resonance spectroscopy) as a green chemistry-based method in sample preparation and characterizing the nutritional qualities of agri-food products from a sustainable perspective	With the power of the NMR technique, sample preparation can have a low impact on the environment, with a focus on sustainability. With the NMR approach, consumer needs can be met in an environmentally friendly way while determining food-quality parameters.	Assessing the challenges that industries will face in the future as cryogen consumption is increasing and resources are limited because sample preparation requires large quantities of cryogen (helium and nitrogen), which has a profound impact on the environment.
Bosona and Gebresenbet, 2023	Addressing the challenges and benefits of primary traceability information, primary information providers, and blockchain technology traceability systems in AFSC	Existing BCT-based traceability systems mainly focus on fruit, vegetables, meat, milk, and dairy products. While research on BCT-based traceability systems in AFSCs has increased since 2016, their application in the business world is still in its infancy and more complex when considering AGTZ.	Conducting further research to design and implement effective BCT-based traceability systems for AFSCs.

Lugo et al., 2023	Addressing AFC's unique features, risk of disruption, high product differentiation, and security requirements	Circularity in the agri-food supply chain is a complex process. Food loss and waste, cost-effective alternatives, policy-making, information and communication technologies, consumer preferences, incentives for stakeholder cooperation, and cooperation between industry, academia and government are all essential considerations when making decisions about circularity.	Examining risk attitudes, preferences, cooperation, and other behavioral characteristics of AFSC stakeholders; considering the links between science, policy, and society, as well as the integration of food loss and waste measurement and applicable policies in future work.
Obonyo et al., 2023	Reviewing of recent literature on knowledge sharing in the context of degradable AFSC in Africa	There is evidence that knowledge sharing in African perishable AFSCs may only be in its infancy.	Conducting quantitative research (e.g. statistical analysis) to test the relationships between mobile phones and factors such as shelf life-shortening, quality monitoring, and sustainability.
Senturk et al., 2023	Explaining the use of IoT and related technologies in AFSCs, great-worldly application examples, and discuss the potential, advantages, and disadvantages of incorporating these technologies into existing processes to ensure sustainability	Today, there is a vast and rapidly growing demand for the integration of IoT-based systems into existing agricultural practices and supply chains.	Testing the field of IoT-based systems in natural agriculture and food systems and proving their success in practical applications.

(*Continued*)

Table 1 (*Continued*)

Research papers	Main contribution - scope	Findings	Future directions
Aldama et al., 2023	Conducting research for a deeper understanding of the production and utilization of different food by-products and food waste streams and industry-specific baseline evaluating existing official guidelines and establishing a more consistent set of recommendations for LCA in line with ISO standards	Many practitioners adopt a zero-load approach to model biomass valorization even when their materials do not meet the ISO 14044 waste definition. A significant number of practitioners do not provide transparency in justifying their methodological choices and fail to consistently present an approach that conforms to the ISO allocation hierarchy.	Researching for a deeper understanding of the production and utilization of different food by-products and food waste streams, conducting industry-specific baseline research evaluating existing official guidelines and establishing a more consistent set of recommendations for LCA in line with ISO standards.
Vignati et al., 2023	Modelling of HSI (hyperspectral imaging) technique and data analysis approaches (data processing and qualitative/quantitative) for fresh-cut produce	Cost-effective and handheld HSI devices are recommended at AFSC. This is because sensor technologies can solve common shelf-life problems such as early decay and spoilage, and improve product standardization in monitoring the fresh-cut production of fruits and vegetables at AFSC. However, the HSI instruments currently available in the market are laboratory instruments and costly.	Researching to increase the flexibility of application areas and develop compact and cost-effective HSI instruments.
Melesse et al., 2023	Addressing the benefits, types, levels of integration, key elements, implementation steps, and challenges of implementing digital twins	The implementation of digital twins shows promising progress in addressing global issues in the agri-food supply chain. Although there are signs of improvement in the industry, real-world applications of this solution are still in their early stages.	Working to develop tools and standards to facilitate the use of digital twins, focusing on proving their technical and economic benefits and investigating how they are technically and economically deployed and operated.

Trevisan and Formentini, 2023	Structuring Industry 4.0 technology adoption at AFSC around the main themes of research design, digital technologies, contextual diversity, governance, and sustainability	The review informs managers in the agriculture and food industry about the potential application of digital technologies to prevent food loss and waste along the agriculture and food supply chain.	Researching differences in the adoption of digital technologies in developing and developed countries, evaluating the impact of digital technologies on management mechanisms along the supply chain and their impact on contextual information, data management, and sustainability
Zhong et al., 2023	Reviewing quality management at AFSC, to achieve the goal of superior quality performance	AFSC evaluates the relationships between quality management practices and SC quality performance from a hierarchy of capabilities perspective.	Conducting research that is needed on behavioral experiments and simulations in addition to empirical research dominated by AFSC quality management studies; researching the adoption of digital technologies such as artificial intelligence, big data, and cloud technology and shifts in organizational learning modes.
Current study	Reviewing current literature in AFSC, assessing progress in the field, digitization, and collaboration, and exploring the relationship between this perspective and infrastructure	The Internet of Things, blockchain technology, big data, sensor technologies and radio frequency identification systems are the most interesting digital technological infrastructures. Areas of application for this technology include product traceability, information sharing among stakeholders, food waste prevention and food quality protection. In the context of collaboration, most studies focus on vertical collaboration. This concept has started to be addressed in the field of AFSC in the last 5 years.	Expand the use of other digital technologies in AFSCs; focus on horizontal partnerships or integrated collaborations in addition to vertical integration; establish integrated research designs that use both qualitative and quantitative methods.

waste that cause four of the seven lean production wastes (defect, unnecessary inventory, overproduction, and inappropriate processing). In agriculture and other food related LCAs related to food production, it is important to work towards allocation decisions on specific food co-products and waste utilization. In addition, food loss and waste materials can be reintroduced into the economy as fertilizers and other products (Aramyan et al., 2021; Berenguer et al., 2023; Aldama et al., 2023; Echegaray et al., 2022; Pearce et al., 2018; Santagata et al., 2021; De Steur et al., 2016).

There are also studies in the literature that examine the pandemic impact for AFSCs. A noteworthy point in these studies is that although there is a prediction of no food deficit during the COVID-19 pandemic, there is an expectation of a recession for India in the post-pandemic period due to disturbances in different farming activities, industrial processing, and related AFSCs. The fact that the concept of 'working from home,' which is relevant to many industries during the pandemic, is unsuitable for agriculture is a significant factor. Although this is a new field, it is estimated that it will take some time to assess the complete impact of the pandemic on the agri-food and allied industries (Adelodun et al., 2021; Hassan and Abbasi, 2021; Sid et al., 2021).

Real-world applications are regarded as essential for future work because practical case studies increase the confidence of AFSC industries to integrate technology and research into their business processes (Hassan and Abbasi, 2021; Lugo et al., 2023; Senturk et al., 2023).

Another common theme in the literature is the issue of AFSC sustainability. The literature demonstrates that there is still a need to develop and model sustainable AFSC, and the importance of case studies in sustainable agriculture is frequently emphasized. Several studies present the social sustainability paradigm as a suggestion for future work, attributing it specifically to the need for development (Aramyan et al., 2009; Matopoulos et al., 2015; Kamble et al., 2020; Melesse et al., 2023).

For developing countries, AFSC is seen as an important issue and an area open to scrutiny. In some countries, food deficits are increasing day by day and this has a negative impact on living standards. Therefore, there is a need to study AFSC, especially for developing countries (Siddh et al., 2015; Stone and Rahimifard, 2018; Trevisan and Formentini, 2023).

Agriculture 4.0 technologies should be adopted to efficiently manage soil and water resources, land and environmental pollution, and ensure less waste is produced. The sector will be highly automated using smart technologies such as robots and smart sensors based on spectroscopy and imaging technology. Convincing cases of blockchain-based agri-food traceability are scarce. The multivariate monitoring process is extended from the maturation/growth process in the greenhouse to various post-harvest product processing steps, presenting the need to develop digitalization studies for a complete supply chain quality control (Echegaray et al., 2022; Melesse et al., 2023; Trevisan and Formentini, 2023).

Research on assessing AFSC sustainability, including the role of coordination between chain stages and its relationship with economic, environmental, and social performance, is an important topic in the literature. The cooperative, which serves the needs of product programming, application, and comparison processes, is recommended for AFSC, especially to protect small producers (Pearce et al., 2018). In terms of agricultural product management and harvest process control, a centralized monitoring is important both at the grower level and at the cooperative level, serving the needs of crop programming and application benchmarking processes (Dania et al., 2018; Stone and Rahimifard, 2018).

To conclude, it is evident that the agri-food sector has a vast scope for research, and by leveraging technical advancements, it has the potential to become a compelling and profitable industry. Based on the analysis conducted, it can be asserted that the significance of digitization and physical aspects in the context of AFSC has not been explicitly examined as infrastructure providers in the existing academic literature. To address this existing gap in the field, the third chapter additionally investigates the roles of digitalization and physical elements as providers of infrastructure, while considering pertinent references in the literature.

3 State of the art in agri-food supply chain network design

Network design decisions are critical in FSCM. Supply chain network design phase includes the role of chain actors, the overall configuration and the institutional arrangements that comprise the network. Supply chain network design requires attention to the issues on which management should focus its attention and resources in order to achieve the chain's goals and success (Van Der Vorst, 2005).

All resources are integrated and play an active role in delivering the product to the customer in the design of the supply chain network. These resources include people, machines, information systems, and technology infrastructures (Van Der Vorst, 2005). In terms of network design, the critical issues are the roles and types of operations performed in the facilities, the locations of the facilities, the capacities allocated to each facility, the markets that the facilities will serve, and the sources that will feed the facilities (Chopra and Meindl, 2010; Soysal et al., 2012). These decisions can be directly related to the technological, macro-economic, political, infrastructural, competitive, and other variables that exist in the supply chain (Chopra and Meindl, 2007).

In agri-food supply chain network design, digital technologies, and physical infrastructure issues have a direct impact on decision-making. In this context, it is important to examine the relevant strategic, operational, and tactical decisions in the design of the agri-food supply chain network.

3.1 Physical infrastructures in agri-food supply chains

Physical infrastructure comprises all the essential elements required for the creation, processing, and distribution of food in urban areas. According to urban cultivators, the top three crucial resources for food production in the city are seeds, land or space, and compost. Water, soil, funding, and seedlings are also important, albeit to a lesser extent. The survey identified the top five resources in short supply as compost, funding or capital, land, staff or volunteers, and soil. These resources were characterized as notably scarce due to accessibility, organic matter, affordability and funding, knowledge, and tools (MacRae et al, 2012).

In order to design a resilient and efficient agri-food supply chain network, careful consideration needs to be given to the physical elements of the infrastructure (Kaimba et al., 2021). The network extends from farms and agricultural production facilities to processing and packaging facilities, storage and warehousing, transportation infrastructure and distribution centres. Technology and information systems, such as tracking, traceability, and inventory management, are essential for optimizing coordination. The cold chain infrastructure guarantees the preservation of perishable goods during transportation and storage. Integral components include safety and quality control facilities, waste management infrastructure, energy and water management systems, and compliance with regulatory standards. To design an efficient agri-food supply chain network, it is crucial to consider sustainable practices, such as the use of renewable energy and environmentally friendly packaging. This approach not only promotes efficiency and reduces costs but also ensures the delivery of safe and high-quality products to consumers.

There are various approaches and frameworks that can be considered from different perspectives to guide the creation and planning of supply chain network design. Factors related to supply chain strategy, infrastructure elements and decision processes are all interrelated and require supply chain design dimensions of different structures (Ivanov, 2010). In this context, Table 2 provides an overview of the relevant technological and physical infrastructure elements identified in the literature review of this study in terms of strategic, operational, and tactical decisions.

AFSC has a complex and multidimensional structure because it involves many actors in different processes. Furthermore, numerous economic, social, and environmental dimensions are essential for the design of the AFSC

Table 2 An exemplar of strategic, tactical, and operational frameworks in agri-food supply chain environment

	Frameworks		
	Strategic	Tactical	Operational
Network Design	Long-term commitments (Obonyo et al., 2023). Value chain integration in corporate success (Senturk et al., 2023). Sustainability factors in supply chain: social sustainability dimension on organization's ability, sustainability analysis for supply chain redesign (Kamble et al., 2020; Matopoulos et al., 2015). Logistics networking system of AFSC (Yadav et al., 2022) Green packaging practices associated with financial and social performance (Lugo et al., 2023). Measuring sustainability performance of organization (Pearce et al., 2018). Supply chain integration on supply chain performance (Hassan and Abbassi, 2021).	Transportation mode selection (Taskiner and Bilgen, 2021) Network optimization decisions (i.e. routing, scheduling, cross-docking, location) (Utomo et al., 2018; Matopoulos et al., 2015; Ahumada and Villalobos, 2009; Yadav et al., 2022) Distribution center design (Sid et al., 2021) Product design and visibility (Silvestri et al., 2022) Traceability and transparency (Silvestri et al., 2022) Capacity considerations (Melesse et al., 2023) Food security (Echegaray et al., 2022) Market linkages (Santagata et al., 2021)	Reuse, remanufacturing, and recycling of goods as part of day-to-day operations (Mehmood et al., 2021; Santagata et al., 2021) Managing significant stages in the supply chain (i.e. production, processing, packaging, handling, storage, distribution) Viral infection risks in daily logistics operations (Sid et al., 2021)
Food Loss and Waste	Restructuring of AFSCs to address unforeseen changes in customer demand (Sid et al., 2021). Assessment of the effectiveness of agri-food chains and supply networks with food loss and waste (Adelodun et al., 2021). Network design reorganization of management and development of new business models (Aramyan et al., 2021). Effectiveness of collaboration in sustainable supply chains (Dania et al., 2018).	Market modernization (i.e. product shelf-life, technical advances, and consumer demand diversification) (Sid et al., 2021)	Considering sustainability in daily operations (Kamble et al., 2020; Van Bussel et al., 2022) Considering government regulations in daily operations (Mehmood et al., 2021; Echegaray et al., 2022)

(Continued)

Table 2 (*Continued*)

	Frameworks		
	Strategic	Tactical	Operational
Infrastructure	The impact of the pandemic on the adoption of a 3D energy model of decarbonization, digitalization and centralization (Adelodun et al., 2021) Circular economy applications in agricultural ecosystem and food supply chain (Adelodun et al., 2021). Sustainability factors in supply chain: social sustainability dimension on organization's ability, sustainability analysis for supply chain redesign (Kamble et al., 2020; Matopoulos et al., 2015)	Actor and workforce training (Santagata et al., 2021; Lugo et al., 2023). Interdisciplinary policy-making (Lugo et al., 2023). Blockchain technology for traceability throughout supply chain (Feng et al., 2020). Central databases (Bosona and Gebresenbet, 2023).	Considering waste reduction in daily operations (De Steur et al., 2016; Santagata et al., 2021). Human rights considerations (Atanasovska et al., 2022; Bosona and Gebresenbet, 2023). IoT tools for automation (Echegaray et al., 2022; Silvestri et al., 2022; Melesse et al., 2023).
Agriculture and Agri-Food Sector	Evidence-based policy creation in the agri-food sector (Adelodun et al., 2021). Restructuring in the AFSCs (production, processing and packaging, transportation and storage, distribution and marketing and consumption practices) (Sid et al., 2021). Selecting suppliers, setting up wineries, bottling plants, and distribution centers under the perishability factor (Taskiner and Bilgen, 2021).	Managing intra/inter-organizational quality-related relationships (Zhong et al., 2023). Livestock management (early disease diagnosis, animal welfare, crop yield, etc.) (Senturk et al., 2023).	Inventory scheduling (Kamble et al., 2020). Scheduling (Ahumada and Villalobos, 2009; Matopoulos et al., 2015). Cost minimization (Taskiner and Bilgen, 2021). Carbon emissions (Taskiner and Bilgen, 2021).
Digitalization	Investment in physical assets (Hu et al., 2023). Development of organizational innovation capabilities (Zhong et al., 2023). The decarbonization effect that emerged with the pandemic process (Adelodun et al, 2021). Network Design Decentralization (Tharatipyakul and Pongnumkul, 2021).	Concerning quality schemes (Pomarici and Sardone, 2020). Digital gaps between stakeholders (Tharatipyakul and Pongnumkul, 2021). Information sharing between members of the supply chain (Obonyo et al., 2023).	Implementation of digital technologies (i.e. IoT, Cloud computing, GPS) for monitoring (Stone and Rahimifard, 2018; Silvestri et al., 2022; Trevisan and Formentini, 2023; Yadav et al., 2022; Kamble et al., 2020).

network. The efficient design of the agri-food network is considered through optimization decisions (i.e. routing, scheduling, cross-docking, location) (Utomo et al., 2018; Matopoulos et al., 2015; Ahumada and Villalobos, 2009; Yadav et al., 2022). Moreover, the factors that include some capacity considerations, market conditions, and transport mode choice provide long-term benefits to all actors in the chain. Network design decisions are directly related to supply chain integration and supply chain performance (Hassan and Abbassi, 2021).

Within the structure discussed in Table 2, food loss and waste are defined as the loss of food and food materials in terms of quantity or nutritional quality. With regard to food loss and waste, changing network design means adding recycling flows to traditional product, information and financial flows. Therefore, food loss and waste are considered to be an issue that influences supply chain network design decisions.

The infrastructure that provides society with essential products and services is important in network design as an indicator of physical, digital, and human resource development. The integration of real-time information flow, data analysis, and network design components require an appropriate infrastructure. A primary logistical risk element is the lack of infrastructure services. Phases relating to inventories, production scheduling, and equipment scheduling, as well as measures for minimizing costs, balancing carbon emissions, and achieving sustainability should be emphasized when making network design decisions.

In a broader sense, within the digital technology elements, digital infrastructure systems offer many benefits, such as optimizing supply chain operations, ensuring data transparency, integrating data into the system and traceability. Ensuring food safety is particularly important in food supply chains. In terms of strategic choices, one could say that investments in digital technology elements have high upfront costs. In the long term, however, these costs will be recouped in the form of many benefits. These include supply chain integration, loss minimization and control of large volumes of data. The relationship between network design and digital technology involves elements of dynamic structure, including data processing, anticipating organizational change, and gaining solid and accurate information to use in these circumstances.

3.2 Digital infrastructure in agri-food supply chains

The last decade has seen a huge increase in the use of digital technologies in SCM. In an ever-changing and fast-paced business environment, managers prefer to make decisions based on data-driven information rather than intuition. Digital technologies, which are seen as valuable in decision-making, are seen as an important step in developing and improving supply chain functions. As a result of the benefits of digital technologies, organizations are motivated to improve their technical and organizational capabilities to obtain various values from these technologies (Arunachalam et al., 2018).

Organizations have to constantly improve their business processes, infrastructures, platforms, and supply chains in response to market conditions. In a business environment that is becoming increasingly complex, the need for an information technology ecosystem grows regularly. In this context, there is a need to discover a new and improved method of storing, transforming, sharing, and utilizing data and information to adapt to the dynamic changes in the business world and benefit from them (Delen and Demirkan, 2013).

From the perspective of agri-food supply chains, it is well known that the increase in the variety of food products and brands has increased the risk of counterfeit products and fraud. This has led to a rapid increase in consumer demand for information about food. This increasing demand for food-specific information in society reveals consumers' desire for product transparency and lack of trust (Ge et al., 2017).

Growing sustainability concerns about food production and consumption have raised public awareness and consumer organizations have influenced government agencies and policymakers to take some decisions on creating sustainable supply chains. Some of the challenges faced and to be addressed in the context of sustainable agri-food supply chains can be discussed in terms of food safety and quality control, lack of industrialization, and lack of information. In order to overcome these challenges and find solutions, it is necessary to take into account sustainability dimensions in a holistic manner. These solutions should cover not only the production stage of food but also the processing, packaging, and distribution of food. In line with these developments, the monitoring and use of newly developing technologies within the scope of agriculture-food supply chains has increased, and it has been observed that digital technology infrastructures have a significant impact on agriculture-food supply chains (Kamble et al., 2020).

The literature review of the agri-food supply chain listed in Table 1 reviews studies that are based on digital technologies or mention these technologies in their research. Table 3 shows the digital technologies employed in the agri-food supply chain and their application areas.

According to Table 3, the most widely used or interesting digital technology infrastructures in the reviewed studies are the internet of things, blockchain technology, big data, sensor technologies, and radio frequency identification systems. When we look at how digital technologies are used in agri-food supply chains in Table 3, certain areas stand out. These include keeping track of products, sharing information among different parties, reducing food waste, and safeguarding food quality. To gain a better understanding of digital

Table 3 Digital technology infrastructures and related Publications

Digital technology infrastructures	Related publications	The use of digital technologies
Internet of Things (IoT)	Kamble et al. (2020), Feng et al. (2020), Demestichas et al. (2020), Echegaray et al. (2022), Taskiner and Bilgen (2021), Silvestri et al. (2022), Bosona and Gebresenbet (2023), Senturk et al. (2023)	Collection and sharing of information in real-time, using positioning systems, providing communication between objects, collecting and sharing information, detecting security vulnerabilities in the system
Blockchain Technology	Kamble et al. (2020), Demestichas et al. (2020), Echegaray et al. (2022), Tharatipyakul and Pongnumkul (2021), Bosona and Gebresenbet (2023)	Preventing information asymmetry, ensuring traceability between actors, measuring process performances in the supply chain, monitoring product quality, ensuring transparency about product originality and quality
Big Data	Kamble et al. (2020), Echegaray et al. (2022), Taskiner and Bilgen (2021), Silvestri et al. (2021)	Ensuring accountability and transparency in information sharing among supply chain actors, accelerating the integration and use of data processes
Sensor Technologies	Kamble et al. (2020), Adelodun et al. (2021), Tharatipyakul and Pongnumkul (2021)	Ability to track the product throughout the entire supply chain, making food traceability systems safer and more efficient, processing of products without exposure to physical contact
Radio Frequency Identification (RFID)	Kamble et al. (2020), Feng et al. (2020), Demestichas et al. (2020)	Ability to track the product throughout the entire supply chain, making food traceability systems safer and more efficient, processing of products without exposure to physical contact
Wireless Sensor Network (WSN)	Feng et al. (2020)	Monitoring and storing information at all stages of production, processing, distribution and consumption of the product, facilitating the development, and implementation of a corporate information system
Cloud Computing	Kamble et al. (2020)	Storage of data in the supply chain within the framework of scalability, usability, data integrity, data transformation, data quality, data heterogeneity, privacy, legal and regulatory issues, development of smart agriculture applications

(*Continued*)

Table 3 (*Continued*)

Digital technology infrastructures	Related publications	The use of digital technologies
Near Field Communication (NFC)	Feng et al. (2020)	Ensuring secure data transfer
QR Code	Feng et al. (2020)	Providing information regarding product names, prices, serial numbers, part numbers, batches, and dates
Information and Communications Technology (ICT)	Demestichas et al. (2020)	Creating an ecosystem that includes supply chain actors and their services using information technology software
Biotechnology	Pearce et al. (2018)	Eco-efficient production and processing of products, ensuring improvements in product quality and efficiency
Remote Sensing Technologies, Satellite Navigation Technologies	Pearce et al. (2018), Adelodun et al. (2021)	Eliminating operational confusion, programming and comparing products in groups, performing data-based product control processes
Artificial Intelligence (AI), Digital Twins	Adelodun et al. (2021), Melesse et al. (2023)	Ensuring flexible, smart and efficient operations, reducing bottlenecks in the supply chain, planning for unexpected situations, improving existing processes and resources
Robotics and Automation, Drones	Adelodun et al. (2021), Echegaray et al. (2022)	Improving working efficiency and reliability, reducing food waste
Nanotechnology	Adelodun et al. (2021)	Improved seeds, genetically modified organisms, biomass accumulation and crop yield improvement
Imaging Technology	Echegaray et al. (2022), Vignati et al. (2023)	Visual evaluation of food on the processing line with minimal human intervention
Cybersecurity	Echegaray et al. (2022)	Ensuring food safety by protecting information and computer technology systems such as networks and computers
Augmented Reality (AR)	Echegaray et al. (2022)	Improving the visual perception of the real world by technological systems, increasing production efficiency, fast, consistent, and more accurate product classification

technology infrastructures and their uses, it would be useful to introduce the definitions and features of these technologies briefly.

The Internet of Things is a technical system that combines different information technologies. It is known as the third innovation of the information industry, after the computer and the Internet. By joining different accesses and technologies through the internet, the Internet of Things constructs a communication model between positioning systems and objects, in other words, a vast intelligence network (Wang et al., 2014). The Internet of Things is widely used throughout the agri-food supply chain, from the farmer to the final consumer. It assists in collecting and sharing information on actors and products, detecting system security weaknesses, and warning of potential food safety concerns. This system ensures that the supply chain's operational processes are in control, preventing issues like food safety, quality uncertainties, and resource waste from occurring on a large scale (Wang and Yue, 2017).

Blockchain technology is acknowledged as a system that provides solutions to issues with transparency and trust. Blockchain technology is distributed and decentralized, consisting of interconnected blocks of information that are distributed over time, eliminating incorrect information transfers and ensuring that the information shared is equivalent. Blockchain can provide a useful solution to supply chain trust problems, enabling the traceability of operations between actors and the accurate measurement of process performance and results. In the agricultural food supply chain, it's a popular technology because of its many benefits, such as verifying product quality during transportation, assuring customers of product authenticity and quality, and reducing costs by eliminating intermediary inspectors (Kshetri, 2018).

Based on big data, we can say that data has four features: volume, variety, speed, and accuracy. Therefore, we need big data analysis to examine the data that has been collected and processed. Although big data analysis is implemented in several fields like art and commerce, it is relatively new in the agri-food industry. This system promotes responsibility, openness in information sharing, and drive among supply chain participants. It allows them to make knowledgeable decisions. For instance, when incorporating a cluster of businesses into the sustainable certification process, the utilization of big data analysis speeds up the integration and use of data. Additionally, expenses related to in-person visits to companies are removed (Shukla and Tiwari, 2017).

The transportation of perishable goods has become one of the most important issues in logistics, with more than 50,000 refrigerated trailers registered in Germany alone. This increased demand for cold chain logistics has created a huge application area for sensor technologies. The large amount of wastage of products transported in the agri-food supply chain and the realization of these losses at the last stages of the supply chain have also created the need for sensor technologies. Significant improvements in product quality and the amount of waste have been achieved with data loggers that allow 'on-the-road' access to sensor data so that the product can be tracked (Jedermann et al., 2006).

Radio frequency identification (RFID) systems are commonly used among sensor technologies and have become a leading technology in agri-food supply chains for traceability operations over the last 20 years. RFID technology is known for its faster reading speed compared to traditional barcode systems, which is due to technical infrastructure such as data logger features and integrated sensor systems. Thanks to these benefits, we can confidently claim that food tracing systems have improved safety and efficiency. In addition, it is crucial to process products without physical contact to ensure the utmost hygiene and quality conditions (Badia-Melis et al., 2015).

As technology becomes increasingly prevalent in the agri-food supply chain context, a wide variety of new technological tools are being developed. These tools demand costly computing hardware due to their capacity for large-scale and intricate operations. Furthermore, processing and analyzing data is seen as a challenging and time-consuming task (Ahumada and Villalobos, 2009). Initially, these technological structures were designed to meet the needs of individual actors and were not open to third parties. But as the sector has progressed and supply chain actors have joined forces, systems that incorporate the entire supply chain into technology have begun to appear (Wolfert et al., 2017).

Customers want accurate product information because of a lack of trust in the food supply chain. Studies show that demands and regulations for sustainable agriculture and food distribution are growing. The use of technology in agri-food supply chains has allowed for better transparency of several operational processes. This includes managing data, recording land and resource usage, tracing products, ensuring product quality, processing, packaging, and distribution. Such technological infrastructure has become necessary for this purpose. In summary, the use of digital technologies in the agri-food supply chain is a result of a need and is developing through the establishment of an organized structure for automation and monitoring (Kamble et al., 2020).

4 Conclusion

Due to the ever-changing nature of the food industry and the specific characteristics of food products, FSCM involves a complex set of activities. In addition, AFSCM faces additional challenges in achieving sustainable development goals, especially in recent years, due to the existing structure and the unique characteristics of agri-food products. Various aspects, such as regulations, customer awareness, and pressure from non-profit organizations, have pushed companies to look for ways to reduce their environmental and social impacts. The increasing development of physical infrastructure and digitalization efforts have made AFSCM's decision-making process more complex and difficult than ever before. The evolution from traditional SCM to FSCM and currently to AFSCM has inevitably impacted food supply systems.

This study provides a detailed review of the literature on AFSCM. To the best of our knowledge, this study is a pioneering work in the field of AFSCM that brings the issues of network design, physical infrastructure, and digitalization to the literature with a deep perspective in the context of its relationship with SCM. Also, this study offers a summary of future research directions for researchers interested in the subject by linking with the existing academic literature (see Table 1).

There has been a significant increase in the number of articles on AFSCM in academic literature, an indication that the importance of SCM in the agricultural sector is increasingly recognized. The need for stakeholders to have a clear view of the supply chain and to be able to track it effectively, the widespread use of digitalization and the fact that sustainability is a concern for companies all suggest that AFSCM is becoming an increasingly important issue.

The distinct features of AFSC are frequently highlighted in various studies. The reviewed articles indicate that AFSCM entails challenging procedures in managing and planning, relying heavily on environmental conditions, and facing intense external intervention. Certifications like EUROPE GAP, BRC, IFS, and LCA are essential in AFSCM due to environmental compliance and high security risks, complying with global standards. Results also show the need for a circular economy approach to secure sustainability in the agriculture and food industry. The issue of food waste is one of the main topics of discussion in the articles. In this regard, we encounter the fact that lost food and waste materials can be recycled into the economy as fertilizer or in other forms.

Food loss and waste, which poses a significant problem for AFSCM, is directly related to infrastructure elements that are evaluated in terms of physical, digital, and human resources when making operational, tactical, and strategic network design decisions. For example, in the context of food loss and waste, network design changes within the framework of recycling structure and redesign decisions. In the context of infrastructure elements, infrastructure investment decisions are made in return for the benefits brought by digital technologies such as fast and transparent information sharing, more efficient coordination of physical assets and more effective use of human resources (see Table 2).

Based on the study's results, we have focused on the relationship between physical and digital infrastructures in supply chain. We acknowledge that sustainability is the most frequently discussed topic in literature. Additionally, while digitalization and physical aspects are regarded as crucial issues in AFSC, they have not been dealt with directly as infrastructure providers. The Internet of Things, blockchain technology, big data, sensors, and radio frequency identification systems emerged as the most widely employed digital technologies. These technologies are primarily used to trace products, share information among stakeholders, prevent food waste, and protect food quality (see Table 3).

The main limitation of the study is that it is limited to review articles. In a similar review, research articles, book chapters, or conference presentations could also be included in the scope of the study. At the same time, the related study can also be conducted in databases other than Web of Science. Sustainability review, standardization features, and product-specific review can also be added to the detailed literature table. The scope of the physical, digital, and human resources infrastructure systems covered in the study can be expanded.

Suggestions for further research are compiled from the articles provided and presented to researchers with detailed information in Table 1. The following general recommendations are worthy of mention:

- Aside from economic and environmental sustainability, there is limited research assessing the social aspects of agricultural chain issues.
- The principles of the circular economy can be applied to the design of AFSC.
- Lean methods need to be applied to manage waste in the design of agricultural chains.
- Advances in reusing wasted food as fertilizer for agriculture are possible in the future.
- There is still a need for mathematical decision support models to ensure sustainable development in the AFSCM design.
- There is a strong need for cooperation between universities and businesses for real-world applications.
- Developments related to Agriculture 4.0 and Agrifood 4.0 applications have great potential.
- Existing digital technologies, such as cloud computing and artificial intelligence, should also be used in AFSCs, in addition to the Internet of Things, blockchain technology, big data, sensor technologies, and radio frequency identification systems.
- In the case of AFSCM, the use of physical and human resources infrastructure systems can be considered and the impact on supply chain performance can be evaluated.
- By considering different approaches and frameworks, network design decisions can be presented in a comprehensive way.

- Integrated research designs, where qualitative and quantitative methodologies are used together, can be introduced into the literature.

5 References

Adelodun, B., Kareem, K. Y., Kumar, P., Kumar, V., Choi, K. S., Yadav, K. K., Yadav, A., El-Denglawey, A., Cabral-Pinto, M., Truong Son, C. T., Krishnan, S. and Khan, N. A. (2021). Understanding the impacts of the COVID-19 pandemic on sustainable agri-food system and agroecosystem decarbonization nexus: a review. *Journal of Cleaner Production* 318, 128451.

Ahmed, H. N., Ahmed, S., Khan, M. A. and Ali, S. M. (2022). Sustainable supply chain in emerging economies during and post COVID-19 pandemic: a systematic literature review and future research directions. *Sustainable Supply Chain in Emerging Economies*, ahead of print, 1-27.

Ahumada, O. and Villalobos, J. R. (2009). Application of planning models in the agri-food supply chain: a review. *European Journal of Operational Research* 196(1), 1–20.

Aldama, D. D., Grassauer, F., Zhu, Y., Ardestani-Jaafari, A. and Pelletier, N. (2023). Allocation methods in life cycle assessments (LCAs) of agri-food co-products and food waste valorization systems: systematic review and recommendations. *Journal of Cleaner Production* 421, 138488.

Aramyan, L., Grainger, M., Logatcheva, K., Piras, S., Setti, M., Stewart, G. and Vittuari, M. (2021). Food waste reduction in supply chains through innovations: a review. In: *Measuring Business Excellence* 25(4), 475–492.

Aramyan, L. H., Meuwissen, M. P. M., Oude Lansink, A. G. J. M., Van der Vorst, J. G. A. J., Van Kooten, O. and Van der Lans, I. A. (2009). The perceived impact of quality assurance systems on tomato supply chain performance. *Total Quality Management and Business Excellence* 20(6), 633–653.

Arunachalam, D., Kumar, N. and Kawalek, J. P. (2018). Understanding big data analytics capabilities in supply chain management: unravelling the issues, challenges and implications for practice. *Transportation Research Part E: Logistics and Transportation Review* 114, 416–436.

Atanasovska, I., Choudhary, S., Koh, L., Ketikidis, P. H. and Solomon, A. (2022). Research gaps and future directions on social value stemming from circular economy practices in agri-food industrial parks: insights from a systematic literature review. *Journal of Cleaner Production* 354, 131753.

Badia-Melis, R., Mishra, P. and Ruiz-García, L. (2015). Food traceability: new trends and recent advances: a review. *Food Control* 57, 393–401.

Berenguer, C. V., Andrade, C., Pereira, J. A. M., Perestrelo, R. and Câmara, J. S. (2023). Current challenges in the sustainable valorisation of agri-food wastes: a review. *Processes* 11(1), 1-36.

Bloemhof, J. M. and Soysal, M. (2017). Sustainable food supply chain design. In: Bouchery, Y., Corbett, C. J., Fransoo, J. C. and Tan, T. (Eds). *Sustainable Supply Chains. Springer Series in Supply Chain Management 4*. Switzerland.

Bosona, T. and Gebresenbet, G. (2023). The role of blockchain technology in promoting traceability systems in agri-food production and supply chains. *Sensors* 23(11), 5342.

Chopra, S. and Meindl, P. (2007). *Supply Chain Management. Strategy, Planning & Operation*. Gabler, New Jersey, pp. 265–275.

Chopra, S. and Meindl, P. (2010). *Supply Chain Management: Strategy, Planning and Operation* (4th edn.). Pearson Education, NJ.

Chopra, S. and Meindl, P. (2012). *Supply Chain Drivers and Metrics: Supply Chain Management, Strategy, Planning and Operation* (5th edn.). Pearson, Upper Saddle River, NJ.

Ciampa, A., Danesi, F. and Picone, G. (2023). NMR-based metabolomics for a more holistic and sustainable research in food quality assessment: a narrative review. *Applied Sciences (Switzerland)* 13(1), 1-14.

Dania, W. A. P., Xing, K. and Amer, Y. (2018). Collaboration behavioural factors for sustainable agri-food supply chains: a systematic review. *Journal of Cleaner Production* 186, 851-864.

Delen, D. and Demirkan, H. (2013). Data, information and analytics as services. *Decision Support Systems* 55(1), 359-363.

Demestichas, K., Peppes, N., Alexakis, T. and Adamopoulou, E. (2020). Blockchain in agriculture traceability systems: a review. *Applied Sciences (Switzerland)* 10(12), 1-22.

Derakhti, A., Santibanez Gonzalez, E. D. R. and Mardani, A. (2023). Industry 4.0 and beyond: a review of the literature on the challenges and barriers facing the agri-food supply chain. *Sustainability* 15(6), 5078.

De Steur, H., Wesana, J., Dora, M. K., Pearce, D. and Gellynck, X. (2016). Applying value stream mapping to reduce food losses and wastes in supply chains: a systematic review. *Waste Management* 58, 359-368.

Echegaray, N., Hassoun, A., Jagtap, S., Tetteh-Caesar, M., Kumar, M., Tomasevic, I., Goksen, G. and Lorenzo, J. M. (2022). Meat 4.0: principles and applications of Industry 4.0 technologies in the meat industry. *Applied Sciences (Switzerland)* 12(14), 6986.

Feng, H., Wang, X., Duan, Y., Zhang, J. and Zhang, X. (2020). Applying blockchain technology to improve agri-food traceability: a review of development methods, benefits and challenges. *Journal of Cleaner Production* 260, 121031.

Food and Agriculture Organization of the United Nations (2017). The future of the food and agriculture. Available at: https://www.fao.org, accessed October 2023.

Food and Agriculture Organization of the United Nations (2019). Technical platform on the measurement and reduction of food loss and waste. Available at: https://www.fao.org, accessed January 2024.

Ge, L., Brewster, C., Spek, J., Smeenk, A., Top, J. and Van Diepen, F. (2017). Blockchain for agriculture and food: *Findings from the pilot study* (No. 2017-112). *Wageningen Economic Research*.

Hassan, N. M. and Abbasi, M. N. (2021). A review of supply chain integration extents, contingencies and performance: a post Covid-19 review. *Operations Research Perspectives* 8, 2214-7160.

Hu, L., Zheng, Y., Woods, T. A., Kusunose, Y. and Buck, S. (2023). The market for private food safety certifications: conceptual framework, review, and future research directions. *Applied Economic Perspectives and Policy* 45(1), 197-220.

Ivanov, D. (2010). An adaptive framework for aligning (re) planning decisions on supply chain strategy, design, tactics, and operations. *International Journal of Production Research* 48(13), 3999-4017.

Jedermann, R., Behrens, C., Westphal, D. and Lang, W. (2006). Applying autonomous sensor systems in logistics–combining sensor networks, RFIDs and software agents. *Sensors and Actuators A: Physical* 132(1), 370–375.

Jia, F., Hubbard, M., Zhang, T. and Chen, L. (2019). Water stewardship in agricultural supply chains. *Journal of Cleaner Production* 235, 1170–1188.

Kaimba, G. K., Mithöfer, D. and Muendo, K. M. (2021). Commercialization of underutilized fruits: baobab pulp supply response to price and non-price incentives in Kenya. *Food Policy* 99, 101980.

Kamble, S. S., Gunasekaran, A. and Gawankar, S. A. (2020). Achieving sustainable performance in a data-driven agriculture supply chain: a review for research and applications. *International Journal of Production Economics* 219, 179–194.

Kshetri, N. (2018). Blockchain's roles in meeting key supply chain management objectives. *International Journal of Information Management* 39, 80–89.

Lugo, S. D. R., Kimita, K. and Nishino, N. (2023). Characteristics of decision process towards circular food economy: a review. *Cleaner Logistics and Supply Chain* 7, 100104.

MacRae, R., Nasr, J., Kuhns, J., Baker, L., Christianson, R., Danyluk, M., Snider, A., Gallant, E., Kaill-Vinish, P., Michalak, M., Oswald, J., Patel, S. and Wekerle, G. (2012). Could Toronto provide 10% of its fresh vegetable requirements from within its own boundaries? Part II, Policy supports and program design. *Journal of Agriculture, Food Systems, and Community Development* 2(2), 147–169.

Matopoulos, A., Barros, A. C. and Van der Vorst, J. G. A. J. (2015). Resource-efficient supply chains: a research framework, literature review and research agenda. *Supply Chain Management: An International Journal* 20(2), 218–236.

Mehmood, A., Ahmed, S., Viza, E., Bogush, A. and Ayyub, R. M. (2021). Drivers and barriers towards circular economy in agri-food supply chain: a review. *Business Strategy and Development* 4(4), 465–481.

Melesse, T. Y., Franciosi, C., Di Pasquale, V. and Riemma, S. (2023). Analyzing the implementation of digital twins in the agri-food supply chain. *Logistics* 7(2), 33.

Moreno-Miranda, C. and Dries, L. (2022). Integrating coordination mechanisms in the sustainability assessment of agri-food chains: from a structured literature review to a comprehensive framework. *Ecological Economics* 192, 107265.

Obonyo, E., Formentini, M., Ndiritu, S. W. and Naslund, D. (2023). Information sharing in African perishable agri-food supply chains: a systematic literature review and research agenda. *Journal of Agribusiness in Developing and Emerging Economies*, ahead-of-print.

Pearce, D., Dora, M., Wesana, J. and Gellynck, X. (2018). Determining factors driving sustainable performance through the application of lean management practices in horticultural primary production. *Journal of Cleaner Production* 203, 400–417.

Pomarici, E. and Sardone, R. (2020). EU wine policy in the framework of the CAP: post-2020 challenges. *Agricultural and Food Economics* 8(1), 1-40.

Ravi, H. K., Degrou, A., Costil, J., Trespeuch, C., Chemat, F. and Vian, M. A. (2020). Larvae mediated valorization of industrial, agriculture and food wastes: biorefinery concept through bioconversion, processes, procedures, and products. *Processes* 8(7), 857.

Santagata, R., Ripa, M., Genovese, A. and Ulgiati, S. (2021). Food waste recovery pathways: challenges and opportunities for an emerging bio-based circular economy. A systematic review and an assessment. *Journal of Cleaner Production* 286, 125–490.

Senturk, S., Senturk, F. and Karaca, H. (2023). Industry 4.0 technologies in agri-food sector and their integration in the global value chain: a review. *Journal of Cleaner Production* 408, 137096.

Shukla, M. and Tiwari, M. K. (2017). Big-data analytics framework for incorporating smallholders in sustainable palm oil production. *Production Planning and Control* 28(16), 1365–1377.

Sid, S., Mor, R. S., Panghal, A., Kumar, D. and Gahlawat, V. K. (2021). Agri-food supply chain and disruptions due to COVID-19: effects and Strategies. *Brazilian Journal of Operations and Production Management* 18(2), 1–14.

Siddh, M. M., Soni, G. and Jain, R. (2015). Perishable food supply chain quality (PFSCQ): a structured review and implications for future research. *Journal of Advances in Management Research* 12(3), 292–313.

Silvestri, C., Silvestri, L., Piccarozzi, M. and Ruggieri, A. (2022). Toward a framework for selecting indicators of measuring sustainability and circular economy in the agri-food sector: a systematic literature review. *The International Journal of Life Cycle Assessment* 2022, 1–39.

Soysal, M., Bloemhof-Ruwaard, J. M., Meuwissen, M. P. M. and Van Der Vorst, J. G. A. J. (2012). A review on quantitative models for sustainable food logistics management. *International Journal on Food System Dynamics* 3(2), 136–155.

Soysal, M., Bloemhof-Ruwaard, J. M. and van der Vorst, J. G. A. J. (2014). Modelling food logistics networks with emission considerations: the case of an international beef supply chain. *International Journal of Production Economics* 152, 57–70.

Soysal, M., Belbağ, S. and Erişkan, S. (2022). Horizontal collaboration among SMEs through a supply and distribution cooperative. *Open Transportation Journal* 16(1), 1-16.

Stone, J. and Rahimifard, S. (2018). Resilience in agri-food supply chains: a critical analysis of the literature and synthesis of a novel framework. *Supply Chain Management* 23(3), 207–238.

Taskiner, T. and Bilgen, B. (2021). Optimization models for harvest and production planning in agri-food supply chain: a systematic review. *Logistics* 5(3), 52.

Tharatipyakul, A. and Pongnumkul, S. (2021). User interface of blockchain-based agri-food traceability applications: a review. *IEEE Access* 9, 82909–82929.

Trevisan, C. and Formentini, M. (2023). Digital technologies for food loss and waste prevention and reduction in agri-food supply chains: a systematic literature review and research agenda. *IEEE Transactions on Engineering Management*, 1-20.

Ünal, V., Ömürgönülşen, M., Belbağ, S. and Soysal, M. (2020). The internet of things in supply chain management. In: Paksoy, T., Koçhan , Ç. and Ali, S. S.(Eds.), *Logistics 4.0*. CRC Press, Florida, , 27–34.

Utomo, D. S., Onggo, B. S. and Eldridge, S. (2018). Applications of agent-based modelling and simulation in the agri-food supply chains. *European Journal of Operational Research* 269(3), 794–805.

Van Bussel, L. M., Kuijsten, A., Mars, M. and Van't Veer, P. (2022). 'Consumers' perceptions on food-related sustainability: a systematic review, *Journal of Cleaner Production* 341, 130904.

Van Der Vorst, J. G. (2005). Performance measurement in agrifood supply chain networks: an overview. *Quantifying the Agri-Food Supply Chain* 15, 13–24.

Van Der Vorst, J. G. A. J., Tromp, S. O. and Van der Zee, D. J. (2009). Simulation modelling for food supply chain redesign; integrated decision making on product quality, sustainability and logistics. *International Journal of Production Research* 47(23), 6611–6631.

Vignati, S., Tugnolo, A., Giovenzana, V., Pampuri, A., Casson, A., Guidetti, R. and Beghi, R. (2023). Hyperspectral imaging for fresh-cut fruit and vegetable quality assessment: basic concepts and applications. *Applied Sciences (Switzerland)* 13(17), 9740.

Wang, H. Z., Lin, G. W., Wang, J. Q., Gao, W. L., Chen, Y. F. and Duan, Q. L. (2014). Management of big data in the internet of things in agriculture based on cloud computing. *Applied Mechanics and Materials* 548, 1438–1444.

Wang, J. and Yue, H. (2017). Food safety pre-warning system based on data mining for a sustainable food supply chain. *Food Control* 73, 223–229.

Wolfert, S., Ge, L., Verdouw, C. and Bogaardt, M. J. (2017). Big data in smart farming – a review. *Agricultural Systems* 153, 69–80.

World Bank (2021). What is cooking: digital transformation of the agrifood systems. Available at: https://openknowledge.worldbank.org, accessed October 2023.

World Bank (n.d.). Agriculture and Food. Available at: https://www.worldbank.org, accessed October 2023.

World Food Programme (WFP) (n.d.). Supply Chain. Available at: https://www.wfp.org, accessed October 2023.

Yadav, V. S., Singh, A. R., Gunasekaran, A., Raut, R. D. and Narkhede, B. E. (2022). A systematic literature review of the agro-food supply chain: challenges, network design, and performance measurement perspectives. *Sustainable Production and Consumption* 29, 685–704.

Zhong, J., Cheng, H., Chen, X. and Jia, F. (2023). A systematic analysis of quality management in agri-food supply chains: a hierarchy of capabilities perspective. *Supply Chain Management* 28(3), 619–637.

Chapter 8

Planning and control in agri-food supply chains

Florian Cramer and Sandra Transchel, Kühne Logistics University, Germany

1 Introduction

Recent events such as the coronavirus disease 2019 (aka COVID-19) pandemic and the Russian-Ukrainian war have chipped away at many supply chains' foundations. Thousands of companies have been left stumbling in the wake of these events and are now refocusing on how to maintain their business activities in the face of a possible recession and increasing labor and material costs. The paradigm shift to more sustainable production and consumption is now competing with the efforts to build more resilient supply chains. At the same time, the issue of sustainability remains of utmost importance on a global scale: In November 2022, the worldwide population reached 8 billion people, and it is estimated that it could grow to up to 9.7 billion by 2050.[1] This increasing amount of population will put an additional strain on already challenged global food systems, which, at their core, rely on a strong agri-food industry. Therefore, we must rethink our current food systems to facilitate the necessary sustainability paradigm shift, which includes an evaluation of existing planning

1 https://www.un.org/en/global-issues/population.

http://dx.doi.org/10.19103/AS.2023.0122.11

and control mechanisms and, eventually, developing novel, innovative systems that enable managers and policymakers to make informed decisions for the planning and control of agri-food supply chains (AFSCs).

Planning and control activities in AFSCs refer to all processes identifying and anticipating the needs and demands of consumers, making strategic, tactical, and operational decisions to meet those needs, and controlling the overall performance of such systems. This includes traditional planning, such as demand planning and forecasting, production planning, inventory management, and distribution, as well as related activities, such as risk management, food quality and safety management, and sustainability assurance activities. A simplified AFSC framework, which describes the interdependencies between various actors in AFSCs, developed by the World Bank is illustrated in Fig. 1. Besides the rather linear flow of physical products from input suppliers via producers, intermediaries, processors, and retailers, the framework also illustrates the diverse range of supporting entities, e.g. logistics service providers, financial services, regulators, and policymakers (Jaffee et al., 2008). These supporting activities encompass various operations directly impacting the performance of AFSC, the quality of products, packaging materials and processes, and warehousing operations.

Managing agri-food products is particularly complex. Their production, processing, handling, and consumption can directly affect human health, well-being, and sustainability and thus require specific safety regulations. In addition, many agri-food products are perishable, and related logistics processes are time and temperature-sensitive, thus requiring cold chain operations throughout the supply chain to minimize loss and to satisfy safety regulations. To better

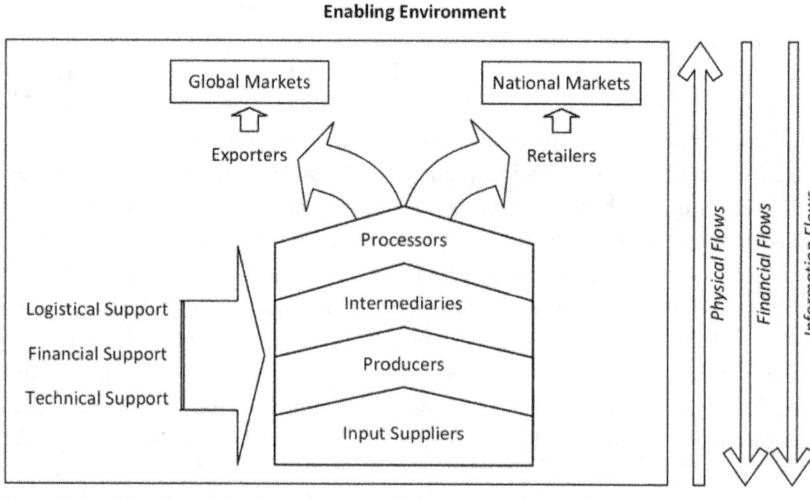

Figure 1 Agri-food supply chain framework. Source: Adapted from: Jaffee et al. (2008).

ensure food safety and transparency in AFSCs, more and more governments and authorities intensify and extend food safety regulations. For example, after a series of food incidents in the 1990s, in 2002, the European Union introduced the General Food Law[2] stating various principles, requirements, and procedures for all stages of the food and feed supply chain. A critical aspect of that law is the traceability requirement of the products along the entire supply chain.

In recent decades, we could observe an increasing development of new technologies and methodologies such as agricultural robotics, imaging technology, and machine-learning-based forecasting that revolutionize operations processes in AFSCs. However, to fully utilize the benefits and adapt to the new technological environment and increasing data availability, decision-makers may require new optimization tools for planning and control activities. Many planning and control activities are based on decision-support mechanisms using planning models from operations management (OM) and quantitative techniques from operations research (OR) and management science (MS).

Consequently, this chapter explores planning and control processes in AFSCs from an OM perspective, mainly considering techniques and methodologies from OR and MS. The subsequent section, Section 2, will provide fundamentals of planning and control in AFSCs. It will highlight essential building blocks and give a brief history of planning and control models in AFSCs in which relevant review articles are highlighted. The following subsections explore integral parts of planning processes in AFSCs, such as forecasting, including price, demand, and shelf life forecasting (Section 3) and production planning and inventory control (Section 4). Furthermore, a brief overview of upcoming challenges and opportunities in these areas is given in Section 5. We finally conclude with brief remarks in Section 6.

2 Fundamentals of planning and control in agri-food supply chains

The planning and control of agri-food systems have a long history, dating back to ancient civilizations, where planers developed structured irrigation and crop management systems. The industrial revolution in the eighteenth and nineteenth centuries spawned new agricultural technologies, such as more effective plows and seed drills, which led to more efficient farming practices and increased crop yields (Rasmussen, 1962). With a growing population in the twentieth century and the further development of agriculture technologies, AFSC planning shifted to increase production to meet the increasing demand.

2 Regulation (EC) No 178/2002 laying down general principles and requirements of food law. https://eur-lex.europa.eu/legal-content/EN/TXT/HTML/?uri=CELEX:32002R0178.

At the same time, the development of chemical fertilizers and pesticides significantly improved yield rates and production output (Gardner, 2009). Consequently, all these technological developments, along with advanced planning and control mechanisms, improved the efficiency and effectiveness at all stages in AFSCs. In recent years, however, the focus has been more and more on sustainable food production and supply chains. Shifting from pure output-oriented production and cost-efficient supply chain planning to an ecologically and socially sustainable planning system of AFSCs that can still feed the growing population is one of the major challenges of our society.

This section first reviews the history of research in the planning and control activities in AFSCs. Although somewhat dated, many of these mathematical models and quantitative techniques still build the foundation of modern decision-support systems. Since many AFSCs span huge distances from the place of origin to the place of final consumption, including lengthy and time-consuming trips with varying modes of transportation and intermediate storage, two important distribution processes in agri-food supply chains must be mentioned: cold chain management (Mercier et al., 2017) and transport mode choice (Ahumada and Villalobos, 2009). In the second step, we briefly explore cold chain management and transport mode choice in the context of AFSCs.

2.1 Planning and control of agri-food systems – a brief literature review

Research on planning and control processes in AFSCs has a long history and has attracted considerable interest over many decades. This development is reflected in the diverse scientific publications studying various planning problems along the entire AFSC. Due to the tremendous amount of publications in this field, instead of presenting a literature review on individual publications, we provide an overview of the individual review articles in particular areas; however, with a specific focus on planning and control problems using quantitative methods of OR and MS.

The first review papers date back to the 1980s when Kennedy (1981) studies applications of dynamic programming formulations[3] as decision-support techniques for forestry, fishery, and agriculture policies. Six years later, Glen (1987) examines OR-based models used for various planning problems, particularly at the farm level. Glen distinguishes two major areas, namely crop and livestock production. He provides a detailed analysis of different modeling and optimization approaches for various fundamental planning problems, such as planning and harvesting operations, evaluating pest and disease control

3 Dynamic Programming is an interactive optimization technique for multistage decision problems (Bellman, 1957).

mechanisms (related to crop production), and breeding and replacement strategies and feeding policies (related to livestock production).

Since 2009, several review papers have been published examining planning tools in food and agriculture supply chains. Lowe and Preckel (2004) follow up on the work by Glen (1987) and study applications of OR-based decision-support methodologies used for crop planning, harvesting, and risk management in AFSCs. Already in 2004, the topic of 'precision production' and the associated changes in agri-food planning (e.g. smaller planning units and the availability of more granular data) was already recognized as one of the emerging issues in agri-food production. In the same year, Lucas and Chhajed (2004) provide a survey of location planning problems in agriculture supply chains. The authors discuss early studies of agriculture location theory, e.g. the seminal work by von Thünen (1826), and present a variety of OR-based modeling approaches to solve location planning problems in agriculture. Ahumada and Villalobos (2009) complement the previous reviews by focusing on food products from crops. The authors illustrate challenges between planning models of nonperishable and fresh (perishable) food products and classify planning tools for strategic, tactical, and operational decision problems successfully implemented in decision-support tools in practice. With a similar focus, however, even more specialized, Zhang and Wilhelm (2011) examine quantitative decision-support models based on OR and MS techniques for particular crops, including fruits, vegetables, grapes, and wine.

Akkerman et al. (2010) take a slightly different view by developing a classification scheme for food distribution systems and discussing related challenges. The authors focus on quantitative OM models facing three important food industry-specific challenges: food quality, food safety, and sustainability. Shukla and Jharkharia (2013) keep a broader perspective and review papers along the entire AFSC utilizing both quantitative and qualitative methodologies. The authors identify that demand planning and forecasting received less attention in AFSCs than in traditional manufacturing. However, the question arises to what extent agri-food products require more specific forecasting techniques or whether the traditional forecasting methods are sufficient (Petropoulos and Carver, 2019).

Tsolakis et al. (2014) provide a comprehensive hierarchical decision-making framework and develop a taxonomy that applies to all stakeholders involved in the design and management of AFSCs. With a similar goal, however, with a focus on decision problems in fruit supply chains, Soto-Silva et al. (2016) review related research that uses mathematical modeling and optimization techniques to support decision-making processes. The authors build a classification along different dimensions: (i) the purpose of the decision (i.e. sowing, production, harvesting, distribution, or inventory planning), (ii) the decision level (i.e. strategic, tactical, or operational decision problems), and (iii) the modeling approach and methodology (e.g. linear programming, nonlinear programming, multiobjective optimization, stochastic optimization, or simulation). Besides these rather generic dimensions, they further identify that only a small fraction of the surveyed papers are based on real-world cases using actual data.

Handayati et al. (2015) investigate collaboration and coordination mechanisms in AFSCs and the interaction of various actors. The authors classify different coordination mechanisms, including information-sharing activities and contract design, based on the interdependencies and challenges to meeting customer service requirements and safety issues. Their review shows that coordination-related problems in AFSCs, where different stakeholders in the supply chain cooperatively plan their decisions, are still at an early stage. With a review period of 30 years ranging from 1983 to 2013, Kusumastuti et al. (2016) examine the literature on crop-related supply chains focusing on the degree to which specific planning decisions (primarily production, harvesting, and processing) are integrated and coordinated throughout the entire chain. Like Handayati et al. (2015), Kusumastuti et al. (2016) also find that most planning models in AFSCs focus on a single decision-making problem and ignore the interdependencies of other planning decisions, e.g. coordinating harvesting and processing or coordinated joint production and replenishment of different actors along the supply chain.[4]

With increasing technological development (sensor and information and communication technology (ICT), companies can continuously monitor various features in food supply chains, such as temperature, humidity, or location. This development affected research on the benefits of novel planning and control mechanisms in agri-food systems. Mercier et al. (2017) assess the current state of time–temperature control mechanisms along food cold chains and review various field studies and their potential for improved supply chain planning. One of their major findings is that continuous monitoring of the temperature in the supply chain and keeping track of the time–temperature history (TTH) of perishable products allows planners to dynamically assess products' remaining shelf lives, which, in turn, may improve inventory levels and service level performance.

Since the beginning of the pandemic in 2020 and the Ukraine war in 2022, supply chain risk management has increasingly become the focus of many companies, not only in AFSCs. However, even before, risks and uncertainties were ubiquitous in AFSCs. Behzadi et al. (2018) provide a literature review of quantitative decision models for agri-food supply chain risk management. Their review provides various implications not only for future research but also

4 While simultaneously considering agri-food characteristics such as yield uncertainty (due to weather) and perishability due to deterioration.

for practical implementation. One of their main findings is that robustness, i.e. the ability to withstand disruption with an acceptable loss of performance, and resilience, i.e. the potential to recover quickly from a disruption are two critical performance measures for managing risk in AFSCs. More recently, Taşkıner and Bilgen (2021) present a structured literature review for optimization models in harvesting and production planning. Like other reviews before, e.g. Soto-Silva et al. (2016), the authors classify the literature regarding the type of decision, the objective, the decision level, and the planning horizon. Besides technical modeling details, the authors also classify the literature regarding the product type (e.g. fruits, grain, or olive oil), level of planning uncertainty (demand and supply uncertainty), perishability, and sustainability aspects such as food loss and waste. Key findings are as follows: The vast majority of papers study planning problems of a single stage in the supply chain and/or a single product/product group; there is little research on multiproduct and integrated multistage decision-making problems. Although such models can provide more sophisticated results, the complexity of such models makes them challenging to analyze, particularly for AFSCs with their unique characteristics, such as random yield and perishability. In 2022, Stüve et al. (2022) present a systematic literature review of modeling approaches and planning tasks supported by advanced planning systems in the food industry. The study reveals that although many scientific planning and decision-support models have been developed, there is little evidence that such models can be successfully used in practice: Empirical studies on applying novel planning models are relatively scarce.

The overview of review papers illustrates the tremendous amount of research studying quantitative techniques for planning and control in AFSCs. Many of the research papers study a particular agri-food product in a particular environment, which is one of the reasons for the enormous amount of literature. A common finding among the most recent review papers is that especially planning problems where multiple (dependent or independent) decision-makers are involved (e.g. mutiechelon production and inventory control, joint pricing, and production/inventory control) are still understudied. A further observation is that there is little research on the impact of governmental policies and regulations on planning and optimization in AFSC. In Section 4.4, we provide a detailed review of some studies investigating this topic.

2.2 Cold chain management in agri-food supply chains

For many AFSCs, cold chain logistics is an integral part of operations. But what exactly is a cold chain, and why is it essential for planning control processes in AFSCs? The answer to this question is quite simple: Cold chains are transport chains with meticulously coordinated and controlled temperatures for all related operations. The need to establish and maintain such supply chains arises from two interconnected challenges, namely the issue of food safety (Mercier et al., 2017), i.e. ensuring the innocuousness of food items, and food security (James and James, 2010), i.e. providing sufficient supply without generating extensive food loss.

2.2.1 Rationale for the use of cold chains in agri-food systems

Many foodborne pathogens and microorganisms, like, e.g. *Campylobacter*, *Salmonella*, *Listeria monocytogenes*, or *Escherichia coli*, are mesophiles, i.e. they thrive at ambient temperatures, while a few others are psychotropic, i.e. they grow at low temperatures (Bhunia, 2018). One way to limit the growth of these organisms is to store agri-food items in a controlled environment with lower temperatures. Decay and ripening processes are often slowed down at lower temperatures; however, they are not completely stopped (Yahia, 2009).

Many countries rely on food safety regulations to reduce the chance of outbreaks of foodborne diseases. A report by the World Health Organization (2010) estimated that 600 million people get infected by foodborne diseases each year, of which they estimate 420 000 people die. In terms of financial loss, the World Bank (2018) estimated that foodborne diseases cause a yearly $110 billion of cost in low- and medium-income countries. Consequently, there are many countries where a cold chain is not a matter of convenience and prevention of lawsuits but is mandated by law. Since there is no such thing as a universal global food safety regulation, companies must consider regulations in each of the markets they are active in carefully. Often these regulations are defined in bilateral treaties and national law (Lin, 2012). As a result, companies often rely on standards that cover a wide array of regulatory aspects. Such standards and regulations include the USA–China Food-Safety Agreement, the Codex Alimentarius, Good Agricultural Practice (GAP), or the ISO 22000 (Gravani, 2009; Lin, 2012; Panghal et al., 2018).

According to estimates by the International Institute of Refrigeration (IIR), around 12% of the food produced in 2017 was lost due to insufficient cooling (Amath et al., 2021). At the same time, cold chains are a major contributor to global energy consumption and emissions (James and James, 2010). While cold chains help to increase food safety and food security, they also pose immense infrastructural challenges due to the required energy and transport infrastructure (Neusel and Hirzel, 2022; Salin and Nayga, 2003). A major challenge that thus arises in evaluating the need and benefits of cold chains is characterized by trade-offs between maintaining food safety and security versus infrastructural challenges, energy consumption, and, of course, costs. Managing these trade-offs can get tricky and may require constant attention, as shown by the recent development of energy markets: A report by Amath et al. (2021) finds that associated energy cost for cold chain operations in the UK has doubled from 2021 to 2022 due to a dramatic increase in energy prices.

2.2.2 Trade-offs in cold chain design decisions

To better understand the trade-offs in agri-food cold chain design, we must look closely at the underlying biological and chemical processes behind spoilage. Two major issues affect food safety and food loss: microbial growth and decay processes (Bintsis, 2018; Prusky, 2011; Raak et al., 2017). Infections of foodstuff can happen at many stages within the supply chain, e.g. during harvesting, production processes, transportation, or storage, either by mechanical injuries such as scratches, blemishes, and bruises or by exposure to contaminated material and environments. Decay processes depend on the storage and handling conditions and can be affected by upsetting and deviating from proper environmental conditions, e.g. by inappropriate storage temperature or the presence of activators such as phytohormones like ethylene. For many of these processes, e.g. ethylene production, ambient temperatures are crucial (Knee et al., 1985). Microbial growth and enzymatic, transpiratory, or respiratory processes such as ethylene respiration greatly depend on ambient temperature; however, decay and ripening processes vary between food products. While some products, like grapes, are highly tolerant to low-temperature storage, others are susceptible and may, like tomatoes, suffer cold damage at low temperatures. Thus, optimally planning cold chains and providing product-specific temperature control is crucial for many agri-food products. Reasons for deviating ambient temperatures include waste heat from operations, outside temperature, and biochemical processes such as ripening.

While storage planning at facilities can be quite challenging, requiring different temperature zones and goods separation based on cross-perishability effects, another bottleneck in cold chain logistics often lies within distribution processes. On the one hand, there is the transition from a field to an initial storage location, the entry point into the cold chain: Depending on the ambient temperature during the harvesting process and the type of products, the products may need to be cooled down to slow deterioration. This initial cooling phase is called precooling (Brosnan and Sun, 2001). On the other hand, there is interfacility transportation of goods between the warehouse locations of various actors along the supply chain. Each of these transitions bears the risk of breaking the cold chain and impeding food quality and safety. Furthermore, logistics efficiency demands the consolidation of goods to reduce less-than-truckload (LTL) transportation. Consolidating goods, however, means that achieving optimal transport temperature for every good within a certain shipment can become challenging. Thus, the cold chain imposes challenges and barriers to the planning and control of related processes in AFSCs.

2.2.3 Cold chain design and management

As described above, the precooling stage is often the entry point into the cold chain. This entry point, which sometimes is omitted due to a lack of infrastructure or the additional cost, can severely impact the shelf life of a product (Brosnan and Sun, 2001). Other elements of a cold chain are cold storages, refrigerated transportation via refrigerator trucks and reefer containers, and distribution centers (Behdani et al., 2019; Neusel and Hirzel, 2022). The resulting challenges related to trade-offs in temperature-controlled transportation and the design of cold chain networks have attracted many researchers from the operations management and operations research community.

As highlighted by Awad et al. (2020) and Shashi et al. (2018), various aspects of cold chain design and management in the agri-food context have been explored. Awad et al. (2020) show that a majority of vehicle-routing-related research in this area has focused on including product quality. The authors identify various factors that have been considered regarding transportation, quality, and environmental constraints. They further discover that the cost associated with product loss is a major concern for including quality parameters. The inclusion of quality parameters is often motivated by a product's loss or waste and associated economic or environmental costs. For instance, some research, e.g. Zhao et al. (2020) or Hsiao et al. (2017), incorporate aspects specific to food safety, such as microbial growth in their optimization models.

In a review of cold chain management, Shashi et al. (2018) categorize four categories associated with cold chain management: root causes for inefficiency, sustainability issues, key metrics measurement, and improvement approaches. The authors find that most research has focused on measuring key metrics, followed by research on root causes for inefficiencies within cold chains. They note that research has increasingly underlined the consumer's role regarding food safety and security issues. Moreover, food quality and safety is the prevalent performance metric among the analyzed papers. From a practitioner's view, food safety is crucial to supply chain operations. Government-mandated laws and regulations, such as hazard analysis and critical control points (HACCP), are essential factors when deciding on supply chain design and the management of related processes. Managers need to incorporate these aspects in their decision-making process when considering how to optimize operations.

2.2.4 Challenges in agri-food cold chains

The presented trade-offs and elements of a cold chain can pose severe entry barriers in some countries. Major issues with cold chain efficiency are related to a lack of infrastructure, insufficient standardization, and the number of involved intermediaries (Shashi et al., 2018). In addition, alternative power systems are required if there is no stable power grid. These systems add an additional layer of investment to the existing unstable power supply problem. However, they also allow to maintain off-grid cold chain facilities. Such facilities can reduce the amount of food loss and, thus, increase food security. For instance, a joint project with InspiraFarms, the Shell Foundation, and the UK's Department for International Development in Africa showed that off-grid cold chain facilities can improve shelf life and help farmers to reduce losses (Shell Foundation, 2019).

Additional to grid infrastructure and stability, the transport infrastructure imposes another constraint. Without proper cooling solutions, such as reefer containers, the resulting break in the cold chain endangers food safety and can increase food loss. Even small fluctuations in temperature can have detrimental effects on food safety. An inadequate road network adds to this by prolonging the transition between warehouses. Additionally, malpractice, poor cooling equipment design, and incorrect storage can also impose additional risks (Ndraha et al., 2018). Consequently, research in this area should aim to support decision-makers to minimize the trade-offs in cold chain operations, design, and management and to increase food safety and security. The major challenges concerning this matter are energy consumption and requisite infrastructure trade-offs. Particularly operations in developing countries need further attention to find sustainable solutions.

2.3 Transport mode choice and collaboration in agri-food supply chains

Global AFSCs often span over multiple countries, including numerous stations at different storage locations and involving various stakeholders such as carriers, customs authorities, and plant inspection. To enable a smooth transition between supply chain tiers, the planning of transportation mode choice and collaboration between stakeholders along the supply chain is essential to avoid unnecessary costs and reduce lead times and product loss.

2.3.1 Multi-, inter-, co- and synchromodal transportation

As Jones et al. (2019) observe, the term 'intermodal transportation' has been defined in numerous ways with varying meanings and emphasis. Agamez-Arias and Moyano-Fuentes (2017) note that 'intermodal transportation' sometimes is used interchangeably with the terms 'multimodal transportation', 'combined transportation', or 'co-modal transportation'. However, according to their literature review, these terms differ slightly in their meaning and scope. While multimodal transportation describes the transportation of goods with at least two different modes of transportation (Pencheva et al., 2022), intermodal transportation requires using the same loading unit throughout the transport chain (Agamez-Arias and Moyano-Fuentes, 2017). Co-modality, in turn, implies the optimal use of various modes of transportation (including multimodal and intermodal transportation) to reduce resource utilization, and synchromodality adds the use of real-time data to flexibly coordinate and optimize transportation (Acero et al., 2022; Agamez-Arias and Moyano-Fuentes, 2017). Consequently, multimodal transportation can be viewed as the basis for intermodal, co-modal, and synchromodal transportation. However, as Acero et al. (2022) point out, the synchronization of transport along the whole supply chain requires a novel set of practices and cannot be viewed as a simple extension of the abovementioned concepts.

Regarding the use of synchromodality in cold chains, Guo et al. (2017) review relevant characteristics and challenges. According to their review, horizontal collaboration and information technology are critical factors for synchromodal transportation in cold chains. The authors find that, in general, synchromodality is still widely unexplored in the context of perishable goods supply chains and cold chains. This is less surprising when considering the challenges and trade-offs imposed by cold chains. At the same time, the combination of synchromodality and cold chains holds many exciting opportunities for future research on operational, tactical, and strategic levels, especially in the area of planning and control of AFSCs.

One essential element for multimodal transportation of agri-food products is the reefer container. Technological advances in tracking and tracing, e.g. by using sensors, radio frequency identification (RFID), or blockchain technology (Tian, 2016), can enhance the functionality of these containers and help to facilitate a transition to co-modal and synchromodal transportation.

2.3.2 Logistics collaboration in agri-food supply chains

As mentioned previously, the collaboration between carriers along the supply chain is crucial in enabling synchromodal transportation and increasing efficiency. Several works have explored logistics collaboration in AFSCs. Fikar and Leithner (2021) develop a simulation-based decision support system to evaluate the trade-off between logistics collaboration and food quality loss for food cooperatives. Using the example of Austrian food cooperatives, the authors show that logistics collaboration can significantly reduce transportation distance and travel time. However, it may also lead to an increasing loss of quality, which may lead to a loss of product. Badraoui et al. (2020) use a case study approach to explore collaboration in Moroccan agri-food industry. While they found that collaboration enabled cost savings, they also identified several barriers to continued collaboration. For example, ideas and goals may diverge over time and thus lead to a termination of collaboration. Furthermore, varying product requirements, demand patterns, and shelf life of agri-food products can significantly impose the complexity of operational planning.

In a study on the antecedents of horizontal logistics collaboration in AFSCs, Badraoui et al. (2022) use confirmatory factor analysis and structural equation modeling to develop a conceptual model and examine the impact of AFSC characteristics. The authors find a link between companies' dependence and the level of commitment to the collaboration, which can be challenging due to seasonal factors or cold chain requirements of agri-food products. This link was not confirmed for the examined nonfood sample. A study by Matopoulos et al. (2007) showed that the power balance between companies does not have to be in favor of large companies, highlighting that small and medium companies also can exert power over large companies in AFSC collaborations. In contrast to previous research, Badraoui et al. (2022) discover that horizontal logistics collaboration instead relies on resource sharing than on long-term investments. The findings highlight the importance of resource sharing in horizontal logistics collaboration. Moreover, the authors deduce that collaborating with companies with a high level of similarity minimizes arising conflicts due to increased logistics complexity. Consequently, further research is necessary to understand dependencies, differences, and ways to bridge arising complexity gaps.

3 Forecasting issues in agri-food supply chains

Forecasting is essential for managers to better understand and anticipate future development, such as the development of demand, price, or product quality. While for many supply chains, forecasting is mainly limited to supply and demand forecasts, in AFSCs, price and shelf-life predictions are critical factors that should not go unnoticed. Thus, the subsequent section briefly explores price (Section 3.1), demand (Section 3.2), and shelf-life forecasting (Section 3.3) in the context of planning and control of AFSCs.

3.1 Price forecasting

Agri-food products can suffer from high price volatility (Zelingher et al., 2021). The major contributors to this volatility are local events such as extreme weather, natural disasters, crop failure, or variations in crop yield (Lestari et al., 2022; Zelingher et al., 2021), but also the political situation as seen by the impact of the Ukrainian war on wheat prices[5]. Therefore, the prospective price of agri-food products is a significant criterion for decision-making that affects crop and field planning (Tatarintsev et al., 2021). Early work by Brandt and Bessler (1983) finds that increased accuracy of price forecasting in agriculture can improve planning and control activities. The authors note that back in the 1980s, forecasts and market projections for agricultural products primarily relied on

5 https://www.economist.com/finance-and-economics/2022/03/12/war-in-ukraine-will-cripple-global-food-markets.

naïve forecasts and domain expert evaluations. Several works (e.g. Bayona-Oré et al., 2021; Fang et al., 2020; Sabu and Kumar, 2020; Şahinli, 2020) have since shown the value of price prediction in AFSCs by employing exponential smoothing, time series, and machine learning models. Empirical work has shown that depending on the product, varying forecast methodologies and parameter settings may be relevant to achieve high forecast accuracy (Xu and Zhang, 2022a).

Applied models in research use time series data as a basis to determine forecast values and include but are not limited to Holt-Winter models, auto-regressive integrated moving average models (ARIMA), vector auto-regressive models (VAR), support vector machine models (SVM), vector error correction models (VECM), seasonal-trend decomposition procedures based on locally weighted scatterplot smoothing (STL), and neural networks (NN) (Bayona-Oré et al., 2021; Fang et al., 2020; Şahinli, 2020; Xu and Zhang, 2021, 2022b). As Zelingher et al. (2021) note, small changes in supply can upset global market prices, highlighting the importance of creating high-quality forecasts to determine market prices and supply better. As such, forecasting prices in AFSCs can help managers to better understand, plan, and control decision-making processes and demand development.

3.2 Demand forecasting

Demand planning and forecasting in AFSCs have not received significantly special attention compared to other industries (Flores et al., 2020; Petropoulos and Carver, 2019; Shukla and Jharkharia, 2013). Consequently, the question arises of how agri-food products require special demand planning methods. One aspect that has been emphasized in previous sections is the aspect of product perishability, which adds additional complexity to forecasting demand in AFSCs. While in supply chains, without this issue, managers will want to avoid lost sales, in AFSCs, high stocks can lead to a loss of product. Taylor (2006) studies demand management in AFSCs and finds that agri-food products often lack on-shelf availability data. Compared to other products, agri-food stock-keeping units can lack a uniform packaging size (as they may be sold by weight rather than as fixed units). Sharing customer demand data accurately can also be challenging. Sales data is often used as a proxy for actual customer demand. However, sales data does not sufficiently account for lost sales, substitution, differences in packaging sizes/weight, or promotional effects. Taylor (2006) notes that often there is little effort to participate in collaborative forecasts along the supply chain. Adding to this is a lack of timely order transmission and data consistency and accuracy. Shukla and Jharkharia (2013) add that agri-food products frequently have a high substitutability, which makes an accurate forecast less critical.

Further complexity arises from volatile customer demand patterns that may be sensitive to weather and seasonal changes. Consequently, managers may instead rely on their instinct when predicting demand and planning supply. Further research should thus not only include new methods of demand forecasting but also capture the status quo of managerial decisions regarding demand forecasting and the underlying complexities of decision-making in AFSCs.

3.3 Shelf-life prediction

Limited shelf life has been identified as one of the most critical characteristics of agri-food products. An accurate prediction of the products' remaining shelf lives can not only improve the performance of planning processes in AFSC and food safety, but it can also reduce loss/waste in the supply chain and increase the freshness and availability of the product. However, predicting shelf life is challenging, especially for ultra-fresh products such as fruits and vegetables, meat, poultry, and dairy products. Moreover, due to various definitions, it is not entirely clear what 'shelf life' precisely means. Depending on the view (e.g. the retailer or the consumer), there may be different interpretations. For example, the 'Directive 2000/13/EC of the European Parliament and of the Council of 20 March 2000 on the approximation of the laws of the Member States relating to the labelling, presentation and advertising of foods' distinguishes between the 'date until which the foodstuff retains its specific properties when properly stored', typically denoted as 'best-before' date (Article 9), and the date after which there is 'an immediate danger to human health' typically denoted as 'use by' date (Article 10) (European Union, 2018). To optimally plan production, distribution, and consumption, understanding the aspects of shelf life is crucial.

Besides packaging, various external factors in cold chains affect a product's (remaining) shelf life, e.g. ambient temperature, the level of sunlight, humidity, oxygen, CO_2, or other chemical substances (e.g. ethylene), and last but not least, handling processes. In recent years, there has been an increasing development in new technologies that can continuously monitor such external factors along the entire cold chain. Furthermore, these technologies collect data that allow for developing methodologies for continuous real-time measurement of (remaining) shelf lives (Albrecht et al., 2020).

Shelf life prediction methods can be classified as follows: (i) *static testing*: the most commonly used testing methodology in practice, which simply counts the number of days until a product reaches its end-of-life state; (ii) *temperature and humidity monitoring*: where the shelf life of a product is predicted by comparing the temperature/humidity profile of a batch with historical data illustrating the relationship between temperature/humidity and shelf life; (iii) *chemical analyses*: where the level of certain chemical substances is used to

Table 1 Business cases for AI-based shelf life prediction technologies

Company/project	Technology
Insignia Technologies	Smart labels based on patented smart pigment that changes the color in response to changing temperature or CO_2 level.
Onethird	AI-powered, near-infrared scanning scanner technology
Clarifruit	Cloud-based quality control App and data analytics platform using imaging machine learning technology
IntelliDate	Intelligent packaging system and smartphone App that allows for real-time sensor-based data collection.

predict the shelf life of a product. For example, ethylene is a phytohormone released during the ripening process of fruits and vegetables. The level of ethylene can be used to predict when products will begin to rot; (iv) *imaging*: where hyperspectral or ultrasonic imaging is used as a shelf-life predictor by reflecting the internal ripening process of fruits and vegetables; and (v) *spectroscopy*: where a unique `spectral fingerprint´ is used to predict the product's remaining (Onethird, 2022).

Shelf-life prediction technologies are still at an early stage. To be practically viable, such technologies must be, on the one hand, economically affordable and, on the other hand, easy to implement and use along the entire cold chain. An example of innovative temperature-monitoring technologies is the so-called smart label, e.g. as offered by Insignia Technologies.[6] These labels provide a time–temperature indicator (TTI) by visually changing colors over time and by a specified temperature range. Using these smart labels is simple and cost-efficient compared to other technologies. Onethird developed a handheld scanner, including a data-sharing platform that uses artificial intelligence to predict shelf lives.[7] Compared to the smart labels that can be used without manual human interventions, the handheld scanner requires manual operation by an employee. By employing imaging technology, Clarifruit offers a quality control application that analyzes real-time data about external and internal factors influencing the shelf life, specifically for fruits and vegetables[8]. The project IntelliDate, funded by the German Federal Ministry of Food and Agriculture, develops an intelligent packaging system for sensor-based prediction of the actual expiration date of food. Instead of a fixed 'best before' date, which food producers primarily set, in this project, researchers develop a method that generates a dynamically changing 'best before' date based on real data in the logistics chain collected by smart sensors.[9] The four business cases and projects are summarized in Table 1.

6 https://www.insigniatechnologies.com/freshtag.
7 https://onethird.io/food-waste-solutions.
8 https://www.clarifruit.com/solutions/technology.
9 https://innovationstage.pageflow.io/intellidate#267150.

Technological innovations have a direct effect on planning and optimization in AFSCs. In recent years, we can observe an increase in research using methods of OM and OR to measure the managerial benefits of such real-time shelf life prediction methods for perishable products. Ketzenberg et al. (2015) show how the TTH can be used to improve the inventory management of perishable products. The authors first develop an inventory control problem as a Markov Decision Process (MDP) that uses the TTH of the product and then establish a heuristic that is easily applicable in practice. Gaukler et al. (2017) have a similar focus and show the value of TTH to set dynamic expiration dates based on a product's exact real-time environmental conditions. Torres et al. (2020) examine a methodology to predict shelf life losses along the supply chain based on sensory and physicochemical quality attributes. To estimate shelf life, the authors use a nonlinear regression model that correlates the theoretical shelf life and the temperature during the entire logistics process, i.e. storage and transport. Albrecht et al. (2020) investigate the applicability of a TTI as a shelf-life predictor in a national poultry chain. The authors develop a model that adapts a label's discoloration to products with different shelf lives, which is integrated into software to support decision-making processes.

However, it must be noted that overall, the integration of real-time information technology and its implication on decision-making processes and the performance of AFSC is still at an early stage. Many food companies struggle with the trade-off between such technologies' investment cost and effort and the long-run benefits. Hence, further research in planning and controlling such processes is necessary.

4 Production planning and inventory control for agri-food products

Production and inventory planning in AFSCs involves complex decision-making problems in various stages along the supply chain. The following section first provides an overview of the most critical building blocks when developing and studying planning models for decision support in AFSCs. Subsequently, we review planning and control models in the context of production, harvesting, and processing (Section 4.2), inventory control of perishable products (Section 4.3), and partnerships and governance programs in AFSCs (Section 4.4).

4.1 Building blocks of planning and control processes

Understanding the planning environment and the relevant building blocks is crucial for developing successful decision-support models for AFSCs. In the following section, we briefly describe some of these building blocks critical for developing optimization models for production planning and inventory control in AFSCs.

4.1.1 Product lifetime (perishability)

Modeling the limited lifetime (perishability) is one of the most fundamental building blocks for perishable agri-food products. The integration of a limited shelf life has a long history in developing decision-support models for production planning and inventory control. Already in the 1950s, Whitin (1953) considered the deterioration of products after a fixed storage time. Around 10 years later, van Zyl (1963) distinguished between age-dependent and age-independent perishability and illustrated the complexity when optimizing inventory levels for perishable products. While age-independent perishability considers products with an a priori fixed lifetime over which the quality remains constant (e.g. products with a specified expiry date), age-dependent perishability refers to products whose quality deteriorates over time (e.g. fresh produce). Following van Zyl's (1963) initial classification, Nahmias (1982) introduces two classes of perishability to which most models can be assigned: (i) 'fixed-lifetime' models and (ii) models where products are subject to 'continuous deterioration'. Under a fixed lifetime, the product expires (perishes) after a fixed predefined lifetime; for example, grocery products with best-before or use-by dates until then, the producer guarantees the same quality. Continuously deteriorating models refer to deterioration modeled as a function of the on-hand inventory. Within these classes, both deterministic and stochastic lifetime models have been studied (Goyal and Giri, 2001). Bakker et al. (2012) provide a structured overview of deteriorating inventory control literature succeeding the work of Goyal and Giri (2001).

4.1.2 Demand modeling

Understanding and modeling demand is key in developing production and inventory control models. Depending on the practical problem, demand can be modeled either *deterministically* or *stochastically*. Assuming a deterministic demand is often a simplifying assumption and not always reasonable (e.g. for slow-moving products in retailing). The inclusion of stochastic elements can significantly impede modeling complexity. Consequently, whether to include demand stochasticity is a choice between increasing model complexity and oversimplifying reality (Powell and Baker, 2016). In a multiperiod context, it is also relevant to model demand processes over time, i.e. as a stationary or nonstationary[10] demand process. Many multiperiod models assume stationary

10 Stationary demand refers to a stochastic demand process over time whose distribution does not change, whereas in nonstationary demand processes, the distribution of the demand changes in time, e.g. due to seasonality, trend,

demand to reduce modeling complexity. However, in retail, stationary demand may not always be suitable since, throughout the week, demand may vary significantly (Broekmeulen and van Donselaar, 2009; van Donselaar et al., 2006).

Moreover, the age of the product may influence the demand for agri-food products. For example, many grocery retailers are concerned that fewer fresh products negatively affect the demand and thus seek to offer markdowns on less fresh products (Buisman et al., 2019; Chen et al., 2018). It has been shown that fresh food plays a critical role in retail assortments, generating up to 40% of the overall revenue (Buck and Minvielle, 2014). Thus, considering the correct level of age dependency when modeling demand in AFSCs may be critical when optimizing inventory levels and production capacities (Chen and Sapra, 2021). Additional relevant features when modeling demand are whether customers, in case of a stockout, are willing to wait or not (*backlog or lost sales*) (Minner and Transchel, 2010), whether demand is price sensitive (relevant when optimizing price and inventory jointly) (Agi and Soni, 2019), how customers pick products (LIFO, FIFO, or hybrid) (Haijema and Minner, 2019), or whether there are cross-price or substitution affects between products (Transchel et al., 2022).

4.1.3 Random supply modeling/yield uncertainty

Yield uncertainty or insufficient supply are unique characteristics in various AFSCs, especially yield uncertainty in harvesting. Research has shown that yield uncertainty can be modeled in several ways depending on the problem context. Although not specific to AFSCs, Yano and Lee (1995) provide an excellent overview of various approaches to modeling yield uncertainty in production systems applicable to different real-life scenarios. For example, one way to model yield uncertainty is to consider a Bernoulli process, i.e. each unit received is 'good' with a certain probability and 'not good' with the complementary probability, where the term 'received' can be broadly interpreted as being produced or delivered by a supplier. Such a model applies to an inventory management problem of a retailer who orders a certain quantity of fresh food products (e.g. boxes of tomatoes or units of pineapples), but the quantity delivered does not entirely fulfill the quality of the retailer. Thus, the yield follows a binomial distribution. Another way is to model the yield rate as a random fraction of the order quantity. The main difference between the two modeling approaches is that in the latter (typically denoted as *stochastically proportional yield*), the yield rate is assumed to be independent of the order quantity, in the first modeling approach following a Bernoulli process, the random yield rate depends on the specified quantity (Yano and Lee, 1995). A

or other external influences.

common feature of both modeling systems is an upper bound of the quantity for the 'good' units.

However, binomial yield and stochastically proportional yield may not be the best choice when modeling the harvesting yield of agri-food products when there is no clear maximum quantity. Alternatively, the harvest yield can also be modeled directly as a random variable with a theoretical or empirical distribution (Bansal and Nagarajan, 2017). More recently, Anderson and Monjardino (2019) investigated the impact of yield uncertainty in designing risk-sharing contracts in AFSCs to better understand growers' fertilizer usage and optimize fertilizer usage for risk-averse growers. The authors propose a novel contract arrangement in a three-echelon supply chain considering yield uncertainty and show that this contract can coordinate the supply chain under specific conditions. Based on the case of an Australian wheat supply chain, the authors further demonstrate the effectiveness of their proposed contract.

4.1.4 Inventory issuing policy

The modeling of inventory issuing policy, i.e. the order in which demand depletes products from stock, is related to demand modeling. Due to the limited lifetime of products, a firm's inventory typically consists of different age levels, and the performance of an inventory control policy depends on the order in which the stock is issued. Most papers in the literature consider one of the two extreme issuing policies, *First-In-First-Out (FIFO)*, i.e. assuming that the oldest products are picked first, or *Last-In-First-Out (LIFO)*, i.e. assuming that the freshest products are picked first (Nahmias, 2011). While in theory, FIFO leads to a lower loss, in practice, there are several reasons why this may not be the case: A new delivery may stem from an older batch, the latest batch may have a lower shelf life (e.g. due to harvesting conditions, handling, etc.), the storage conditions may not be the same for all products, etc. Thus, in practice, especially for fresh produce, FIFO is not always preferred by companies. Often, selecting the appropriate merchandise for an order is therefore relinquished to on-site quality assurance and warehouse staff.

Furthermore, in B2C markets, pure FIFO policies are challenging to implement since retailers must trade off availability and the order in which products a purchased. In practice, a hybrid FIFO–LIFO approach is more likely, which is also studied in the literature, e.g. Haijema and Minner (2019). However, the terms 'First-In' and 'Last-In' assume that later-arriving inventory is always fresher, which is not always the case. A more flexible wording that also handles cases such as uncertain quality in supply or random lead times is *First-Expired-First-Out (FEFO)* and *Last-Expired-First-Out (LEFO)* (Hansen et al., 2022). In addition, many retailers apply the so-called stock rotation, i.e. resorting items on the shelf, to facilitate FEFO in-store by placing items with shorter expiry dates and lower remaining shelf life at the front rows of the shelf.

4.1.5 Objective

Many papers have focused on a single objective, such as profit maximization or cost minimization problems. However, recent works have increasingly featured multi-objective optimization models in light of the triple bottom line and supply chain design trade-offs (see, e.g. Ugarte et al., 2016). In regard to cost objectives, typically considered costs are fixed production setup or ordering costs, variable production/procurement costs, waste/disposal costs in case of unsalable overstocks, and backlog/lost sales costs in case of stockouts, typically not all at once. Since backlog and lost sales costs are challenging to estimate in B2C markets, retailers tend to meet predetermined target service levels that indirectly trade off the costs of overstocking and understocking, e.g. Hansen et al. (2022) and Minner and Transchel (2010).

4.1.6 Planning horizon and review policy

Production and inventory control problems typically boil down to two fundamental questions: When and how frequently should be produced or inventory be reordered? How much should be produced or reordered? (Silver et al., 2016). Many decision problems in OM can be formulated by a single-period, e.g. Newsvendor-type models where a planner makes decisions (e.g. production capacity, workforce, inventory, or size of land) at the beginning of a single period (e.g. a day, month, year) without knowing all data with certainty. Typically, demand is the unknown parameter in such systems, whereas the demand distribution is given. These single-period models assume there is no future decision that can be affected. In practice, however, the decision in one period may directly affect decisions in future periods, and to take this interaction into account, multi-period planning models must be used. In multi-period inventory control, the 'review policy' is critical when building the model and finding the most suitable mathematical tool for the analysis. Most models either belong to the class of *continuous-review* or *periodic-review* policies. While a periodic-review approach evaluates the inventory level only at specific time points, a continuous-review policy assumes that the inventory can be monitored continuously (e.g. by using scanning technologies such as barcode or RFID) (Silver et al., 2016).

4.1.7 Lead time

It is well known that an increasing lead time in perishable production and inventory control problems significantly complicates planning models and makes them challenging to solve optimally (Broekmeulen and van Donselaar, 2009). Thus, many problems have first been studied under the assumption of zero lead time (Nahmias, 2011). To cope with the increasing complexity in perishable-product inventory control due to lead time, various heuristics have been developed, e.g. order-up-to policies (Broekmeulen and van Donselaar, 2009; Haijema and Minner, 2016, 2019) or service-level-constraint models (Hansen et al., 2022; Minner and Transchel, 2010).

4.1.8 Number of products and locations

Most papers study planning problems of a single product in a single location. Multiproduct and multilocation problems are typically more complex since they must coordinate multiple interacting decisions. Multiproduct models are relevant when such products, e.g. share a common resource, e.g. limited shelf space in a retail store (Hübner et al., 2020) or transportation capacity and fixed costs (Ambrosino and Sciomachen, 2007). Planning problems across multiple locations, e.g. centralized versus decentralized inventory management, transshipment and distribution problems, or replenishment and allocation problems (Ahumada and Villalobos, 2011; Karaesmen et al., 2011), also consider the dependency between the decision at individual locations and, thus, add additional complexity to the optimization model.

4.2 Production, harvesting and processing

For agri-food products, production and harvesting are seasonal activities, negatively impacting farming equipment and technology utilization. Therefore, much OM- and OR-related research in this area has focused on capacity investments and planning problems for agriculture products, specifically focusing on the impact of yield uncertainty and perishability. On that note, capacity planning must be viewed in the broader sense and can, e.g. be related to farmland planning (e.g. size of the farmland and allocation of farmland to different crops), machinery investments (e.g. planting and harvesting equipment), the planning of farmworkers (e.g. harvesters), and planning the harvesting patterns. The following sections briefly explore examples of relevant production and processing stages in AFSCs.

4.2.1 Seed production

Planning in AFSCs already starts with seed production. The global seed market was valued at US$77.4bn in 2021 and is expected to grow at an annual rate of 4.7% to nearly US$111.8 billion in 2029.[11] Producing high-quality and resilient

11 https://www.maximizemarketresearch.com/market-report/global-seeds-market/111623/.

seeds is the first step to ensuring food security. However, developing such seeds can take up to a decade before they reach the quality to be marketed (Singh et al., 2013). Moreover, climate change and the increasing weather volatility require more precise planning processes on, e.g. which seeds to cultivate and where to produce. However, surprisingly little research studies the benefits of decision-support models from OM and OR in seed production.

Bansal and Nagarajan (2017, 2022) study a production planning problem of an agribusiness firm that produces and sells a large variety of hybrid seeds from cross-pollination between two different parent seeds. The firm can plant the seeds in two subsequent seasons, first in the Northern hemisphere and subsequently, with the information of the yield rate a second time in the Southern hemisphere. Furthermore, the firm produces multiple types of such hybrids, which share common parent seeds with limited inventories. This inventory limitation of parent seeds leads to the problem of optimally allocating the parent seeds to multiple hybrids produced in different regions. Given this decision, the firm must decide on the land size to grow each seed in the two different growing periods. Hence, the overall decision problem is a joint 'parent-allocation' problem and a sequential production planning problem with yield uncertainty and information update after the first production period. Bansal and Nagarajan (2017) use a decomposition-based exact solution approach to address the complexity of the stochastic and dynamic nature of the problem and show the benefits of operational flexibility for an actual data set from the firm.

Burer et al. (2008) examine different contracts relevant to the agricultural industry between a seed supplier and various retailers. In the seed industry, seed suppliers traditionally are allowed complete returns of unsold products from retailers with a 100% buyback price. Anticipating the inefficiency of such a 100%-buyback price contract, the authors study bonus-based contracts and the so-called reduced-buyback-price contracts. Under a bonus-based contract, the retailer gets a bonus (or rebate) on every unit sold as long as the sales quantity exceeds a prespecified percentage of the retailer's order quantity. Under a reduced-buyback-price contract, the retailer gets a buyback price lower than the wholesale price for a unit of seed exceeding a certain percentage of the amount ordered. The authors study whether and when the contract coordinates the supply chain. It is important to note that their model does not consider yield uncertainty at the seed supplier. The authors demonstrate that the existing contracting models achieving supply chain coordination[12] may not be suitable in seed supply chains. They present modified contracts (and contract

12 Roughly described: Supply chain coordination seeks to align all decentralized plans and decisions as if they would have been optimized by a central decision-making unit.

parameters) that coordinate decision-making processes in the commercial seed industry.

4.2.2 Crop-planning

Motivated by the case of an olive oil producer in Turkey, Kazaz (2004) studies a two-stage planning problem of a food producer under yield and demand uncertainty. Before knowing how much yield is produced, the farmer must first decide how much farmland to lease and till (first-stage decision). Then, after the exact olive crop is known, they must decide how much of the harvest to use for their own production and how much to purchase from or sell to other growers (second-stage decision). The second-stage decision highly depends on the realized yield, such that the yield uncertainty - taking this into account also affects the first-stage farmland leasing decision. Boyabatli et al. (2019) examine a farmland allocation problem, where farmland is allocated between crops utilizing crop rotation benefits across the growing season, an instrument in agriculture planting a series of different crops on the same land in sequential seasons. The authors characterize the optimal dynamic farmland allocation policy of a farmer who periodically allocates farmland between two crops in the presence of revenue uncertainty. Flores and Villalobos (2020) examine a joint capacity decision problem (acres to plant) and technology decision problem of a more complex agriculture network consisting of several farmers in several regions. The authors develop a two-stage stochastic model and validate their model through a case study that estimates the production potential for three vegetable items, bell pepper, romaine lettuce, and tomato (plum type), in the southwest region of Arizona and New Mexico in the United States.

4.2.3 Harvesting

Harvest planning involves various types of decisions on a strategic, tactical, and operational level. Among others, it encompasses the investment, selection, and allocation decisions of farming technologies, the timing and scheduling of harvest operations, and effective planning of limited resources, such as fertilizers, workforces, and in recent years more and more water consumption (Iakovou et al., 2016b). Allen and Schuster (2004), e.g. examine an investment problem of a grape farmer who must decide on the effort to devote to the harvest (indirectly the capital investment in equipment), which directly affects the pressing rate of the grape harvest (the tons of grape that can be pressed per day). The uncertainty the farmer faces is the length of the harvest season (unforeseen frost may abruptly end the harvest) and the crop size. The decision-support model addressing this problem has the structure of the classical Newsvendor problem with underage and overage costs. The overage cost (corresponding to an overinvestment) is the cost of harvesting too quickly and ending up with an excessive amount of unused resources. The underage cost is reflected in the cost of harvesting too slowly and ending up with missed sales and food loss. Ahumada and Villalobos (2011) developed a comprehensive operational planning model (mixed-integer problem) that incorporates various short-term planning issues such as postharvest behavior of crops, weather uncertainty, postharvest decay (yield loss), transportation cost and time, and market uncertainty. Since the problem is NP-hard[13] and thus difficult to solve for realistic problem instances, the authors present a heuristic solution algorithm.

4.2.4 Processing and production

Food processing transforms fresh agricultural products such as milk, meat, fruits and vegetables, and grains into ingredients and direct consumable food products. The food processing market was valued at $143.51 bn in 2020 and is projected to increase to $235.67 bn by 2028,[14] one of the largest markets globally. Food processing and production are often handled by a different entity in the food supply chain than growing and harvesting and thus require coordination processes with both the upstream production/farming stage and the downstream distributor/retail stage, which often lead to inefficiencies in the supply chain. Kusumastuti et al. (2016), investigating applications of OR and MS to harvesting and processing problems until 2015, reveal that only 40 papers have been published studying planning problems related to food processing. Most of them investigate optimal production plans or schedules and inventory control policies under different types of constraints and levels of uncertainty, mainly in demand and supply.

Since 2015, we can observe an increase in research on advanced planning problems in agri-food production and processing, which includes a higher level of uncertainty in these systems and integrates sales-oriented decisions such as pricing. For example, Boyabatli et al. (2017) study the impact of spot price and yield uncertainty on the optimal capacity decisions of an agricultural processing firm. The authors develop a planning model calibrated based on the palm industry, an industry in focus in recent years. More recently, Noparumpa et al. (2022) study an advanced selling problem of a fine wine producer through selling wine futures, a practice in the wine industry to sell wine (particularly fine wine) before it is bottled, finally priced, and released to the market. The authors develop a mathematical model that provides decision support regarding questions like how winemakers should allocate production between futures and retail distribution in the presence of an uncertain bottle rating that indicates

13 NP-hard is a term from computational complexity theory which describes a set of problems that cannot be solved in polynomial time.
14 https://www.verifiedmarketresearch.com/product/food-processing-market/.

the final quality of the wine and provides answers regarding the impact of risk aversion and market characteristics on the winemakers' allocation decision.

Review papers focusing on agri-food production are Amorim et al. (2013), who review the literature with a specific focus on integrating production and distribution planning for perishables, and Pahl and Voß (2014), who provide a comprehensive survey of literature on how deterioration and lifetime constraints are integrated with production and supply chain planning.

Many OM- and OR-based planning tools have been developed in the past 50 years. Especially the past decade has yielded tremendous development in innovative farming technologies, such as new sensing technologies allowing yield and quality mapping[15] that enable a shift toward more sustainable production in AFSCs. Fountas et al. (2016) provide a comprehensive overview of innovative technologies in precision farming and the need for novel planning and optimization tools in AFSCs.

4.3 Inventory control of perishable food products

When planning inventory levels, it is crucial to understand that inventory serves different organizational functions. Among others, it can serve as a cycle, pipeline, anticipation, and safety stock (Silver et al., 2016). Cycle stock, e.g. occurs when products are ordered or produced in batches to utilize economies of scale or benefit from quantity discounts in purchase prices or freight rates. Inventory between two stages in a supply chain (e.g. on trucks, ships, or railways) is also referred to as pipeline stock. The longer transportation lead times, the more inventory is in the pipeline, and the higher effects on companies' financials (in the balance sheet and profit-and-loss report). Inventory may also serve as *safety stock,* which is used to create a buffer against all types of uncertainty in the system (e.g. demand, supply, and price uncertainty). It should optimally balance the costs of having too much inventory (causing holding, disposal, and waste costs), also denoted as overage cost, and the costs of having too little inventory (causing missed current and future sales, backlog costs, and eventually penalty for not meeting customer demands), also denoted as underage cost.

Estimating underage and overage costs can sometimes be tricky for agri-food products. For example, retailers struggle with estimating the correct underage costs since future missed sales due to stockouts are hard to predict. The exact overage cost depends on whether a product can be stored and carried over to the next period. In this case, inventory holding costs (capital cost and cost of managing inventory) occur. However, perishable products need to be removed from stock after a finite shelf life, which may cause disposal

15 Sensing technologies attached to harvesting machines allow a yield monitoring during harvest.

or waste cost). In exceptional cases, secondary sales opportunities may exist; however, typically, with lower revenue than production and ordering costs.

The literature on inventory control models for perishable products is very rich, and depending on the product type, and how products perish over time, different modeling and optimization techniques are needed. In particular, the type of perishability and the way how it is modeled (see Section 4.1) has led to different literature sub-streams. Nahmias (1982, 2011) and Karaesmen et al. (2011) provide comprehensive literature reviews of inventory control models with fixed-lifetime (shelf life) products, and Raafat (1991), Goyal and Giri (2001), Bakker et al. (2012), and Janssen et al. (2016) review the literature studying continuously deteriorating inventory models.

Besides the general review papers, Nahmias (2011) and Karaesmen et al. (2011) present various approaches to modeling inventory systems for perishable products and illustrate specific challenges. Borodin et al. (2016) provide an extensive overview of OR methodologies for handling uncertainty in agricultural supply chain management problems. They classify the literature concerning modeling approaches, e.g. stochastic programming, robust optimization, or simulation.

4.4 Impact of partnerships and governance programs

The increasing pressure on farmers and food producers both from the supply side (e.g. the growing effects of climate change and the rising cost of equipment, labor, and material costs) and from the demand side (e.g. the increasing demand and customer expectations regarding, e.g. quality and transparency) forces farmers to seek out and identify valuable partnerships in the supply chain. We observe an increase in research studying the benefits and obstacles of collaborative planning and the value of partnerships and cooperatives in AFSCs. Within that research area, there is an increasing number of modeling and optimization studies analyzing the fundamental effects of partnerships and governance on the overall performance of AFSCs.

4.4.1 Contract farming

Contract farming describes a partnership between different stakeholders in an AFSC, typically a farmer and a buyer, characterized by an *ex-ante* agreement (contract). Such an agreement typically encompasses (among others) the type and the quality of the product (specifying production and processing conditions), the quantity supplied, and the price (Eaton and Shepherd, 2001). The advantages of such contractual agreements for farmers are, e.g. guaranteed prices and quantities, access to reliable markets, and potentially better access to financial support. However, contract farming may also bear certain risks. The prospect of higher revenues may be outweighed by, e.g. higher yield risk or an increased dependency from the buyer (Eaton and Shepherd, 2001). Chen and Chen (2021) investigate the effect of a farmer's production efficiency on joining a contract farming agreement for high-value agriculture products. Their main finding is that production cost is a driving force for the execution of a contract and severely affects the impact on both stakeholders. Dutta and Kaur (2022) study advanced-payment contracts where the farmer receives parts of the buyer's payment preharvest and the remaining payment is paid after the supply of the products. Such agreements are discussed as socially responsive since they benefit poorer smallholder farmers by allowing them better investments, especially in emerging economies.

4.4.2 Farmer cooperatives

Other forms of partnerships in AFSCs are farmer cooperatives, business associations owned and operated by their members (farmers) to pool resources and help with processing, financing, and marketing activities.[16] Especially in emerging economies, farmer cooperatives can enable smallholder farmers to operate more cost-efficiently and productively. However, organizing such cooperatives requires significant coordination and transaction costs. An et al. (2015) study five effects related to smallholder farmers participating in an agriculture cooperative: lower cost, stable process yield, stronger brands, shorter supply chains, and exclusive direct sales with guaranteed selling prices. The authors show that farmers do not always benefit from participating in a cooperative. One critical factor determining the value of a membership is the size of the cooperative. Shi et al. (2019) study a selling problem of farmer cooperatives facing random supply and random prices. They develop optimal pricing and inventory policies for different holding cost structures and conduct a numerical study using the data of a farmer cooperative in Kenya. More recently, Neupane et al. (2022) examined to what extent agricultural cooperatives assist farmers in improving production performance and efficiency. Based on the case of smallholder goat farmers in Nepal, one of their main findings is that cooperative memberships improve the skill and motivation of farmers through increased accessibility to information, capacity development, access to finance, and access to inputs and services.

4.4.3 Governance programs

To encourage investments in technological innovations, various policy instruments (e.g. taxes or subsidies) seek to support farmers in adopting innovative concepts and technologies. In OM and SCM, we observe increasing

16 https://www.usda.gov/topics/rural/co-ops-key-part-fabric-rural-america.

research to support policymakers in understanding the effects of specific policy instruments on planning and decision-making processes in AFSCs. For example, motivated by an enterprise in South Asia cooperating with smallholder buffalo farmers to produce semifinished dairy products such as Mozzarella cheese, Hsu et al. (2019) examine the impact of innovative partnerships between a production enterprise and smallholder farmers on quantity and quality aspects (in this case the milk). The authors study the impact of government subsidies that reward farmers for improving quality on the performance and profitability of the farmers and the enterprise. Through a stylized model, the authors find that subsidies are not necessarily beneficial to all actors in such a partnership. Alizamir et al. (2019) develop a decision-support model guiding farmers to enroll in subsidy programs. They examine two subsidy programs offered by the U.S. government (i) the Price Loss Coverage (PLC) program and (ii) the Agriculture Risk Coverage (ARC) program. Their results provide guidelines for farmers enrolling their crops in these subsidy programs and show that farmers' enrollment statistics support these guidelines. Akkaya et al. (2021) develop a quantitative model to investigate to what extent taxes and subsidies encourage farmers to experiment with innovative production methods when facing yield uncertainty. The underlying case is the experimentation with hybrid seeds. Hybrid seeds are more costly and thus a barrier to investing in. However, they are more robust against extreme weather events and may increase the farmers' yields and income. Chintapalli and Tang (2021, 2022) examine the effectiveness of crop 'minimum support prices' (MSPs), specific governmental subsidy schemes in agriculture to support farmers and to guarantee crop availability. The efficacy of MSPs is influenced by several factors, among others, yield uncertainty of the crop, market conditions (size and prices), and production cost. The authors show that a higher MSP may not always be beneficial, mainly when the cost of offering MSPs exceeds the overall surplus.

5 Future challenges and opportunities for planning and control of agri-food supply chains

In light of global climate change and the accompanying extreme weather events, such as droughts, floodings, and storms, and an increasing scarcity of natural resources, such as water and arable land, global agri-food production chains are under pressure to become more efficient and effective than ever. Combined with a growing global population and the quest for more sustainable agri-food systems, this has inspired new technologies, primarily based on industry 4.0 technologies (Oltra-Mestre et al., 2021), and emergent issues for planning and control processes in AFSCs. This section explores several of these emerging topics by briefly highlighting current challenges and opportunities related to the planning and control processes of AFSCs.

5.1 Precision farming and smart agriculture

Precision agriculture uses precision sensing technology such as positioning satellites or yield, irrigation, and nutrition sensors to increase crop yield and land use and to reduce resource consumption. In addition, the combination of precision agriculture with information and communications technology (ICT) has led to the development of smart agriculture (Inoue, 2020). The resulting increase in data granularity calls for more efficient planning and control mechanisms. However, depending on the sensing technology employed, the accuracy may vary significantly (Daponte et al., 2019).

Smart agriculture comprises sensing technology, ICT, data analytics, artificial intelligence, and robotics (Inoue, 2020; Puri et al., 2017; Spanaki et al., 2022). For example, drones can be used to collect remote sensing data (Daponte et al., 2019), and machine learning can be applied to analyze such data (Ali et al., 2015). Such nearly real-time information allows for more effective use of agrochemicals, better utilization of labor, and helps to better predict yield rates. Another example of smart agriculture is the use of blockchain technology to improve traceability in agri-food systems (Feng et al., 2020). Furthermore, smart agriculture can also include using ICT to share farming equipment, such as expensive machinery (Sengupta et al., 2019). Agriculture equipment is one of the most significant expenses for farmers, which led to the development of the sharing economy in farming, e.g. Agrishare initiated by the Welthungerhilfe is a mobile phone-enabled platform linking farmers in developing countries without assets with service providers for production, processing and transporting.[17] However, the development of appropriate collaboration policies for such systems poses a significant challenge for the planning and control in AFSCs, which needs further exploration.

To provide adequate decision support for practitioners, researchers interested in smart agriculture thus can focus on developing new decision-support models and tools, e.g. by including sophisticated data analytics techniques, machine learning, artificial intelligence, or heuristic approaches (Ali et al., 2015). On the one hand, researchers may focus on operational aspects, e.g. by developing heuristics for drone swarms (Spanaki et al., 2022) or optimization models to determine the number of drones and the placement of recharging facilities. Furthermore, managerial aspects such as collaboration policies and risk management strategies also need further attention.

17 https://www.welthungerhilfe.org/agrishare.

5.2 Controlled environment agriculture and circular economy solutions

The popularity of Controlled environment agriculture (CEA) has increased globally. CEA describes the production of agricultural produce in a highly controlled environment, such as greenhouses and insect farms, as well as hydroponics, aeroponics, or aquaponics systems (Ojha et al., 2020; Palm et al., 2018; Shamshiri et al., 2018). Compared to open-field agriculture, CEA can offer tremendous advantages in land and water usage or yield rates, and it can be built where traditional open-field agriculture may not be feasible (Benke and Tomkins, 2017; van Delden et al., 2021). However, at the same time, the management of such systems can be quite challenging. For example, the biodiversity in the vicinity of a greenhouse can contribute to phytosanitary health and pest control (Messelink et al., 2021).

Recent trends and developments have included the consideration of vertical and urban farming. As Butturini and Marcelis (2019) point out, several terms exist for vertically stacked agricultural production systems. While in Europe, the term 'vertical farming' is prevalent, in Asia, the term 'plant factory' is more commonly used. Furthermore, the authors identify four different types of vertical farming systems: vertical farms with artificial lighting, container farms, in-store farms, and appliance farms. While vertical farming has many benefits, such as a low environmental impact, investment and operational costs are still quite high, and as van Delden et al. (2021) highlight, many vertical farming technologies need further improvement in optimizing the various control parameters. Furthermore, due to the higher cost involved in the production, many crops are not financially viable for vertical farming (Benke and Tomkins, 2017). Thus, challenges include the choice of profitable crop varieties and optimal planning and scheduling of harvesting cycles. However, there is almost no scientific literature on the economic feasibility of vertical farming concepts, which makes it highly speculative to what extent vertical farming will develop in the future (van Delden et al., 2021). Moreover, such new concepts require new and innovative planning methods for strategic and tactical decision-making problems (van Delden et al., 2021) and detailed operational problems for optimizing the management of vertical farms (Santini et al., 2021). A specialized form of vertical farming is in-store farming solutions like those provided by Infarm.[18] While the concept has been experimented with in practice by various companies (Butturini and Marcelis, 2019), the in-store farming concept remains widely unexplored in the scientific literature and offers exciting new planning and control-related challenges in the field of operations research.

18 https://www.infarm.com/.

Two streams of CEA systems that are bridging the gap toward the circular economy are aquaponics, i.e. the integration of aquaculture and hydroponics (Folorunso et al., 2021), and insect farms, i.e. the valorization of food waste to grow insects and compostable mass (Ojha et al., 2020). Insect farms can help to convert waste into reusable materials such as fodder; however, these farms require optimization regarding insect choice, waste presorting, and waste distribution. Aquaponic systems require great care regarding the selection of plants and fish. Some plant additives, such as pesticides, can adversely affect the fish population; thus, aquaponics similarly offers research opportunities regarding operations and management decisions. The German retail chain REWE experiments with an aquaponic system in one of its markets. As a showcase for a sustainability concept in food retail, the supermarket cultivates herbs (basil) and fish (cichlids) on its roof, then sold downstairs in the store.[19]

5.3 Supply chain shortening and farm-to-fork concepts

Another upcoming area is the focus on the so-called farm-to-fork or farm-to-table concepts, in which consumers and distributors omit intermediaries and source agri-food products directly from the farmer. A related area of research that has gained attention lately is 'short food supply chains' (SFSC) (Barbosa, 2021). The rationale behind SFSC often lies within social aspects (Corsi et al., 2018), e.g. in local self-sufficiency or supporting the local community. However, these supply chains build on the backs of many small and medium-sized enterprises, which often do not have the same level of logistics efficiency as global supply chains.

As discussed in previous sections, trade-offs related to distribution processes of perishable agri-food products require the development of novel and optimized distribution processes to empower the actors in SFSC. One example could be using ICT to offer platform services like The Food Assembly.[20] According to Zhu et al. (2018) the sustainability aspect holds opportunities for further model-based research in SFSC. Furthermore, research into logistics-related optimization of SFSC also requires further study (Paciarotti and Torregiani, 2021).

6 Conclusion

This chapter has given a brief overview of the development of scientific research on planning and control processes in AFSCs and their related challenges. Our

19 https://www.rewe-group.com/de/presse-und-medien/newsroom/stories/aquakultur-mit-barsch-und-basilikum-die-dachfarm-versorgen/.

20 The Food Assembly has been developed in France in 2011, and was introduced in 2014 in UK, Spain, Germany, and other European countries. https://thefoodassembly.com/.

review focused on quantitative modeling approaches considering techniques and methodologies from OR and MS. We have shown that there is a tremendous amount of research developing decision-support tools for planning processes in AFSCs. Although some models are somewhat dated, their mathematical foundation can still be used as technical building blocks for developing new innovative planning tools, e.g. by developing new methodologies based on artificial intelligence and machine learning procedures such as deep reinforcement learning and neural networks.

As highlighted at the outset of this chapter, several factors and events have led to further challenges to the sustainability paradigm shift, which is necessary to achieve global sustainability as framed by the United Nations' Sustainable Development Goals. Several areas for future research were portrayed, and some building blocks of AFSCs were explored. New technologies and the integration of ICT play an essential role in the future development of the management of AFSCs; however, understanding their exact benefits and how they can support managerial decision-making processes is as essential. Thus, integrating new concepts, such as artificial intelligence, has to be grounded in practice and, consequently, must be explored according to their relevance to practitioners. This also means that future research will have to address multiobjective solutions that can not only further sustainability efforts but also cater to the needs of companies by, for example, including supply chain resilience.

The rising interest in the topic of AFSCs is generally reflected in the number of academic articles and books published in recent years. For an extended overview in regard to planning and control processes in sustainable food supply chains, we would like to refer the interested reader to the works of Accorsi and Manzini (2019) and Iakovou et al. (2016a), who provide comprehensive overviews of critical drivers behind the design, planning, and control of complex AFSC, structural support for risk management strategies, and sustainability concepts.

7 Where to look for further information

The following articles provide a good overview of the subject:

- Accorsi, R. and Manzini, R. (Eds). (2019). *Sustainable Food Supply Chains. Planning, Design, and Control through Interdisciplinary Methodologies*. 1st edition. Academic Press, London.
- Ahumada, O. and Villalobos, J. R. (2009). Application of planning models in the agri-food supply chain: a review. *European Journal of Operational Research* 196(1): 1–20. DOI: 10.1016/j.ejor.2008.02.014.
- Bakker, M., Riezebos, J. and Teunter, R. H. (2012). Review of inventory systems with deterioration since 2001. *European Journal of Operational Research* 221(2): 275–284. DOI: 10.1016/j.ejor.2012.03.004.

- Mercier, S., Villeneuve, S., Mondor, M. and Uysal, I. (2017). Time-temperature management along the food cold chain: a review of recent developments. *Comprehensive Reviews in Food Science and Food Safety* 16(4): 647–667. DOI: 10.1111/1541-4337.12269.

Key research in this area is frequently published in the following journals:

- *Advances in Agronomy*;
- *British Food Journal;*
- *Computers and Electronics in Agriculture;*
- *European Journal of Operational Research'*
- *Food Control;*
- *Global Food Security;*
- *International Journal of Production Economics;*
- *International Journal of Production Research;*
- *Journal of Forecasting;*
- *Journal of Food Safety*;
- *Nature Food*; and
- *Production Planning and Control.*

8 Acknowledgement

This chapter was funded by dtec.bw – Digitalization and Technology Research Center of the Bundeswehr which we gratefully acknowledge. dtec.bw was funded by the European Union – NextGenerationEU.

9 References

Accorsi, R. and Manzini, R. (Eds). (2019). *Sustainable Food Supply Chains. Planning, Design, and Control through Interdisciplinary Methodologies.* 1st edition. Academic Press, London.

Acero, B., Saenz, M. J. and Luzzini, D. (2022). Introducing synchromodality: one missing link between transportation and supply chain management. *Journal of Supply Chain Management* 58(1): 51–64. DOI: 10.1111/jscm.12269.

Agamez-Arias, A. and Moyano-Fuentes, J. (2017). Intermodal transport in freight distribution: a literature review. *Transport Reviews* 37(6): 782–807. DOI: 10.1080/01441647.2017.1297868.

Agi, M. A. N., and Soni, H. N. (2019). Joint pricing and inventory decisions for perishable products with age-, stock-, and price-dependent demand rate. *Journal of the Operational Research Society* 71(1): 85–99. DOI: 10.1080/01605682.2018.1525473.

Ahumada, O. and Villalobos, J. R. (2009). Application of planning models in the agri-food supply chain: a review. *European Journal of Operational Research* 196(1): 1–20. DOI: 10.1016/j.ejor.2008.02.014.

Ahumada, O. and Villalobos, J. R. (2011). Operational model for planning the harvest and distribution of perishable agricultural products. *International Journal of Production Economics* 133(2): 677–687. DOI: 10.1016/J.IJPE.2011.05.015.

Akkaya, D., Bimpikis, K. and Lee, H. (2021). Government interventions to promote agricultural innovation. *Manufacturing and Service Operations Management* 23(2): 437–452. DOI: 10.1287/msom.2019.0834.

Akkerman, R., Farahani, P. and Grunow, M. (2010). Quality, safety and sustainability in food distribution: a review of quantitative operations management approaches and challenges. *OR Spectrum* 32(4): 863–904. DOI: 10.1007/s00291-010-0223-2.

Albrecht, A., Ibald, R., Raab, V., Reichstein, W., Haarer, D. and Kreyenschmidt, J. (2020). Implementation of time temperature indicators to improve temperature monitoring and support dynamic shelf life in meat supply chains. *Journal of Packaging Technology and Research* 4(1): 23–32. DOI: 10.1007/s41783-019-00080-x.

Ali, I., Greifeneder, F., Stamenkovic, J., Neumann, M. and Notarnicola, C. (2015). Review of machine learning approaches for biomass and soil moisture retrievals from remote sensing data. *Remote Sensing* 7(12): 16398–16421. DOI: 10.3390/rs71215841.

Alizamir, S., Iravani, F. and Mamani, H. (2019). An analysis of price vs. revenue protection: government subsidies in the agriculture industry. *Management Science* 65(1): 32–49. DOI: 10.1287/mnsc.2017.2927.

Allen, S. J. and Schuster, E. W. (2004). Controlling the risk for an agricultural harvest. *Manufacturing and Service Operations Management. INFORMS* 6(3): 225–236. DOI: 10.1287/MSOM.1040.0035.

Amath, J. P., Dupont, J. L. and Guilpart, J. (2021). *The Carbon Footprint of the Cold Chain, 7th Informatory Note on Refrigeration and Food*. DOI: 10.18462/iir.INfood07.04.2021.

Ambrosino, D. and Sciomachen, A. (2007). A food distribution network problem: a case study. *IMA Journal of Management Mathematics* 18(1): 33–53. DOI: 10.1093/IMAMAN/DPL012.

Amorim, P., Meyr, H., Almeder, C. and Almada-Lobo, B. (2013). Managing perishability in press-distribution planning: a discussion and review. *Flexible Services and Manufacturing Journal* 25(3): 389–413. DOI: 10.1007/s10696-011-9122-3.

An, J., Cho, S. H. and Tang, C. S. (2015). Aggregating smallholder farmers in emerging economies. *Production and Operations Management* 24(9): 1414–1429. DOI: 10.1111/POMS.12372.

Anderson, E. and Monjardino, M. (2019). Contract design in agriculture supply chains with random yield. *European Journal of Operational Research* 277(3): 1072–1082. DOI: 10.1016/J.EJOR.2019.03.041.

Awad, M., Ndiaye, M. and Osman, A. (2020). Vehicle routing in cold food supply chain logistics: a literature review. *International Journal of Logistics Management*. DOI: 10.1108/IJLM-02-2020-0092.

Badraoui, I., van der Vorst, J. G. A. J. and Boulaksil, Y. (2020). Horizontal logistics collaboration: an exploratory study in Morocco's agri-food supply chains. *International Journal of Logistics Research and Applications* 23(1): 85–102. DOI: 10.1080/13675567.2019.1604646.

Badraoui, I., van der Lans, I., Boulaksil, Y. and van der Vorst, J. G. A. J. (2022). Antecedents of horizontal logistics collaboration in agri-food supply chains. *International Journal of Logistics Management* 33(1): 239–260. DOI: 10.1108/IJLM-09-2020-0362.

Bakker, M., Riezebos, J. and Teunter, R. H. (2012). Review of inventory systems with deterioration since 2001. *European Journal of Operational Research* 221(2): 275–284. DOI: 10.1016/j.ejor.2012.03.004.

Bansal, S. and Nagarajan, M. (2017). Product portfolio management with production flexibility in agribusiness. *Operations Research* 65(4): 914–930. DOI: 10.1287/opre.2017.1608.

Bansal, S. and Nagarajan, M. (2022). Portfolio management issues in the commercial seed industry: a modeling framework and industry implementation. In: Boyabatli, O., Kazaz, B. and Tang, C. S. (Eds) *Agricultural Supply Chain Management Research: Operations and Analytics in Planting, Selling, and Government Interventions*. Springer Series in Supply Chain Management, pp. 53–67.

Barbosa, M. W. (2021). Uncovering research streams on agri-food supply chain management: a bibliometric study. *Global Food Security* 28(February): 100517. DOI: 10.1016/j.gfs.2021.100517.

Bayona-Oré, S., Cerna, R. and Hinojoza, E. T. (2021). Machine learning for price prediction for agricultural products. *WSEAS Transactions on Business and Economics* 18: 969–977. DOI: 10.37394/23207.2021.18.92.

Behdani, B., Fan, Y. and Bloemhof, J. M. (2019). Cool chain and temperature-controlled transport: an overview of concepts, challenges, and technologies. *Sustainable Food Supply Chains: Planning, Design, and Control. through Interdisciplinary Methodologies* (October 2022): 167–183. DOI: 10.1016/B978-0-12-813411-5.00012-0.

Behzadi, G., O'Sullivan, M. J., Olsen, T. L. and Zhang, A. (2018). Agribusiness supply chain risk management: a review of quantitative decision models. *Omega* 79: 21–42. DOI: 10.1016/j.omega.2017.07.005.

Bellman, R. (1957). *Dynamic Programming*. Princeton University Press, Princeton, NJ.

Benke, K. and Tomkins, B. (2017). Future food-production systems: vertical farming and controlled-environment agriculture. *Sustainability: Science, Practice and Policy* 13(1): 13–26. DOI: 10.1080/15487733.2017.1394054.

Bhunia, A. K. (2018). *Foodborne Microbial Pathogens*. Springer, New York.

Bintsis, T. (2018). Microbial pollution and food safety. *AIMS Microbiology* 4(3): 377–396. DOI: 10.3934/microbiol.2018.3.377.

Borodin, V., Bourtembourg, J., Hnaien, F. and Labadie, N. (2016). Handling uncertainty in agricultural supply chain management: a state of the art. *European Journal of Operational Research* 254(2): 348–359. DOI: 10.1016/j.ejor.2016.03.057.

Boyabatli, O., Nguyen, J. and Wang, T. (2017). Capacity management in agricultural commodity processing and application in the palm industry. *Manufacturing and Service Operations Management* 19(4): 551–567. DOI: 10.1287/MSOM.2017.0624.

Boyabatli, O., Nasiry, J. and Zhou, Y. H. (2019). Crop planning in sustainable agriculture: dynamic farmland allocation in the presence of crop rotation benefits. *Management Science* 65(5): 2060–2076. DOI: 10.1287/MNSC.2018.3044.

Brandt, J. A. and Bessler, D. A. (1983). Price forecasting and evaluation: an application in agriculture. *Journal of Forecasting* 2(3): 237–248. DOI: 10.1002/for.3980020306.

Broekmeulen, R. A. C. M. and van Donselaar, K. H. (2009). A heuristic to manage perishable inventory with batch ordering, positive lead-times, and time-varying demand. *Computers and Operations Research* 36(11): 3013–3018. DOI: 10.1016/J.COR.2009.01.017.

Brosnan, T. and Sun, D. W. (2001). Precooling techniques and applications for horticultural products – a review. *International Journal of Refrigeration* 24(2): 154–170. DOI: 10.1016/S0140-7007(00)00017-7.

Buck, R. and Minvielle, A. (2014). Fresh take on food retailing | McKinsey. Available at: https://www.mckinsey.com/capabilities/growth-marketing-and-sales/our-insights/fresh-take-on-food-retailing (accessed 3 January 2023).

Burer, S., Jones, P. C. and Lowe, T. J. (2008). Coordinating the supply chain in the agricultural seed industry. *European Journal of Operational Research* 185(1): 354–377. DOI: 10.1016/j.ejor.2006.12.015.

Butturini, M. and Marcelis, L. F. M. (2019). Vertical farming in Europe: present status and outlook. *Plant Factory: An Indoor Vertical Farming System for Efficient Quality Food Production* (2nd edition) (February 2021): 77–91. DOI: 10.1016/B978-0-12-816691-8.00004-2.

Chen, J. and Chen, Y. J. (2021). The impact of contract farming on agricultural product supply in developing economies. *Production and Operations Management* 30(8): 2395–2419. DOI: 10.1111/POMS.13382.

Chen, L. M. and Sapra, A. (2021). Inventory renewal for a perishable product: economies of scale and age-dependent demand. *Naval Research Logistics (NRL)* 68(3): 359–377. DOI: 10.1002/NAV.21956.

Chintapalli, P. and Tang, C. S. (2021). The value and cost of crop minimum support price: farmer and consumer welfare and implementation cost. *Management Science* 67(11): 6839–6861. DOI: 10.1287/MNSC.2020.3831.

Chintapalli, P. and Tang, C. S. (2022). The implications of crop minimum support price in the presence of myopic and strategic farmers. *European Journal of Operational Research* 300(1): 336–349. DOI: 10.1016/j.ejor.2021.09.034.

Corsi, A., Barbera, F., Dansero, E., Orlando, G. and Peano, C. (2018). Multidisciplinary Approaches to Alternative Food Networks. In: Corsi, A., Barbera, F., Dansero, E., Peano, C. (eds) Alternative Food Networks. Palgrave Macmillan, Cham. https://doi.org/10.1007/978-3-319-90409-2_2

Daponte, P., De Vito, L., Glielmo, L., Iannelli, L., Liuzza, D., Picariello, F. and Silano, G. (2019). A review on the use of drones for precision agriculture. *IOP Conference Series: Earth and Environmental Science* 275(1). DOI: 10.1088/1755-1315/275/1/012022.

Dutta, N. and Kaur, A. (2022). A socially responsible decision-making model for firms contracting with constrained farmers. *International Transactions in Operational Research*: 1–28. DOI: 10.1111/ITOR.13124.

Eaton, C. and Shepherd, A. W. (2001). Contract farming - partnerships for growth. FAO Agricultural Services Bulletin. Available at: https://www.fao.org/3/Y0937E/y0937e00.htm#toc (accessed 5 January, 2023).

European Union. (2018). Regulation (EU) No 1169/2011 of the European Parliament and of the Council of 25 October 2011 on the provision of food information to consumers, amending Regulations (EC) No 1924/2006 and (EC) No 1925/2006 of the European Parliament and of the Council. *Official Journal of the European Communities* 54(L304): 18–63. DOI: 10.3000/19770677.L_2011.304.eng.

Fang, Y., Guan, B., Wu, S. and Heravi, S. (2020). Optimal forecast combination based on ensemble empirical mode decomposition for agricultural commodity futures prices. *Journal of Forecasting* 39(6): 877–886. DOI: 10.1002/for.2665.

Feng, H., Wang, X., Duan, Y., Zhang, J. and Zhang, X. (2020). Applying blockchain technology to improve agri-food traceability: a review of development methods, benefits and challenges. *Journal of Cleaner Production* 260: 121031. DOI: 10.1016/j.jclepro.2020.121031.

Fikar, C. and Leithner, M. (2021). A decision support system to facilitate collaborative supply of food cooperatives. *Production Planning and Control* 32(14): 1179–1190. DOI: 10.1080/09537287.2020.1796135.

Flores, H. and Villalobos, J. R. (2020). A stochastic planning framework for the discovery of complementary, agricultural systems. *European Journal of Operational Research* 280(2): 707–729. DOI: 10.1016/J.EJOR.2019.07.053.

Flores, R., Lázaro, E., Ramos, E., Provost, K. and Habib, H. (2020). Demand management in the supply chain: a focus on agribusiness. In: Mrugalska, B., Trzcielinski, S., Karwowski, W., et al. (Eds) *Advances in Manufacturing, Production Management and Process Control.*. Springer International Publishing, Cham, pp. 333–340.

Folorunso, E. A., Roy, K., Gebauer, R., Bohatá, A. and Mraz, J. (2021). Integrated pest and disease management in aquaponics: a metadata-based review. *Reviews in Aquaculture* 13(2): 971–995. DOI: 10.1111/raq.12508.

Fountas, S., Aggelopoulou, K. and Gemtos, T. A. (2016). Precision agriculture: crop management for improved productivity and reduced environmental impact or improved sustainability. In: Iakovou, E., Bochtis, D., Vlachos, D. and Aidonis, D. (Eds). *Supply Chain Management for Sustainable Food Networks*. John Wiley & Sons, Ltd, UK, pp. 41–65. DOI: 10.1002/9781118937495.CH2.

Gardner, B. L. (2009). *American Agriculture in the Twentieth Century: How It Flourished and What It Cost*. Harvard University Press, Cambridge, Mass.

Gaukler, G., Ketzenberg, M. and Salin, V. (2017). Establishing dynamic expiration dates for perishables: an application of RFID and sensor technology. *International Journal of Production Economics* 193: 617–632. DOI: 10.1016/j.ijpe.2017.07.019.

Glen, J. J. (1987). Mathematical models in farm planning: a survey. *Operations Research* 35(5): 641–666. DOI: 10.1287/opre.35.5.641.

Goyal, S. K. and Giri, B. C. (2001). Recent trends in modeling of deteriorating inventory. *European Journal of Operational Research* 134(1): 1–16.

Gravani, R. B. (2009). The role of good agricultural practices in produce safety. In: Fan, X., Niemira, B. A., Doona, C. J., Feeherry, F. E. and Gravani, R. B. (Eds) *Microbial Safety of Fresh Produce (Institute of Food Technologists Series)*. Wiley-Blackwell, Singapore.

Guo, W., van Blokland, W. B. and Lodewijks, G. (2017). Survey on characteristics and challenges of synchromodal transportation in global cold chains. In: Bektaş, T., Coniglio, S., Martinez-Sykora, A., and Voß, S. (Eds) *Computational Logistics. ICCL 2017*. Lecture Notes in Computer Science. Springer, Cham, pp. 420–434. DOI: 10.1007/978-3-319-68496-3_28.

Haijema, R. and Minner, S. (2016). Stock-level dependent ordering of perishables: a comparison of hybrid base-stock and constant order policies. *International Journal of Production Economics* 181: 215–225. DOI: 10.1016/j.ijpe.2015.10.013.

Haijema, R. and Minner, S. (2019). Improved ordering of perishables: the value of stock-age information. *International Journal of Production Economics* 209: 316–324. DOI: 10.1016/j.ijpe.2018.03.008.

Handayati, Y., Simatupang, T. M. and Perdana, T. (2015). Agri-food supply chain coordination: the state-of-the-art and recent developments. *Logistics Research* 8(1): 1–15. DOI: 10.1007/s12159-015-0125-4.

Hansen, O., Transchel, S. and Friedrich, H. (2022). Replenishment strategies for lost sales inventory systems of perishables under demand and lead time uncertainty. *European Journal of Operational Research*. DOI: 10.1016/J.EJOR.2022.11.041.

Hsiao, Y. H., Chen, M. C. and Chin, C. L. (2017). Distribution planning for perishable foods in cold chains with quality concerns: formulation and solution procedure. *Trends in Food Science and Technology* 61: 80–93. DOI: 10.1016/j.tifs.2016.11.016.

Hsu, V. N., Lai, G. and Liang, G. (2019). Agricultural partnership for dairy farming. *Production and Operations Management* 28(12): 3042–3059. DOI: 10.1111/poms.12992.

Hübner, A., Schäfer, F. and Schaal, K. N. (2020). Maximizing profit via assortment and shelf-space optimization for two-dimensional shelves. *Production and Operations Management* 29(3): 547-570. DOI: 10.1111/POMS.13111.

Iakovou, E., Bochtis, D., Vlachos, D. and Aidonis, D. (2016a). *Supply Chain Management for Sustainable Food Networks*. John Wiley & Sons, Ltd, Chichester.

Iakovou, E., Bochtis, D., Vlachos, D. and Aidonis, D. (2016b). Sustainable agrifood supply chain management. In: Iakovou, E., Bochtis, D., Vlachos, D. and Aidonis, D. (Eds) *Supply Chain Management for Sustainable Food Networks*. John Wiley & Sons, Ltd, UK, pp. 1–39.

Inoue, Y. (2020). Satellite- and drone-based remote sensing of crops and soils for smart farming – a review. *Soil Science and Plant Nutrition* 66(6): 798–810. DOI: 10.1080/00380768.2020.1738899.

Jaffee, S., Siegel, P. and Andrews, C. (2008). *Rapid Agricultural Supply Chain Risk Assessment*. World Bank, Washington DC. (Commodity Risk Management Group).

James, S. J. and James, C. (2010). The food cold-chain and climate change. *Food Research International* 43(7): 1944–1956. DOI: 10.1016/j.foodres.2010.02.001.

Janssen, L., Claus, T. and Sauer, J. (2016). Literature review of deteriorating inventory models by key topics from 2012 to 2015. *International Journal of Production Economics* 182: 86–112. DOI: 10.1016/j.ijpe.2016.08.019.

Jones, W. B., Cassady, C. R. and Bowden, R. O. (2019). Developing a standard definition of intermodal transportation. *Kemampuan Koneksi Matematis (Tinjauan Terhadap Pendekatan Pembelajaran Savi)* 53(9): 1689–1699.

Karaesmen, I. Z., Scheller-Wolf, A. and Deniz, B. (2011). Managing perishable and aging inventories: review and future research directions. *International Series in Operations Research and Management Science* 151: 393–436. DOI: 10.1007/978-1-4419-6485-4_15.

Kazaz, B. (2004). Production planning under yield and demand uncertainty with yield-dependent cost and price. *Manufacturing and Service Operations Management* 6(3): 209–224. DOI: 10.1287/msom.1030.0024.

Kennedy, J. O. S. (1981). Applications of dynamic programming to agriculture, forestry and fisheries: review and prognosis. *Journal of Gender. Agriculture and Food Security* 49(3): 141-173. DOI: 10.22004/ag.econ.12234.

Ketzenberg, M., Bloemhof, J. and Gaukler, G. (2015). Managing perishables with time and temperature history. *Production and Operations Management* 24(1): 54–70. DOI: 10.1111/poms.12209.

Knee, M., Proctor, F. J. and Dover, C. J. (1985). The technology of ethylene control: use and removal in post-harvest handling of horticultural commodities. *Annals of Applied Biology* 107(3): 581-595.

Kusumastuti, R. D., Donk, D. P. and Teunter, R. (2016). Crop-related harvesting and processing planning: a review. *International Journal of Production Economics* 174: 76–92. DOI: 10.1016/j.ijpe.2016.01.010.

Lestari, E. P., Prajanti, S. D. W., Wibawanto, W. and Adzim, F. (2022). ARCH-GARCH analysis: an approach to determine the price volatility of red chili. *AGRARIS* 8(1): 90-105. DOI: 10.18196/agraris.v8i1.12060.

Lin, C.-F. (2012). SPS-plus and bilateral treaty network: a 'global' solution to the global food safety problem? *Wisconsin International Law Journal* 29(4): 694–734.

Lowe, T. J. and Preckel, P. V. (2004). Decision technologies for agribusiness problems: a brief review of selected literature and a call for research. *Manufacturing and Service Operations Management* 6(3): 201–208. DOI: 10.1287/MSOM.1040.0051.

Lucas, M. T. and Chhajed, D. (2004). Applications of location analysis in agriculture: a survey. *Journal of the Operational Research Society* 55(6): 561–578. DOI: 10.1057/palgrave.jors.2601731.

Matopoulos, A., Vlachopoulou, M., Manthou, V. and Manos, B. (2007). A conceptual framework for supply chain collaboration: empirical evidence from the agri-food industry. *Supply Chain Management: An International Journal* 12(3): 177–186. DOI: 10.1108/13598540710742491.

Mercier, S., Villeneuve, S., Mondor, M. and Uysal, I. (2017). Time–temperature management along the food cold chain: a review of recent developments. *Comprehensive Reviews in Food Science and Food Safety* 16(4): 647–667. DOI: 10.1111/1541-4337.12269.

Messelink, G. J., Lambion, J., Janssen, A. and van Rijn, P. C. J. (2021). Biodiversity in and around greenhouses: benefits and potential risks for pest management. *Insects* 12(10): 1–16. DOI: 10.3390/insects12100933.

Minner, S. and Transchel, S. (2010). Periodic review inventory-control for perishable products under service-level constraints. *OR Spectrum* 32(4): 979–996. DOI: 10.1007/s00291-010-0196-1.

Nahmias, S. (1982). Perishable inventory theory: a review. *Operations Research* 30(4): 680–708. DOI: 10.1287/opre.30.4.680.

Nahmias, S. (2011). *Perishable Inventory Systems* (Vol. 160). Springer, New York.

Ndraha, N., Hsiao, H. I., Vlajic, J., Yang, M. and Lin, H. V. (2018). Time-temperature abuse in the food cold chain: review of issues, challenges, and recommendations. *Food Control* 89: 12–21. DOI: 10.1016/j.foodcont.2018.01.027.

Neupane, H., Paudel, K. P., Adhikari, M. and He, Q. (2022). Impact of cooperative membership on production efficiency of smallholder goat farmers in Nepal. *Annals of Public and Cooperative Economics* 93(2): 337–356. DOI: 10.1111/APCE.12371.

Neusel, L. and Hirzel, S. (2022). Cleaner logistics and supply chain energy efficiency in cold supply chains of the food sector: an exploration of conditions and perceptions. *Cleaner Logistics and Supply Chain* 5: 100082. DOI: 10.1016/j.clscn.2022.100082.

Noparumpa, T., Kazaz, B. and Webster, S. (2022). A prescriptive model for selling wine futures to mitigate quality uncertainty. In: *Springer Series in Supply Chain Management 12*. Springer Nature, pp. 123–139. DOI: 10.1007/978-3-030-81423-6_8.

Ojha, S., Bußler, S. and Schlüter, O. K. (2020). Food waste valorisation and circular economy concepts in insect production and processing. *Waste Management* 118: 600–609.

Oltra-Mestre, M. J., Hargaden, V., Coughlan, P. and Segura-García del Río, B. (2021). Innovation in the agri-food sector: exploiting opportunities for Industry 4.0. *Creativity and Innovation Management* 30(1): 198–210. DOI: 10.1111/caim.12418.

Onethird. (2022). The ultimate guide to fresh produce shelf life prediction. Available at: https://onethird.io/ultimate-guide-to-fresh-produce-shelf-life-prediction/ (accessed 3 January 2023).

Paciarotti, C. and Torregiani, F. (2021). The logistics of the short food supply chain: a literature review. *Sustainable Production and Consumption* 26: 428–442. DOI: 10.1016/j.spc.2020.10.002.

Pahl, J. and Voß, S. (2014). Integrating deterioration and lifetime constraints in press and supply chain planning: a survey. *European Journal of Operational Research* 238(3): 654–674. DOI: 10.1016/J.EJOR.2014.01.060.

Palm, H. W., Knaus, U., Appelbaum, S., Goddek, S., Strauch, S. M., Vermeulen, T., Haïssam Jijakli, M. and Kotzen, B. (2018). Towards commercial aquaponics: a review of systems, designs, scales and nomenclature. *Aquaculture International* 26(3): 813–842. DOI: 10.1007/s10499-018-0249-z.

Panghal, A., Chhikara, N., Sindhu, N. and Jaglan, S. (2018). Role of food safety management systems in safe food production: a review. *Journal of Food Safety* 38(4). DOI: 10.1111/jfs.12464.

Pencheva, V., Asenov, A., Sladkowski, A., Ivanov, B. and Georgiev, I. (2022). Current issues of multimodal and intermodal cargo transportation. In: *Modern Trends and Research in Intermodal Transportation*. Springer, Cham, pp. 51–124. DOI: 10.1007/978-3-030-87120-8_2.

Petropoulos, F. and Carver, S. (2019). Forecasting for food demand. In: Accorsi, R. and Manzini, R. (Eds). *Sustainable Food Supply Chains*. Elsevier Inc., pp. 237–248. DOI: 10.1016/B978-0-12-813411-5.00016-8..

Powell, S. G. and Baker, K. R. (2016). *Business Analytics: The Art of Modeling With Spreadsheets*. 5th edition. John Wiley & Sons, Ltd, *Hoboken, NJ*.

Prusky, D. (2011). Reduction of the incidence of postharvest quality losses, and future prospects. *Food Security* 3(4): 463–474. DOI: 10.1007/s12571-011-0147-y.

Puri, V., Nayyar, A. and Raja, L. (2017). Agriculture drones: a modern breakthrough in precision agriculture. *Journal of Statistics and Management Systems* 20(4): 507–518. DOI: 10.1080/09720510.2017.1395171.

Raafat, F. (1991). Survey of literature on continuously deteriorating inventory models. *Journal of the Operational Research Society* 42(1): 27–37. DOI: 10.1057/jors.1991.4.

Raak, N., Symmank, C., Zahn, S., Aschemann-Witzel, J. and Rohm, H. (2017). Processing- and product-related causes for food waste and implications for the food supply chain. *Waste Management* 61: 461–472. DOI: 10.1016/j.wasman.2016.12.027.

Rasmussen, W. D. (1962). The impact of technological change on American agriculture, 1862–1962*. *The Journal of Economic History* 22(4): 578–591. DOI: 10.1017/S0022050700066778.

Şahinli, M. A. (2020). Potato price forecasting with holt-winters and Arima methods: a case study. *American Journal of Potato Research* 97(4): 336–346. DOI: 10.1007/s12230-020-09788-y.

Salin, V. and Nayga, R. M. (2003). A cold chain network for food exports to developing countries. *International Journal of Physical Distribution and Logistics Management* 33(10): 918–933. DOI: 10.1108/09600030310508717.

Santini, A., Bartolini, E., Schneider, M. and Greco de Lemos, V. (2021). The crop growth planning problem in vertical farming. *European Journal of Operational Research* 294(1): 377–390. DOI: 10.1016/j.ejor.2021.01.034.

Sengupta, T., Narayanamurthy, G., Moser, R. and Hota, P. K. (2019). Sharing app for farm mechanization: gold Farm's digitized access based solution for financially constrained farmers. *Computers in Industry* 109: 195–203. DOI: 10.1016/j.compind.2019.04.017.

Shamshiri, R. R., Kalantari, F., Ting, K. C., Thorp, K. R., Hameed, I. A., Weltzien, C., Ahmad, D. and Mojgan Shad, Z. (2018). Advances in greenhouse automation and controlled environment agriculture: a transition to plant factories and urban agriculture. *International Journal of Agricultural and Biological Engineering* 11(1): 1–22. DOI: 10.25165/j.ijabe.20181101.3210.

Shashi, S., Cerchione, R., Singh, R., Centobelli, P. and Shabani, A. (2018). Food cold chain management: from a structured literature review to a conceptual framework and research agenda. *International Journal of Logistics Management* 29(3): 792–821. DOI: 10.1108/IJLM-01-2017-0007.

Shi, J., Zhao, Y., Kiwanuka, R. B. K. and Chang, J. (2019). Optimal selling policies for farmer cooperatives. *Production and Operations Management* 28(12): 3060–3080. DOI: 10.1111/POMS.13091.

Shukla, M. and Jharkharia, S. (2013). Agri-fresh produce supply chain management: a state-of-the-art literature review. *International Journal of Operations and Production Management* 33(2): 114–158. DOI: 10.1108/01443571311295608.

Silver, E. A., Pyke, D. F. and Thomas, D. J. (2016). *Inventory and Production Management in Supply Chains*. 4th edition. CRC Press. DOI: 10.1201/9781315374406.

Singh, R. P., Prasad, P. V. V. and Reddy, K. R. (2013). Impacts of changing climate and climate variability on seed production and seed industry. *Advances in Agronomy* 118: 49–110. DOI: 10.1016/B978-0-12-405942-9.00002-5.

Soto-Silva, W. E., Nadal-Roig, E., González-Araya, M. C. and Pla-Aragones, L. M. (2016). Operational research models applied to the fresh fruit supply chain. *European Journal of Operational Research* 251(2): 345–355. DOI: 10.1016/J.EJOR.2015.08.046.

Spanaki, K., Karafili, E., Sivarajah, U., Despoudi, S. and Irani, Z. (2022). Artificial intelligence and food security: swarm intelligence of AgriTech drones for smart AgriFood operations. *Production Planning and Control* 33(16): 1498–1516. DOI: 10.1080/09537287.2021.1882688.

Stüve, D., van der Meer, R., Ali Agha, M. S. and Lütke Entrup, M. (2022). A systematic literature review of modelling approaches and implementation of enabling software for supply chain planning in the food industry. *Production and Manufacturing Research* 10(1): 470–493. DOI: 10.1080/21693277.2022.2091057.

Taşkıner, T. and Bilgen, B. (2021). Optimization models for harvest and production planning in agri-food supply chain: a systematic review. *Logistics* 5(3): 52. DOI: 10.3390/logistics5030052.

Tatarintsev, M., Korchagin, S., Nikitin, P., Gorokhova, R., Bystrenina, I. and Serdechnyy, D. (2021). Analysis of the forecast price as a factor of sustainable development of agriculture. *Agronomy* 11(6): 1–15. DOI: 10.3390/agronomy11061235.

Taylor, D. H. (2006). Demand management in agri-food supply chains: an analysis of the characteristics and problems and a framework for improvement. *The International Journal of Logistics Management* 17(2): 163–186. DOI: 10.1108/09574090610689943.

Tian, F. (2016). An agri-food supply chain traceability system for China based on RFID & blockchain technology. 2016 13th International Conference on Service Systems and Service Management, ICSSSM 2016. DOI: 10.1109/ICSSSM.2016.7538424.

Torres, R., Martínez-Zafra, M. T., Castillejo, N., Guillamón-Frutos, A. and Artés-Hernández, F. (2020). Real-time monitoring system for shelf life estimation of fruit and vegetables. *Sensors* 20(7): 1860. DOI: 10.3390/s20071860.

Transchel, S., Buisman, M. E. and Haijema, R. (2022). Joint assortment and inventory optimization for vertically differentiated products under consumer-driven substitution. *European Journal of Operational Research* 301(1): 163–179. DOI: 10.1016/j.ejor.2021.09.041.

Tsolakis, N. K., Keramydas, C. A., Toka, A. K., Aidonis, D. A. and Iakovou, E. T. (2014). Agrifood supply chain management: a comprehensive hierarchical decision-making framework and a critical taxonomy. *Biosystems Engineering* 120: 47–64. DOI: 10.1016/J.BIOSYSTEMSENG.2013.10.014.

Ugarte, G. M., Golden, J. S. and Dooley, K. J. (2016). Lean versus green: the impact of lean logistics on greenhouse gas emissions in consumer goods supply chains. *Journal of Purchasing and Supply Management* 22(2): 98–109. DOI: 10.1016/j.pursup.2015.09.002.

van Delden, S. H., Sharath Kumar, M., Butturini, M., Graamans, L. J. A., Heuvelink, E., Kacira, M., Kaiser, E., Klamer, R. S., Klerkx, L., Kootstra, G., Loeber, A., Schouten, R. E., Stanghellini, C., van Ieperen, W., Verdonk, J. C., Vialet-Chabrand, S., Woltering, E. J., van de Zedde, R., Zhang, Y. and Marcelis, L. F. M. (2021). Current status and future challenges in implementing and upscaling vertical farming systems. *Nature Food* 2(12): 944–956. DOI: 10.1038/s43016-021-00402-w.

van Donselaar, K., van Woensel, T., Broekmeulen, R. and Fransoo, J. (2006). Inventory control of perishables in supermarkets. *International Journal of Production Economics* 104(2): 462–472. DOI: 10.1016/J.IJPE.2004.10.019.

van Zyl, G. (1963). *Inventory Control for Perishable Commodities*. North Carolina State University. Available at: https://repository.lib.ncsu.edu/handle/1840.4/2387(accessed 5 January 2023).

von Thünen, H. (1826). *Der Isolirte Staat in Beziehung auf Landwirthschaft und Nationalökonomie*. Perthes, Hamburg.

Whitin, T. M. (1953). *The Theory of Inventory Management*. Princeton University Press, Princeton, NJ.

World Bank. (2018). Food-borne illnesses cost US$ 110 billion per year in low- and middle-income countries. Available at: https://www.worldbank.org/en/news/press-release/2018/10/23/food-borne-illnesses-cost-us-110-billion-per-year-in-low-and-middle-income-countries (accessed 3 January 2023).

World Health Organization. (2010). Estimating the burden of foodborne diseases. Available at: https://www.who.int/activities/estimating-the-burden-of-foodborne-diseases (accessed 3 January 2023).

Xu, X. and Zhang, Y. (2021). Corn cash price forecasting with neural networks. *Computers and Electronics in Agriculture* 184: 106120. DOI: 10.1016/j.compag.2021.106120.

Xu, X. and Zhang, Y. (2022a). Canola and soybean oil price forecasts via neural networks. *Advances in Computational Intelligence* 2(5): 1–12. DOI: 10.1007/s43674-022-00045-9.

Xu, X. and Zhang, Y. (2022b). Commodity price forecasting via neural networks for coffee, corn, cotton, oats, soybeans, soybean oil, sugar, and wheat. *Intelligent Systems in Accounting, Finance and Management* 29(3): 169–181. DOI: 10.1002/isaf.1519.

Yahia, E. M. (2009). *Modified and Controlled Atmospheres for the Storage, Transportation, and Packaging of Horticultural Commodities*. CRC Press, Boca Raton, FL.

Yano, C. A. and Lee, H. L. (1995). Lot sizing with random yields: a review. *Operations Research* 43(2): 311–334. DOI: 10.1287/OPRE.43.2.311.

Zelingher, R., Makowski, D. and Brunelle, T. (2021). Assessing the sensitivity of global maize price to regional productions using statistical and machine learning methods. *Frontiers in Sustainable Food Systems* 5: 1–11. DOI: 10.3389/fsufs.2021.655206.

Zhang, W. and Wilhelm, W. E. (2011). OR/MS decision support models for the specialty crops industry: a literature review. *Annals of Operations Research* 190(1): 131–148. DOI: 10.1007/s10479-009-0626-0.

Zhao, Z., Li, X. and Zhou, X. (2020). Optimization of transportation routing problem for fresh food in time-varying road network: considering both food safety reliability and temperature control. *PLoS ONE* 15(7): e0235950. DOI: 10.1371/journal.pone.0235950.

Zhu, Z., Chu, F., Dolgui, A., Chu, C., Zhou, W. and Piramuthu, S. (2018). Recent advances and opportunities in sustainable food supply chain: a model-oriented review. *International Journal of Production Research* 56(17): 5700–5722. DOI: 10.1080/00207543.2018.1425014.

Chapter 9

Information and communication technology in agri-food supply chains

Gerhard Schiefer, University of Bonn, Germany

1 Introduction

Food production involves one of the most complex organizational models of enterprise cooperation. The core is constituted by the actors involved in the flow of food products, reaching from agricultural production to food processing, trade and retail. The actors involved in the production flow are summarized as members of an agricultural or *agri-food supply chain*. An agri-food supply chain (or shorter: *food chain*) is a theoretical construct for characterizing the flow of products from agriculture to retail and consumers. However, a food chain with a fixed group of enterprises with continuous trade relationships is more the exception than the rule. A typical food chain is comprised of a network of enterprises with dynamically changing trading relationships reaching from agriculture to consumers. Such chains involve partners with a regional, national or even international base, resulting in a dynamically changing organization of food chains and participating enterprises. An example is shown in Fig. 1 where grain farms ('farmer') serve farms engaged in the production of milk, meat or eggs ('MME') which, in turn, serve processors or retailers. All are embedded in a complex network of different suppliers, processors and distributors.

Within this network of interrelated enterprises, it is possible to separate out individual food chains where enterprises are linked through a primary product

http://dx.doi.org/10.19103/AS.2023.0122.12

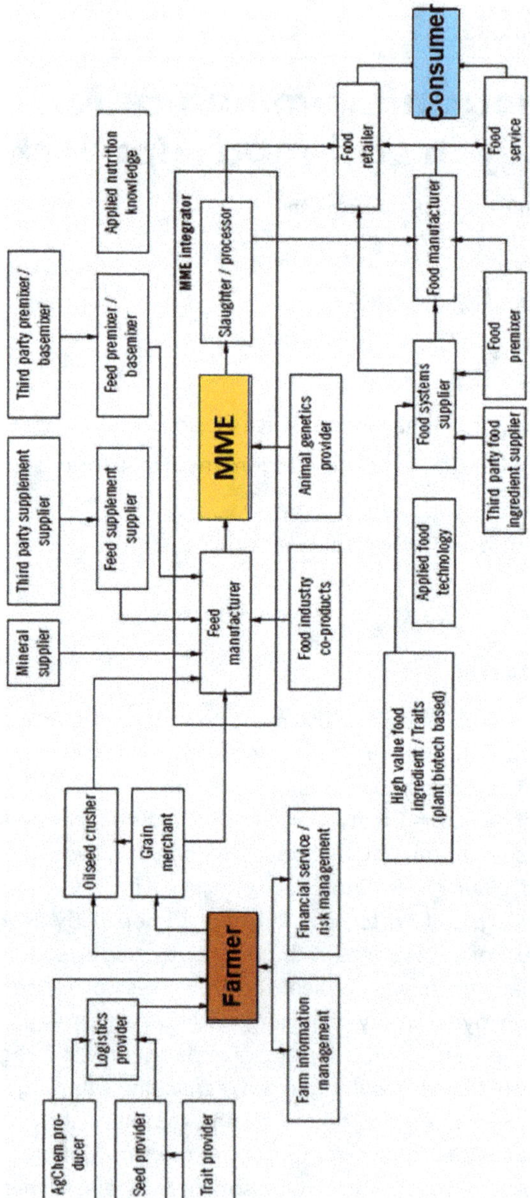

Figure 1 Model of a food network. Source: Author.

flow, focused on a single product that is produced, processed and eventually consumed or integrated into another product (Fig. 2). This is the traditional view taken in food chain analysis which, however, disregards the complexity of the food chain network.

However, whatever the flow of products, coordination is needed to assure that production and distribution processes along the chain deliver certain product qualities required by consumers or meet environmental requirements, depending on government or wider societal requirements. These 'classical' coordination needs have been discussed among stakeholders or interested groups in business, policy and research for a long time (Moreno-Miranda and Dries, 2022). The foundation of all coordination activities is the availability and exchange (communication) of data and information derived from data which

- support coordination of decisions and actions along the chain; and
- serve the information interests of consumers, primarily focusing on product quality characteristics and of society in general with a focus on e.g. environmental impacts of food processes.

This underlines the crucial need for availability and communication of data as well as their transformation into information for food chain organization and success. This chapter outlines challenges in communicating information across food chains, discusses alternative solutions and identifies issues that require further research and development.

2 The coordination challenge

2.1 Coordination approach

The need for data is determined by interdependencies within the food chain. To deal with interdependencies and coordinating actions between individual partners in the food chain, food chains could agree on a so-called 'chain captain', usually a strong partner in the chain who organizes and coordinates interdependencies in decisions and activities across the food chain and assures appropriate availability of data for all partners as well as to meet consumer and societal interests (Chaddad and Rodriguez-Alcala, 2010). This approach is traditionally favored in situations where one partner in a chain has a very dominant role such as a regionally dominant dairy company or slaughterhouse and where partners have close and maybe even exclusive trade relationships. The drawback of this model is its limited flexibility of aligning agricultural production variations with the need for serving market agreements.

The alternative to agreeing a chain captain is an open network situation where trade relationships along the value chain are dynamically evolving within

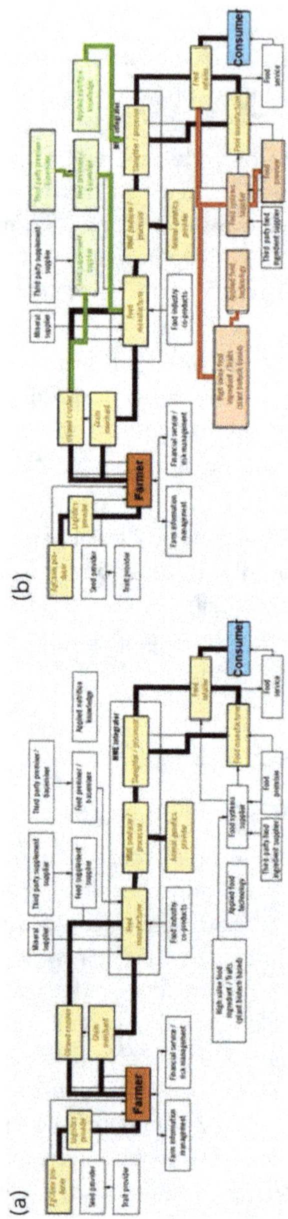

Figure 2 A single (a, black connections) and three interrelated (b, black, green and red connections) food chains within the complex food chain network.

a food (chain) network situation depending on evolving production and market developments, i.e. where enterprises change trading partners in response to production and market developments. Coordination in an open network situation is hard to achieve. The sector usually needs an external facilitator to move forward. Governments might assume a facilitator role through regulatory activities. Alternatives are new organizational chain models where partners have agreed on specific management rules, possibly formulated by an external organization such as a quality board. Examples are discussed in Sections 5.1 and 5.2.

2.2 Communication challenge for coordination

The coordination of data collection and exchange depends on prior agreements on the following:

- Focus;
- Content;
- Technology/formats for data transmission; and
- Abilities of enterprises to meet data agreements.

These agreements must also take account of the complexity of data collection and exchange.

Agreements on common objectives (*focus*) is the baseline for any further coordination engagement. Agreements on objectives need to be followed by agreements on how objectives can best be served through decisions and activities by the various trade partners which determines the need for data and data exchange (*content*). Agreements on responsibilities and timing in the collection, processing and exchange of data assure the *operability* of chain coordination. As the point of data use might differ from the point of data collection, enterprises have to agree on what data to collect, where to collect it, when to collect, what to communicate, how to communicate, where to use the data, by whom and what for.

In the end, the collection, processing and communication of data are dependent on *technologies* dealing with data formats, communication standards and data carriers, including the broad range of ICT technologies. However, as food chains usually involve a range of types of companies with different expertise, the *abilities* of companies in enabling decision-making, performance of key activities, provision of data and understanding of technology (sometimes expressed by the term 'e-readiness'; Bui et al., 2003) are crucial issues in coordination efforts.

In an open network situation, it cannot be guaranteed that all enterprises commit to agreements across all dimensions that are compatible in level and direction. Apart from regulatory initiatives by government or powerful actors in the sector such as dominant retail groups, compatibility of agreements is usually confined to smaller groups of enterprises. If those groups trade with each other, emerging product characteristics and the related need for data exchange and use will usually be determined by the lowest level of compatibility.

3 Data exchange across food chains

3.1 Infrastructure model

The exchange of data between enterprises can build on any of three different models which represent different levels of cooperation between enterprises as well as different levels of data exchange efficiency:

- Level 1: In this level, enterprises keep their local databases and exchange data with their direct suppliers and customers.
- Level 2: Enterprises keep their local databases but allow suppliers and customers to connect with their databases for retracting or providing specified datasets.
- Level 3: Groups of enterprises agree on a joint database.

There is not a clear separation of models. Enterprises might be engaged on all three levels, keeping certain data with relevance across the value chain in a joint database while still having local databases for their own use.

3.2 Direction of exchange

The direction of data exchange is a critical issue as it is closely related to market power. The 'traditional' view is that data are collected from the source, especially from agricultural settings, and communicated through to the end of the chain and even toward consumers (forward exchange). This approach supports enterprises toward the end of the chain in their operations and marketing of their products. However, there is an increasing interest of enterprises at the beginning of the chain including agricultural enterprises to receive data from enterprises further down the chain and especially from retail and consumers as the final customers. Such a backward exchange allows better and earlier adjustments to 'downstream' needs and changing consumer requests. This fits in with the increasing interest of consumers in production processes in agriculture and facilitates a combination of forward and backward exchange of data.

A combination of forward and backward exchange constitutes a balance between the provision and the reception of information at all stages of the chain and contributes to stable chain relationships. The communication of data along the chain for serving chain decisions could

- be linked to the products that are being exchanged; or
- communicated independently of the flow of products.

Data linked to products focus on the communication of characteristics that have relevance for product quality or authenticity. Data that are not linked to products are usually confined to data other than measurable quality characteristics of products. Examples are outlined in Section 4.

3.3 New technologies and standards

The organization of food chain communication systems now has access to an increasing range of technologies. It can be argued that deficiencies in chain communication systems are no longer due to deficiencies in available technology but primarily due to deficiencies in their use, combined with deficiencies in organization and coordination agreements.

Dynamic developments in technology provide a broad range of options to support chain data communication. These include:

- Blockchain technology (Galvez et al., 2018) contributes to data security and trust.
- Sensor technology facilitates data collection and communication (Weston et al., 2021).
- Cloud technology (Rahul et al., 2022) contributes to data accessibility and exchange.
- Artificial intelligence contributes to data analysis and decision support (Rahul et al., 2021).

The use of all of these technologies has been widely discussed. Difficulties are not in their use but in their suitable integration into IT systems for utilization in chain management and communication.

3.4 Data standards in communication and content

Apart from technology, communication requires standards in communication infrastructure. Examples include the Electronic Product Code Information Services (EPCIS) standards developed by the global data standards organization GS/1 (www.gs1.org) governing data exchange and standards governing communication of content and in assuring its reliability. Standards usually try to cover all these areas.

The most common standards relating to content are food safety standards focusing on products and food processing and their combination with auditing schemes for reliability. They may be initiated and managed by

- certain groups within chains (as demonstrated by the example of GlobalG.A.P., a standardization activity evolving from retail groups; www.globalgap.org);
- jointly by all groups of the value chain (as demonstrated by the example of QS, a standard management group initially evolving from agriculture but incorporating representatives from all stages of food chains; www.q-s.de/en/); and
- independent standardization organizations (as e.g. the example of the ISO standards involving the ISO9001 quality management standards; www.iso.org).

4 Data carriers

4.1 Data exchange linked to products

Following a product forward from the source to the end of the chain and from the end of the chain to the source is usually referred to as 'tracking and tracing'. 'Tracking and tracing' was initially discussed in connection with the assurance of food safety and food quality (Fritz and Schiefer, 2009). The process is outlined in Fig. 3.

- Figure 3a describes a forward search (identified by red circles with arrows symbolizing movement along the product flow) to identify and respond to a deficient product sample discovered in a stage of the chain (here: trade).
- Figure 3b describes a backward search (identified by blue circles and arrows moving against the product flow) to identify the source of a food safety issue (here: trade) discovered in the final product.

The 'tracking and tracing' feature has enormous potential for data exchange. It can pick up information at a certain stage and deliver it for use in subsequent stages or to inform customers and other societal stakeholders (e.g. governments). It can focus not just on specific product characteristics but also on processing characteristics or the environmental conditions (as e.g. temperature) affecting the product. It could also be utilized in situation where data are being accumulated during a product's path along the chain. An example involves the collection of the range of data at different stages of the food the chain needed to calculate a product's carbon footprint.

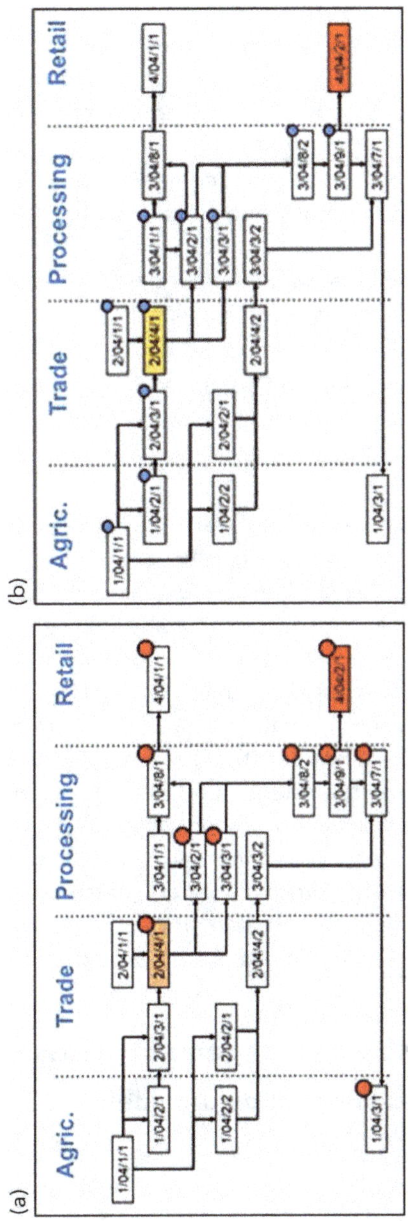

Figure 3 (a) The forward search and (b) the backward search. Source: Author.

4.2 Data exchange with no product relevance

The consideration of data without direct relevance for measurable characteristics of a specific product allows the decoupling of data flows from product flows. These data flows will have increasing future relevance especially in connection with environmental issues which do not affect measurable characteristics of products (e.g. measuring the environmental impact of producing and distributing a product).

A simplified version of product-independent data communication has been established in the 'book and claim' concept (Mol and Osterveer, 2015) that has been used in soybean production and sales (https://greenpalm.org/about-greenpalm/why-greenpalm-makes-a-difference/book-and-claim-supply-chain-model). It is summarized in Fig. 4:

- Figure 4a shows the traditional sales approach where retail does not separate between products produced under different conditions.
- Figure 4b shows the 'book and claim' approach.

In production of high-quality soybeans, the quality information captured in a quality label (representing a data cluster) is separated from the actual products. Products and labels are then marketed separately. The products are channeled into the regular soybean market without any link to the initial label information.

At the end of the chain, buyers of soybean can purchase the required number of labels and attach them to the products, selling them as 'high-quality' products. This approach in separating data from products works well if data refer to product or process characteristics that are not linked to measurable product characteristics but refer to credence attributes such as *fair trade* or *environmental-conscious production* claims that cannot be verified by consumers from product characteristics (such as quality attributes or price). It makes it easier to ensure that the sales quantity of 'high-quality products' matches the production quantity. In cases where data refer to measurable product characteristics, data markets have to assure that traded data remain linked to the specific products they refer to.

5 Simplifying data management across food chains

5.1 Certificates as information providers

Certificates related to product and/or processing characteristics can be viewed as a kind of condensed data communication approach. A certificate represents data linked to a product which can be used by others in the chain. As an example, if certificates linked to all products state 'free of pesticides' then the product delivered to its customers can be classified as 'free of pesticides' even

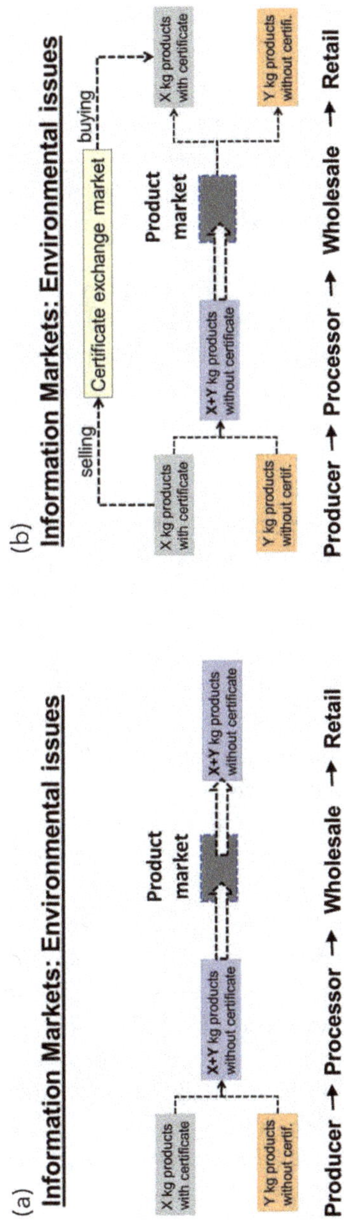

Figure 4 Traditional sales (a) and 'book and claim' concept (b). (Source: Author.

if an enterprise does not know anything about the production process of its suppliers. A chain can build on a series of certificates that covers the food chain from agriculture to retail. Along the chain, certificates might add up to a bundle of certificates or might be integrated subsequently into a single certificate at the end of the chain. Specific examples include quality systems such as the QS system (https://www.q-s.de) where each stage of the chain relies on the quality criteria fulfilled in the preceding stage. A drawback of using certificates in communication is their focus on certain standards which are certified as being fulfilled. Higher levels of achievements are not documented. Therefore, a system building on certificates describes a specified minimum level of achievements.

Traditionally, certificates are provided in paper format. However, certificates are increasingly provided in digital format which allows extraction of the data behind the certificate and linking them to the datasets of the receiving enterprise, allowing combination with other data collected by the receiving enterprise. A recent development is transmission of certificate data to consumer smartphones to give consumers more information on product and process characteristics (Kalinowska et al., 2021). The transformation could be based on data stored in the smartphone or by linking the certificates with cloud-based data linked to the certificate.

There are a number of well-developed communication models implemented in the food sector that build on the certificate approach. A basic approach is realized through external quality management audit schemes as exemplified by the schemes QS in Germany (https://www.q-s.de/en/) and IKB in The Netherlands (http://www.prnp.nl/public/doc/PRLT_4230_IKBdiervoeder.pdf?PHPSESSID=cb8cc97f8e6db655ee51 8ce10a1cf5fb). The coordination is not between individual enterprises but through an external organization. It specifies production requirements for each individual stage of the food value chain which assure certain agreed levels of food quality and environmental protection.

With this approach, companies do not have to exchange data on food quality or environmental issues as they 'know' what requirements have been fulfilled by its suppliers (Fig. 5). In addition, they are not bound to a certain supplier but could choose between all suppliers that follow the scheme. This simplifies the implementation of a coordinated quality and environmental quality management scheme and the exchange of the related data clusters not only along a chain but also within the whole food sector. It is obvious that such a scheme depends on a reliable monitoring and audit activity that assures adherence of all companies to the scheme .

This principle approach could be extended to include data beyond quality and environmental issues depending on the focus of the organization managing the scheme. However, the scheme also has drawbacks. As its focus

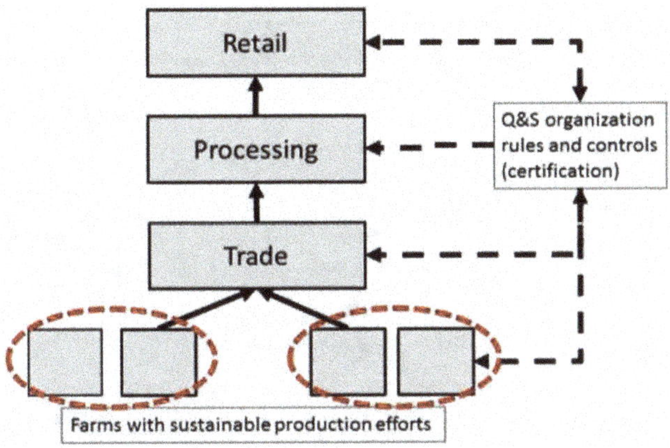

Figure 5 A basic quality coordination scheme. Source: Author.

is on assuring certain levels in quality and environmental protection across a whole sector, improvements in quality and environmental protection and, in turn, in the exchange of data, have to assure that the majority of enterprises are prepared to follow the scheme. The scheme is, therefore, most suitable for situations where a sector-wide basic level of coordination needs to be implemented in a particular area.

5.2 Organizing an information circle

There have been efforts to offset the drawbacks of certification-based systems through coordinated data collection and exchange throughout the food chain, covering forward and backward data exchanges. Examples include attempts to achieve higher levels of quality assurance and environmental protection by developing a flexible internal data communication and exchange system within food chains which could both improve visibility of products in markets and flexible farm production to market developments.

An example has been implemented by a major retailer (https://www.cscp.org/our-work/rewe/). Here final consumer products must fulfill higher levels of quality assurance and environmental protection based on appropriate requirements on all stages of the chain. These higher levels will be communicated to consumers through specific brand names which allow higher prices (Fig. 6). The additional gains are partly channeled back to agriculture which carries the major burden in reaching quality and environmental requirements.

The scheme involves setting a range of levels of environmental performance that are captured in an index which allows some flexibility in production. The final products must reach a certain index level to be able to carry the exclusive

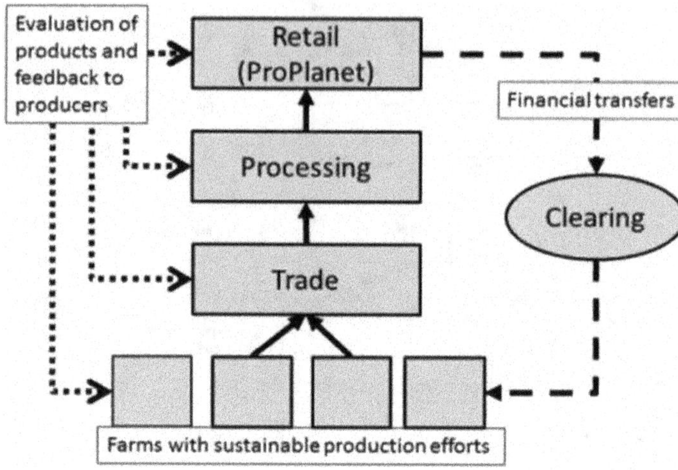

Figure 6 An advanced coordination scheme. Source: Author.

brand name. It is obvious that the formulation of such an index requires intensive data collection and data use at each stage of the food chain. The advantage of an index-driven system is its flexibility in adapting to new requirements and its transparency. It provides

- flexibility in defining an index level that participating enterprises need to reach;
- flexibility in the composition of the index; and
- transparency of the contribution of enterprises to the index along the chain.

However, as in the preceding example, the assurance of quality and environmental protection depends not only on information exchange between enterprises but also on the establishment of a robust monitoring and audit scheme. This highlights data coordination difficulties and still poorly developed schemes for assuring reliability in data collection, processing and communication. New developments in data collection and communication technology including sensors, blockchain technology, smart contracts linked to blockchain technology and others will change the picture in the future and reduce the need for external monitoring and auditing activities.

5.3 Data carriers in logistics

It is most common to communicate data linked to products through data communication technology using product or batch identification (ID) aligning products with data. However, products and packaging are increasingly being used as data carrier themselves. Radio frequency identification (RFID) as integral parts of packaging may carry the information or incorporate links to cloud-based databases. It is now possible to print RFID tags on packages or directly on products (Fathi et al., 2020). The combination of RFID tags with sensor technology opens opportunities for monitoring a product's path through the logistics system and capturing relevant data from its environment such as temperature, climatic conditions or location.

This approach has been implemented in food packaging as an offline and online solution or a combination of both. Online connection of RFID tags allows continuous monitoring of environmental data collected by sensors and, among other things, the continuous tracking of products (Bouzembrak et al., 2019). However, apart from cost issues, it is dependent on the accessibility of a telecommunication network such as the internet. Offline collection of data is independent of telecommunication networks, works everywhere and is (due to its the price advantage) the choice for systems where online monitoring is not relevant.

5.4 Sales and procurement platforms

Sales and procurement business platforms (https://www.bito.com/en-gb/expert-knowledge/article/the-future-of-the-food-trade/) offer an opportunity to establish a data exchange 'on the fly'. By connecting to a platform with an appropriate standardized interface, an enterprise may pick up the relevant product related data in standardized format whenever a product is being bought or sold. The definition of the interface is still primarily based on mutual agreements between enterprises. In data connectivity, it may follow the GS/1 EPCIS standards and, in data content, it may follow one of the quality assurance systems. However, it is frequently based on requests by dominant market participants. Large retail chains may invite procurement offers from smaller providers such as farms and food processors in line with their proprietary platform interface. Similarly, large producers of production inputs may offer proprietary data interfaces on their sales platform. Independent marketing platforms are in development and in experimental use but have not yet reached mainstream business application. For example, see https://www.foodcircle.com/magazine/b2b-procurement-platforms.

6 Specific challenges in data management across food chains

6.1 Dealing with diversity in the sector

The sector is comprised of enterprises with a broad range of diversity in, e.g. size, ability to invest, expertise in data communication, etc. This diversity is hampering the integration of higher levels of data communication and exchange systems. It is in the interest of the sector to reduce variation especially in investment ability and data communication expertise.

One of the options to deal with this problem is networking small- and medium-sized enterprises who, together, could then afford to operate on a level achieved by bigger and more powerful enterprises. Such networks do not need to be organized as formal networks but build on informal cooperation. A specific example has been implemented by the GlobalG.A.P quality management scheme. It allows groups of farms to act as a single unit in GlobalG.A.P.s auditing scheme and distribute the costs among members of the group.

Examples linked to marketing and sales, especially by farms and small food enterprises, include the so-called e-communities and virtual enterprises (Camarinha-Matos et al., 2004). Virtual enterprises build on individual enterprises that might be far away from each other but present themselves jointly to their customers and suppliers. This is supported by market access through online presentations and efficient internal digital communication systems. E-communities are less rigid and are based on digital networking of whatever kind, including the establishment of joint databases or an integrated exchange of data with suppliers and customers.

Alternatives to networking include

- training through industry associations or public institutions that could improve e-competence within the sector; or
- investment support through joint industry efforts or policy engagement that could eliminate investment deficiencies in especially small and medium sized enterprises.

6.2 Distribution of costs and benefits

Major difficulties in reaching agreements on data exchange along food chains are related to the uneven data collection and data use. This uneven situation is due to the fact that, while most information that might be relevant for markets and consumers is to collected by farmers the main beneficiary is the retail sector. It is usually the case that the costs for collecting data about product characteristics occur at stages that are different from the stages were the benefits are realized. Increasing market demands for more environmentally sustainable and high-quality production processes, which might lead to higher returns for retailers, raises the issue of balancing costs and returns along the chain. This, of course, requires transparency on costs and benefits. A typical example 'animal welfare' where a higher level of animal welfare is provided by livestock farmers while higher prices with consumers are realized by retailers. This is the basis for coordination needs which do not just have to agree on data collection and date use but also on agreements regarding the allocation of costs and benefits.

Most of today's discussions on food chain developments focus on agreements regarding data collection and use and omit a discussion of the allocation of costs and benefits. However, the coordination of costs and benefits is of crucial relevance for realizing developments that need to find acceptance with all enterprises along the chain and especially with farms in the long run. Solving this problem is difficult to realize as long as chains do not have full transparency not only on costs and added value of specific market relevant investments but also on profits in different product categories.

6.3 Data ownership

Closely related to the uneven situation in data collection and data use within the food chain is the issue of data ownership. Initially, data are owned by those who collect them and not by those who use them. However, in present systems for data exchange, farmers are usually no longer considered as owners of the data they have collected as soon as they have left their premises. As a consequence, farmers can no longer influence what their data are being used for, a major obstacle for improvements in data communication. Farms are 'asked' to provide data which processing and retail companies use according for their own interest. The transfer of data ownership from farms to other enterprises is, in principle, an act that requires permission in the first place and, if given, should clarify a transfer price.

That is a situation where new data collection and communication approaches have to be identified and evaluated by all concerned. Sensor-based technology might help cutting costs for data collection and transfer. An appropriate allocation of costs, benefits and ownership requires more radical solutions. One of the options discussed in the literature that could help in solving the cost and benefit problem is the establishment of 'data markets'. Farmers would offer data (e.g. through an online platform) and trading partners could make offers for receiving data. This alternative could easily deal with differences in types, quality and volumes of data. Such approaches could lead to differences in product offers in retail shops, products with less and products with more information on issues of interest to consumers and in consequence, with differences in prices. Groups with intensive data collection initiatives in agriculture such as GlobalG.A.P. have long discussed the need for differing products at the point of retail subject to availability of product information. A simplified version of a data market has been established in the 'book and claim' concept (see earlier).

6.4 Data manipulations

The elimination of data manipulation in sourcing or along the chain and the assurance of data reliability have been a central issue in the development of communication systems. Classical approaches involved auditing systems. The sector has developed varieties of auditing schemes, especially in connection with the auditing of quality assurance and environmental management systems. However, auditing systems can only look at samples and do not include continuous monitoring activity. Artificial intelligence may reduce the need for human auditing but cannot replace it completely. However, technology developments have provided means that could reduce the problem. Data sourcing can be improved through emerging *sensor technologies* (Ruiz-Garcia et al., 2009) and data communication through *blockchain concepts* (Galvez et al., 2018).

6.5 E-readiness

In data management, enterprises are confronted with problems that reach beyond data coordination problems. They refer to the differences in enterprises' ability to collect, communicate and use data, known as e-readiness (Bryceson, 2006). While one enterprise might be able to rely on a sophisticated digital data processing and communication ability, another one might still rely on paper-based data management. The level of data communication along the chain is determined by the lowest level of e-readiness in any of the participating enterprises. In the identification of suitable alternatives for communication improvement, one needs to integrate costs and efforts for bringing all enterprises through training and technological investments to the level all members of the chain would like to realize.

6.6 Staying in line with dynamic developments

The food sector is one of the most dynamic sectors in the economy. Changing consumer demands and technology developments require enterprises to continuously monitor developments, to adapt where necessary and, in turn, to adapt the data and communication systems in line with their suppliers and customers. Trends toward vegan food, new product varieties with changes in nutritional characteristics, new plant-based meat alternatives, new regulations on environmental or nutritional transparency, emerging consumer requests on product characteristics, etc. require continuous innovation in the sector's food production and offers. The sector needs to be continuously aware of new requirements and opportunities.

This is why a regular scanning of business and sector activities regarding emerging inventions (sometimes referred to as *environmental scanning*, Aguilar,

1967; Morrison, 1992) is important in assuring that appropriate inventions are not overlooked and that innovation support is focusing toward 'best' development. Environmental scanning is described as an activity dealing with a '... *process that systematically surveys and interprets relevant data to identify external opportunities and threats that could influence future decisions...*' (https://businessmantalk.com/environmental-scanning/).

Environmental scanning is supported by networks where enterprises communicate with each other, with experts or with customers and consumers. This has been acknowledged by policy and business services that support networking, especially among SMEs, through network initiatives and meeting platforms of various kinds, such as the European Enterprise Network (EEN; http://een.ec.europa.eu/, 2020). The EEN had been set up with the explicit mission of helping SMEs to innovate.

These discussions lead to two different and conflicting approaches for adapting data and communication systems to market needs:

- As the development of inventions (e.g. new products) and enablers (e.g. data and communication systems) might not always perfectly match, innovators need to realize a forward-looking approach that avoids dismissing an invention with available enablers (e.g. an established data and communication system) as infeasible (and deciding on a less attractive alternative). At the same time, one might already be able to imagine that supportive enablers (e.g. higher levels of IT technology) could support the realization of a new solution.
- Developments in dynamically changing food environments might be so dynamic that any adoption of inventions (e.g. new product alternatives), even with a forward-looking approach, might be already outdated by the time investments in adoption of enablers have taken place. This view is especially critical in a competitive environment when an adopter might expect a latecomer in innovation to build on newer technology and gain a competitive advantage. It constitutes an innovation barrier that can only be overcome in the long run if enterprises in a sector agree on developments. Networking is a crucial success factor for solving this problem.

Many requirements, e.g. by consumers, on product and production characteristics such as quality cannot be dealt with by decisions of an individual company. It is dependent on its suppliers for providing appropriate information and guarantees and (if not at the end of the chain) its customer on utilizing the information on the level of characteristics reached within the company. If not embedded in an agreement with suppliers and customers, an individual investment would not reach consumers and meet their requirements.

This constitutes a barrier for individual investments and initial initiatives by a first mover as the investment costs could not generate the expected return. In its efforts to innovate, the sector is dependent on coordinated agreements within food chains which, in turn, limits the innovation dynamics of the sector.

7 Conclusion

The food sector and its food chains are continuously challenged to improve in meeting changing consumer needs relating to products, processes and transparency while following ever increasing sustainability and environmental expectations of society. Keeping in line with such developments raises challenges for the dynamic adaptation of data and communication systems within enterprises, along food chains, and toward consumers and the general public.

These challenges cannot be solved by individual enterprises due to the interdependencies within the food chains. Looking at both ends of food chains, data gathered by farms might be relevant for informing consumers while retail or consumer data might be relevant for supporting farms in adapting levels of production or production processes.

Due to the variability in production yields, at one end, and changes in consumer demand, at the other end of the food chain, coordination of data and communication systems along the chain cannot be limited to companies that are constituting a food chain at a certain point in time but needs to incorporate enterprises that are potential suppliers and customers in dynamically changing food chain relationships. This is the challenge which sets food chain coordination apart from coordination efforts in non-food industries and has initiated various models for coordination and coordination management within the sector through external partners or governmental initiatives. This challenge is further aggravated by the global sourcing of food chains which have to cope with global differences in the suitability of climatic conditions for varieties of products with specific climatic needs.

8 Where to look for further information

Good sources for further information are the European IT projects dealing with farms and the food sector. Examples include "Internet of Food and Farm 2020" (https://cordis.europa.eu/project/id/731884) and "Farmtopia" (https://farmtopia.eu).

9 References

Aguilar, F. (1967). *Scanning the Business Environment*. New York: Macmillan.

Bouzembrak, Y., Klüche, M., Gavai, A. and Marvin, H. J. P. (2019). Internet of things in food safety: literature review and a bibliometric analysis. *Trends in Food Science and Technology* 94, 54–64. https://doi.org/10.1016/j.tifs.2019.11.002.

Bryceson, K. P. (2006). *E Issues in Agribusiness, the What, Why, How*. Oxford: Cabi Publishers. https://doi.org/10.1079/9781845930714.0000.

Bui, T. X., Sankaran, S. and Sebastian, I. M. (2003). A framework for measuring national e-readiness. *International Journal of Electronic Business* 1(1), 3–22.

Camarinha-Matos, L. M., Afsarmanesh, H. and Ollus, M. (2004). Virtual Organizations: Systems and Practices. Springer, New York.

Chaddad, F. and Rodriguez-Alcala, M. E. (2010). Inter-organizational relationships in Agri-food systems: a transaction cost economics approach. In: Fischer, Ch. and Hartmann, M. (Eds) *Agri-Food Chain Relationships*. Oxfordshire: CAB, pp. 45–60.

Fathi, P., Karmakar, N. C., Bhattacharya, M. and Bhattacharya, S. (2020). Potential chipless RFID sensors for food packaging applications: a review. *IEEE Sensors Journal* 20(17), 9618–9636. https://doi.org/10.1109/JSEN.2020.2991751.

Fritz, M. and Schiefer, G. (2009). Tracking, tracing, and business process interests in food commodities: a multi-level decision complexity. *International Journal of Production Economics* 117(2), 317–329.

Galvez, J. F., Mejuto, J. C. and Simal-Gandara, J. (2018). Future challenges on the use of blockchain for food traceability analysis. *TrAC Trends in Analytical Chemistry* 107, 222–232.

Kalinowska, K., Wojnowski, W. and Tobiszewski, M. (2021). Smartphones as tools for equitable food quality assessment. *Trends in Food Science and Technology* 111, 271–279.

Mol, A. P. J. and Osterveer, P. J. M. (2015). Certification of markets, markets of certificates: tracing sustainability in global agro-food value chains. *Sustainability* 7(9), 12258–12278.

Moreno-Miranda, C. and Dries, L. (2022). Integrating coordination mechanisms in the sustainability assessment of agri-food chains: from a structured literature review to a comprehensive framework. *Ecological Economics* 192, 107265. https://doi.org/10.1016/j.ecolecon.2021.107265.

Morrison, J. L. (1992). Environmental scanning. In: Whitely, M. A., Porter, J. D. and Fenske, R. H. (Eds) *A Primer for New Institutional Researchers*. Tallahassee, FL: The Association for Institutional Research, pp. 86–99.

Rahul, K., Banyal, R. K., Kumar, V. and Hooda, D. (2022). Cloud computing: technological innovations in the food industry. In: Mor, R. S., Kamble, S. S. and Sangwan, K. S. (Eds) *Lecture Notes in Management and Industrial Engineering*. Springer, New York, pp. 127–141.

Rahul, K., Mor, R. S. and Malik, M. (2021). Sustainable innovations in the food industry through artificial intelligence and big data analytics. *Logistics* 5(4), 66–81.

Ruiz-Garcia, L., Lunadei, L., Barreiro, P. and Robla, J. I. (2009). A review of wireless sensor technologies and applications in agriculture and food industry: state of the art and current trends. *Sensors* 9(6), 4728–4750. https://doi.org/10.3390/s90604728.

Weston, M., Geng, S. and Chandraeati, R. (2021). Food sensors: challenges and opportunities. *Advanced Materials Technologies* 6(5), 2001242.

Part 3

Performance improvement in agri-food supply chains: Case studies

Chapter 10

Improving agri-food supply chains in North America: the case of the sugarcane agro-industry in Mexico

José Luis Solleiro and Rosario Castañón, National University of Mexico (UNAM), Mexico

1 Introduction

Mexico's agricultural modernization began in the 1940s, instigated by the Mexican government's goal of driving a broader transition of the nation's economy. This involved agriculture moving beyond subsistence small-holder farming to large-scale, commercial enterprises using modern plant breeding and technology. The government also intended to produce crops that could be exported to the rest of the world, in order to generate export income and bring in foreign currency. This model led to an agricultural development characterized by the following (Mandel, 1987):

http://dx.doi.org/10.19103/AS.2023.0122.14

- The increasing conversion of agricultural production into an industrial process (e.g. characterized by selecting high-yielding varieties, developing monocultural cultivation, a greater reliance on synthetic inputs and mechanization);
- Increasing intensification of capital investment to create large-scale, capital-intensive farming enterprises;
- Mechanization of all agricultural tasks;
- Intensive use of science and technology;
- Accelerated division of labour; and
- Increasing productivity.

In Mexico, smallholders have not been an obstacle to this process of development since a large number of smallholdings are leased, giving landowners the power to consolidate landholdings and drive through a program of modernization (González-Estrada, 2009).

The last two decades of the twentieth century saw considerable changes in Mexican economic policy, giving priority to open markets and integration into global value chains. Despite the predominance of smallholder farms, Mexico has become one of the major food-exporting countries in the world. By the beginning of 2010, Mexico was the eighth largest producer of agricultural produce and the main exporter of food to the USA. It had a network of 10 free-trade agreements with 45 countries. According to SAGARPA (2012), in 2012 global agri-food trade in Mexican produce reached U$50.579 billion, 2.8 times higher than in 2000. Exports totalled U$22.805 billion and imports U$27.774 billion. Mexico was the fourth and fifth largest producer of chicken meat and eggs, respectively. It has become the eighth largest producer of beef and annually exports more than one million calves to the USA. It is the top exporter of beer to the USA and third for orange juice. It is one of the three largest exporters of avocados, onions, raspberries, asparagus, cucumbers, tomatoes, pumpkins, chilli peppers and honey. This demonstrates the achievements of the new model, but many socio-economic problems remain as a consequence of the heterogeneity of this sector.

At the beginning of this decade, Mexican agriculture faced fresh challenges arising from new obligations under the USA-Mexico-Canada Free-Trade Agreement (USMCA) and the coronavirus disease 2019 (COVID-19) pandemic. Some of these are discussed in the chapter using sugarcane as an example. In response to the pandemic, a project was conducted in 2020 and 2021 to developing an Innovation Agenda targeting the sugarcane industry in Mexico. The aim was to create efficient, inclusive and resilient agri-food systems to achieve objectives such as food security and sovereignty. The Agenda was developed through broad consultation with the main actors in the value chain, including producers, processors, business organizations and government authorities .

The research began with reports, academic articles and interviews with the leaders of the main trade organizations representing producers and processors. Subsequently, consultation workshops were organized in three production regions:

- A high production region (Veracruz as the leading producer);
- An area of medium production volume (San Luis Potosí); and
- An area of low production (Oaxaca).

These three regions represent 50% of national sugarcane production. Three workshops were held in each region, focusing on three main topics:

- Sustainability, productivity, technological development and innovation;
- Regulatory, social and labour issues; and
- The market and value chain.

The workshops were held between August and November 2020 and were attended by 560 stakeholders in the sugarcane agro-industrial chain. All the workshops were attended by experts who introduced the main topics of discussion, both to share up-to-date information with the participants and to help guide an informed debate. The study results were presented in 2021 to stakeholders as a foundation for the Innovation Agenda. This chapter summarizes problems in the sector and considers customized solutions for the states of Oaxaca, San Luis Potosí and Veracruz, according to the vision articulated by the different actors in the value chain.

2 Sugarcane production in Mexico

In Mexico, sugarcane occupies an important position in relation to other crops in terms of area planted and volume of production (Figs 1 and 2). Indeed, sugarcane can be considered Mexico's most important agro-industrial crop. About 70% of sugarcane production is still from small- and medium-sized producers (Sader, 2020). The sugarcane value chain ranges from suppliers of input, e.g. machinery, equipment and professional services for its cultivation, to sugar production in mills and its subsequent processing by industries such as the food and beverages sector. It has been estimated that the sugarcane sector supports around 500 000 families. The sector supports more than 180 000 sugarcane producers, 155 000 day labourers, 70 000 cutters and 32 000 involved in transportation. The sector employs 35 000 unionized workers, 7689 trust workers and 11 822 retirees. It is estimated that the sugarcane value chain generates 500 000 direct jobs and 2.4 million indirect jobs (SADER, 2020).

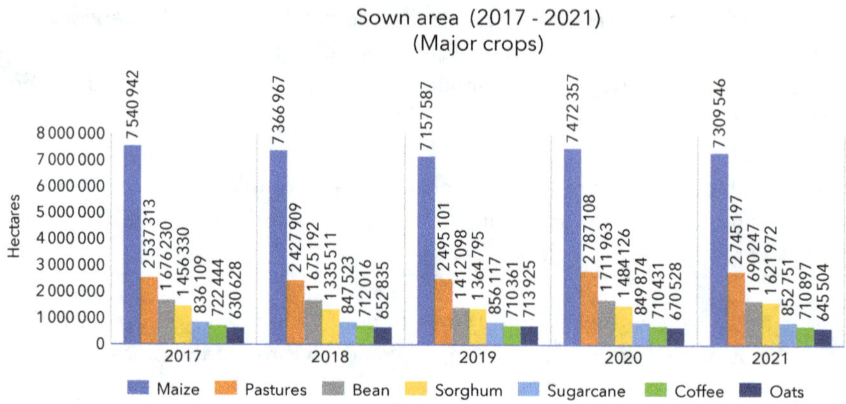

Figure 1 Five crops with the largest planted area in Mexico (2017–21). Source: Elaboration of data from Sistema de Información Agroalimentaria de Consulta (SIACON). https://www.gob.mx/siap/documentos/siacon-ng-161430.

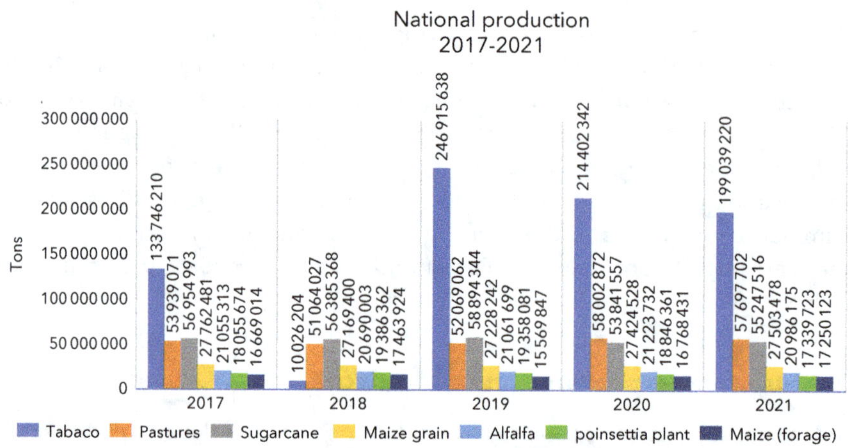

Figure 2 Five of the largest production crops in Mexico (2017–21). Source: Elaboration of data from Sistema de Información Agroalimentaria de Consulta (SIACON). https://www.gob.mx/siap/documentos/siacon-ng-161430.

Over the last 5 years, sugarcane production has shown sustained growth in the top-producing states. For the 2019–20 harvest, the nationwide cultivated area was approximately 912 968 ha (García, 2020):

- 62% corresponds to rainfed areas;
- 18% corresponds to irrigated areas;
- 13% corresponds to partially irrigated areas; and
- 7% corresponds to sowing and sugarcane under development.

Between 2015 and 2019, Veracruz was the leading producer with the largest cultivated area, followed by Jalisco, San Luis Potosí, Oaxaca and Tamaulipas. In terms of crop field yields for the 2018–19 harvest, the leading states were Puebla, Morelos, Jalisco, Chiapas and Michoacán. However, yields across farms vary due to the interaction of multiple variables that can be grouped into three types (SAGARPA-Conadesuca, 2015):

- Human and economic;
- Environmental; and
- Agronomic.

Morelos, Puebla, Nayarit, San Luis Potosí and Jalisco reported the highest yields for the 2018–19 harvest.

In Mexico, sugarcane is mainly used to produce sugar by extracting juice from its stalks (SIAP, 2018). However, it is also used as fertilizer, animal feed or to make paper. To a lesser extent, the juice is also used to produce industrial alcohol and alcoholic beverages.

There are significant differences in the milling capacity of the country's sugar mills:

- 4 have a capacity greater than 10 000 tons;
- 22 have a capacity of between 6000 tons and 10 000 tons; and
- 24 have a capacity of fewer than 6000 tons.

The 50 mills are located in 15 states: Campeche, Chiapas, Colima, Jalisco, Michoacán, Morelos, Nayarit, Oaxaca, Puebla, Quintana Roo, San Luis Potosí, Sinaloa, Tabasco, Tamaulipas and Veracruz. The state with the largest number of mills is Veracruz with 18, followed by Jalisco with 6 and San Luis Potosí with 4.

Mexico has the capacity to both supply domestic demand and generate surplus that has historically been exported to North America, mainly to the USA. While sugar production has grown at an average annual rate of 1.4% between the 2007/08 and 2018/19 cycles, domestic consumption in the same period reduced at an average annual rate of -0.6%. According to García (2020), the reduction in sugar consumption is primarily due to:

- impact of the new labelling on food and beverage products;
- changes in the consumption habits of the population in favour of healthier products; and
- substitution of caloric sweeteners for non-caloric sweeteners.

This has put the focus on increasing exports. However, this alternative is less profitable because prices in the global market are low (García, 2020). It is

Table 1 Factors limiting the competitiveness of the sugarcane industry

Crop field related	Processing related
Low use of biological controls Low use of soil improvers Low degree of crop renewal Low rate of strain substitution Low yields in the field Soil degradation Varietal imbalance Sacre national research for technology development Insufficient resources for the science and technology systems of the sugarcane sector Scarce promotion of public bodies to develop national technology Excessive use of agrochemicals Excessive extraction of nutrients from the soil Chemical fertilization without considering the specific needs of the crop, cycle and climate Negative environmental impact of cane burning Limited economies of scale Monocropping Technological lag in transport equipment Lengthy duration of moving sugar cane	Heterogeneous raw material (content of sucrose and impurities) Dependence on foreign technology for the manufacture of process equipment Incipient linkage of sugar mills with universities and research centres High factory production costs due to low energy efficiency (equipment operates at pressures and temperatures outside the ideal range) Low financing for modernization and diversification Limited use of new technologies to mitigate the environmental impact on water, soil and air caused by agrochemical emissions or spills Economies of scale cannot be achieved in several mills Low product storage capacity Technological lag in milling, condensing, casting and crystallizing equipment, as well as in steam plants Little encouragement to boost the process equipment industry Insufficient research to develop equipment and processes

Source: Elaboration based on Conadesuca (2018).

evident that the competitiveness of the sugarcane industry is affected by numerous factors that influence different stages, from cultivation to processing (shown in Table 1).

The current situation faced by the sugarcane agro-industry derives from a complex historical context in which the role of the state has been significant. Table 2 shows some of the most important elements that have influenced its development.

3 Factors determining the competitiveness of the sugarcane agro-industry

Intensification of international competition through globalization requires new approaches to understand and promote competitiveness. Adopting a practical approach (Müller, 1992), *Canada's Task Force on Competitiveness in the Agri-food Sector* defined competitiveness is the sustained capacity to

Table 2 Key historical elements to understand the current state of the Mexican sugarcane agro-industry

Year/period	Element	Implications for the sugarcane industry
1522		Introduction of sugarcane to Mexico
Early twentieth century	Organization of production in systems known as 'Haciendas'	Large estates where production was concentrated to a single owner who owned the land and the mills. Decisions on production were centralized
1910–17	Mexican Revolution	Abandonment of land and destruction of infrastructure (sugar mills) due to the war. Reduction in sugarcane and sugar production
Years after 1920	Agrarian reform	Disappearance of large estates, distribution of land. Small plots for sugarcane (a phenomenon that persists to this day[1]). Disorganized link between sugarcane production and sugar production
1932	The company Azúcar S.A. is created, integrating sugar-producing companies from several Mexican states	The first effort to associate sugar producers (after the Mexican Revolution) as a measure to face the economic, financial and market crisis in the industry
1938	Azúcar S.A. is restructured to form Unión Nacional de Productores de Azúcar, S.A. (UNPASA)	During the creation of UNPASA, the state intervened to try to organize producers and deal with the problem of sugar surpluses and prices
1943	The government creates the Financiera Nacional Azucarera (National Sugar Financing Agency)	State intervention to finance the industry
1944	First Sugarcane Decree	Established a mechanism to organize cane producers and processors[2]. Forced producers in the sugarcane zones to produce sugarcane and the mills to buy it from them
1948	Creation of the institute for the improvement of sugar production	Provide the sector with more productive varieties (closed in 1991)
1951	Decree to fix sugar prices	Price control by the state for the different forms of sugar
1953	Sugarcane Decree	Increases the percentage to the farmer in the net value of the final honey product (50%)
1970	The National Commission of the Sugar Industry is created (closed in 1983)	New government mechanism to restructure the industry, manage state mills and try to improve the competitiveness of the industry. Price control and greater state control of the industry

(*Continued*)

Table 2 (*Continued*)

Year/period	Element	Implications for the sugarcane industry
1971	The National Sugar Mill Operator is created (liquidated in 1979)	Manage parastatal mills (49 of 66 existing)
1975	Cane Decree	The sowing, cultivation, harvesting and industrialization of sugarcane is declared of public interest. The price of cane is regulated in relation to its sucrose content
1991	Public research centres focusing on sugarcane were closed	Research programmes involving agri-business are slowed down, which causes a lag in the introduction of improved varieties
1994	Signing of the North America Free-Trade Agreement (NAFTA)	Gradual deregulation of the market. Elimination of export tariffs. Impact of the world market on the national industry. Adjustments in the quality of export products
2000	Substitution of sugar by other products, particularly high fructose corn syrup	Decrease in sugar consumption in various industries such as soft drinks. Increase in sugar surpluses
2001	Expropriation of 27 mills	Involvement of the state to rescue the processing industry again
2005	Extinction of sugarcane decrees	The price of sugar is set by supply and demand. The diversification of products derived from cane (sugar and fuel) begins
2010	The Sugarcane Sustainable Development Commission (CONADESUCA) is created	One more effort to co-ordinate activities involving the sugarcane agro-industry. Information systems were created with the purpose of supporting public policies that concern the sector
2013	Special Law on Production and Services	A special tax on sugar-sweetened beverages was implemented. With this measure, the food and beverage industry began searching for sugar substitutes
2020	Frontal warning labelling	Consumers are informed of high content of calories and sodium so they can decide to choose alternative products
2020	National Sugarcane Agro-industry Programme 2021–24	Cane producers with a surface area of up to 4 hectares are given direct support of 7300 pesos, regardless of the harvested area. It is likely that the producer does not use that resource for activities that the incentive is intended for

Source: Elaboration based on: (i) Aguilar et al. (2011). Competitividad de la Agroindustria Azucarera de la Huasteca México. Doctoral Thesis Ciencias Ambientales. Universidad Autónoma de San Luis Potosí, México; (ii) Diario Oficial de la Federación (1975). Decreto por el que se reforman diversos artículos del Decreto que crea el Organismo Descentralizado denominado Comisión Nacional de la Industria

Azucarera. 27 de octubre de 1975. https://dof.gob.mx/nota_detalle.php?codigo=4822416&fecha=27/10/1975#gsc.tab=0. Downloaded on 4 March 2023; (iii) Perdomo, J. (2010). Punto de acuerdo para la Comparecencia del Secretario de la SAGARPA en relación a la expropiación de 27 ingenios azucareros. Gaceta del Senado. https://www.senado.gob.mx/65/gaceta_del_senado/documento/24270. Downloaded on 4 March 2023; (iv) Pérez, P. (2022). La agroindustria cañera en México y los factores que han determinado su historia reciente (1961-2021). Revista Inclusiones 9, 41-71; (v) Secretaría de Agricultura y Desarrollo Rural - Comité Nacional para el Desarrollo Sustentable de la Caña de Azúcar (sf). Programa Nacional de la Agroindustria de la Caña de Azúcar 2021-24. https://www.gob.mx/cms/uploads/attachment/file/709503/PROGRAMA_PRONAC_2021-2024vf_web.pdf. Downloaded on 4 March 2023; (vi) Zuale, L.E. (2022). Impacto de políticas públicas en el desarrollo y evolución del campesinado cañero. Masters thesis. Facultad de Ciencias Biológicas y Agropecuarias. Universidad Veracruzana. https://cdigital.uv.mx/bitstream/handle/1944/52277/ZualeJuarezLuz.pdf?sequence=1. Downloaded on 2 March 2023.[1]This situation hinders the modernization of the field in its different aspects by not achieving economies of scale.
[2]Sugarcane producers had a 25% share of the net value of the final molasses.

gain and maintain a good market share. It is clear from this definition that competitiveness must therefore be understood as a relationship between business organizations and markets, where the various expressions of the power structures of both governments and interest groups, which determine the context in which companies compete, play a determining role.

In the specific case of the sugarcane agro-industry, according to Aguilar et al. (2011) competitiveness is associated with

- the incorporation of new mill processing methods to produce standard, white or refined quality sugar;
- significantly increased sugarcane and sucrose yields;
- identifying areas to reduce costs on farms, in processing and distribution; and
- diversification so as not to depend on a single market niche.

In contrast, the National Sugarcane Agro-industry Program (PRONAC is the Spanish acronym) during 2014–18 (Sagarpa-Conadesuca, n.d.) identified productivity as a key element for competitiveness, involving technological innovation and technology transfer in areas such as milling. According to Santoyo (2020), the critical factors for increasing mill productivity are

- energy efficiency (insufficient energy production and very high consumption make energy self-sufficiency and co-generation difficult);
- use of wastes [e.g. cachaza (a distilled spirit made from fermented sugarcane juice), use of wastewater and stillage];
- time lost (e.g. due to fluctuations in supply from farms);
- determining actual efficiency (above 82.37%); and
- ethanol production (to produce a biofuel co-product).

Given decreasing global demand for sugar resulting from changes in consumption to reduce obesity and diseases, such as diabetes, diversification has been identified as another element of competitiveness. As an example, farms could move into production and marketing e.g. of panela (unrefined whole cane sugar used e.g. in baking or manufacture of confectionery), aguardientes (schnapps), trapiche honey, edible mushrooms and livestock feed, as well as supplying the mill with agricultural residue from the harvest. For their part, mills could move into production of sucrose, ethanol and electric energy, as well as returning by-products such as ashes and cachaza to crop fields for use as biofertilizers. However, global warming, caused by climate change, is a major threat to competitiveness, reducing water availability which particularly affects rainfed crop yields. Climate change also affects proliferation of pests (insects, fungi, viruses and bacteria) affecting crop production.

Recent policies and regulations that affect competitiveness include reform of the official Mexican standard NOM-051 on food labelling, the provisions of the USMCA Treaty between Mexico, the USA and Canada regarding child and forced labour, and sanitary measures to combat the COVID-19 pandemic. They have introduced new restrictions, new technological and operational challenges, as well as increased costs for the industry.

4 Technological issues related to sugarcane agri-business development

Different organizations involved in the sugarcane sector (Osoyo et al., 2017)[1], as well as the participants in the various study workshops (Flores, 2020; Santoyo, 2020; Aguilar, 2020), have highlighted the main technological challenges facing the industry (Table 3).

Incorporating technologies addressing the eight areas identified in Table 3 will help achieve a more competitive industry by reducing costs, increasing productivity and protecting non-renewable natural resources. The Innovation Agenda resulting from the workshops includes specific actions. Examples include developing new varieties

- that are disease, pest and drought resistant;
- with a higher sucrose content;
- that are early maturing or late maturing to extend the harvest season;
- that are suitable for different agro-ecological regions; and
- that can adapt to climate change.

1 Comité Nacional para el Desarrollo Sustentable de la Caña de Azúcar-CONADESUCA (National Committee for the Sustainable Development of the Sugar Cane Industry); Unión Nacional de productores de Caña-UNPC (National Union of Sugarcane Producers); Coordinación de Fundaciones Produce-CUFUPRO (Coordination of Produce Foundations); Instituto Nacional de Investigaciones Forestales Agrícolas y Pecuarias-INIFAP (National Institute of Forestry, Agricultural and Livestock Research); Universidad Autónoma Chapingo-UACH (Chapingo Autonomous University); and Colegio de Postgraduados-COLPOS (postgraduate college).

Table 3 Main technological challenges facing the sugarcane agro-industry

Area	Challenges identified at the roundtables
Crop varieties	• Using a small number of varieties increases susceptibility to pests and diseases • Old varieties lead to meagre yields and do not meet the current needs of the industry • Vegetative material not coming from seedbeds • Drought-resistant varieties with higher sucrose, and pest and disease resistance are required • Some varieties that are incorporated into the crop field are not sufficiently validated
Diseases and pests	• The climatic conditions in which sugarcane grows are conducive to multiple pests and diseases • The indiscriminate use of chemical products to combat pests and diseases
Soil nutrition and management	• High consumption of chemical fertilizers impacts costs and soil erosion • Inadequate fertilizer application methods • Low use of composts and biofertilizers • Soils with salinity, sodicity, acidity and compaction problems • Inadequate use of machinery that damages soils
Irrigation	• Most sugarcane-growing areas are rainfed, which makes them dependent on weather conditions • The crop is subjected to water stress • Most of the plots use gravity irrigation, which is a highly inefficient and unsustainable system • Shallow soils with low water retention capacity • Soil moisture is not retained
Harvest and post-harvest	• Poor cut scheduling • Loss of cane during transport • Loss of cane freshness due to delays in sugarcane entry and processing at the mills • Burning the residues generated during harvesting pollutes the soil and air • Low levels of mechanization and lack of capacity to design equipment adapted to sugarcane field conditions
Diversification	• It is undesirable to place the sugar production surplus in foreign markets because of the low price • After sugar, ethanol is the other best-known product derived from sugarcane. However, the variety of products that can be obtained from sugarcane is vast, including those from waste (e.g. fodder and compost)
Production process	• Low level of automation • Low energy efficiency (low-pressure boilers, steam leaks, fossil fuel consumption) • Highly polluting effluents (high COD and BOD values)
Characterization of the plots	• The lack of information on the plots leads to general solutions with particular characteristics due to soil and climatic conditions

Source: Elaboration.

A related issue is to establish synergies with experimental crop fields, seedling production centres and certified seedbeds to fully evaluate varieties before marketing, in order to ensure the availability of quality genetic material. Another example is increasing mill productivity through better planning of cutting of sugar.

In the area of disease and pest control (mainly yellow aphids, rodents, screwworm and pin flies), integrated pest management is recommended to allow more sustainable and eco-friendly methods of control, including production and use of biological pest and disease control agents (e.g. Trichogramma, Beauveria bassiana, Metarhizium, Tetrastichus and Lixophaga) that can effectively target pests in different regions.

In terms of nutrition and soil management, priority needs to be given to use of biofertilizers such as humus processed by earthworms, promotion of mycorrhizae and other biostimulant microbial consortia, the incorporation of harvest residues into the soil and production of compost from the sizeable organic waste generated by the sugarcane industry. In addition, chemical fertilizers should be applied in precise doses using precision agriculture techniques to target the right place, time and dose. More targeted application will reduce input costs and potential soil degradation. To improve soil, physico-chemical studies are required to identify nutrient composition and needs; as well as inputs and techniques to conserve moisture and prevent erosion. Regarding irrigation, precision systems are needed for more efficient use of water resources. In the case of harvesting and post-harvesting operations, mechanization should be encouraged where soil conditions permit, as well as the use of remote-control systems, software for production control and more efficient equipment suitable for soils.

A wide variety of products can be obtained from sugarcane, which are marketable if quality and safety requirements are met. Diversification includes production of ethanol and biodiesel, as well as potential supply to the chemical industry where sucrose and lignin, present in sugarcane, can be used to create other products such as organic acids and other high-value compounds. The paper industry is another niche, where bagasse can be used more intensively as a raw material. Finally, the importance of efficient equipment to extract the maximum amount of juice and thus increase sugar production and yield, as well as the efficient use of water through wastewater treatment for recycling, is emphasized.

In relation to production, the key is to achieve energy efficiency using highly efficient equipment to reduce energy consumption and co-generating energy using organic material from sugarcane, which can be a source of income if the surplus energy is marketed. Finally, for plot characterization, satellite images (hyperspectral) are recommended to provide detailed and specific information on sugarcane growing areas at plot level to meet the particular needs of each plot. These technologies allow precision agricultural agronomic management according to the potential of each area which can vary within several metres.

However, the fact that the average size of sugarcane smallholder farms is less than 5 hectares complicates technological development such as use of harvesting equipment or installation of irrigation systems, particularly on cost grounds. The Agenda therefore encourages partnerships at different levels:

- Between producers to make shared purchases and achieve economies of scale by working together; and
- Between the public, private, academic and research sectors, for financing, generation and transfer of technological packages, establishing demonstration plots and ensuring the incorporation of technological developments in all links of the production chain.

It is also recognized that a specific innovation policy is needed for the sector. Under the current conditions of withdrawal of government support for scientific and technological activities, such a policy will need to be based on the co-ordinated efforts of the various private actors in the value chain.

5 The impact of the coronavirus disease 2019 pandemic on sugarcane agri-business development

A recent problem that the sugarcane sector has had to address is the COVID-19 pandemic, which has had negative health and economic effects worldwide. The sugarcane sector (from the crop field to the factory) has work practices that must be changed to minimize, as much as possible, the spread of COVID-19, especially during the sugarcane harvest season. Different strategies to prevent the spread of infection are required for the main work environments along the value chain:

- Crop fields;
- Factories and processing plants; and
- Distribution and supporting service providers.

Of these, the field is the riskiest environment, particularly for cutting tasks. The elements that make the cutting activity dangerous in the context of COVID-19 are shown in Fig. 3.

If preventive actions are not taken, the number of infections will probably increase, with a consequent impact on sugarcane cutting, delivery and sugar production. For this reason, Conadesuca, in collaboration with other organizations

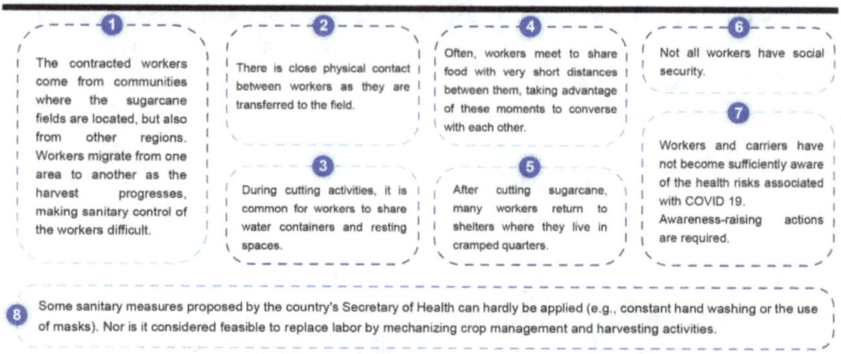

Figure 3 Elements that make the cutting activity dangerous in the context of COVID-19. Source: Elaboration.

such as World Vision, has established a strategy for COVID prevention. This strategy recommends, in addition to the usual sanitary measures applicable worldwide, provisions on *ad hoc* social distancing and promoting awareness among workers not only on procedural aspects in the workplace but also on the precautions they should take in their daily lives, since it is likely that workers will be the main source of infection being brought onto farms.

In addition to the above, Agenda recommendations include calculating the investment needed to carry out risk assessment and protection of workers' health and, based on this, making decisions on how to prioritize and finance particular actions. Other recommendations include forming a Health Committee, implementing Conadesuca guidelines (2020) for responsible recruitment and treatment of worker, and establishing isolation zones for workers who have been infected and paying part of the salary for workers who become ill so that their families are not left unprotected. In training and communication, strategies need to include cutters' families and should be designed with the right target audience in mind (mostly people with low socio-economic and educational levels, and who may come from indigenous groups with a language other than Spanish and with different cultural traditions. These precautions are a sensible investment on financial grounds, to avoid infringing health and safety regulations, to reduce absenteeism and staff turnover that can jeopardize production, and to attract more favorable investment from those looking to reduce investment risks.

6 Regulatory issues related to sugarcane agri-business development

From a regulatory point of view, the sugarcane agri-business has special status due to the legal instrument, the *Law for the Sustainable Development of*

Sugarcane. The purpose of this Law, according to Article 2, is: (...) to regulate the activities associated with contract farming and the sustainable integration of sugarcane, of the processes of planting, cultivation, harvesting, industrialization, and commercialization of sugarcane, its products, subproducts, co-products and derivatives (Cámara de Diputados del H. Congreso de la Unión, 2005).

The law is a legal instrument that covers aspects such as the establishment of contractual conditions between sugarcane suppliers and mills, prices and dispute resolution mechanisms. It also defines the responsibilities of the Secretariat of Agriculture and Rural Development (Sader), including those related to the public policy framework. The National Sugarcane Agro-industry Program (PRONAC) 2021–2024 was published on 29 December 2021, and covered recent developments such as the provisions of the USMCA regarding child labour and forced labour. However, it did not address the possible consequences of reforming NOM-051 for the labelling of processed foods and special taxes for sugar-sweetened beverages. It is a matter of urgency that the sugarcane agro-industry takes a position on these public policy issues.

The reform of NOM-051 was prompted by health problems in the Mexican population associated with overweight and obesity. This led the National Institute of Public Health to develop guidance on choosing healthy foods through a Food and Beverage Front Labelling System (Sefab). A particular objective was to identify products containing sugars, fats, sodium and other components in amounts that may be harmful to health. Identification was derived from the Frontal Warning Labelling standard, published in the Official Journal of the Federation on 27 March 2020, and which came into force in October 2020 (Vázquez, 2020). The standard requires that, if a product has excess amounts of critical compounds, it must use the labels: EXCESS CALORIES, EXCESS SODIUM, EXCESS TRANS FATS, EXCESS SUGARS, EXCESS SATURATED FATS.

The main effect of the modifications to NOM-051 for the sugarcane agro-industry has been to reduce demand for products containing excess sugars, regardless of their source. In 2016, the World Health Organization (WHO) recommended the 'implementation of an effective tax on sugar-sweetened beverages' as a key measure to decrease childhood obesity. The recommendation was adopted by several countries, which created special tax charges for these products. In Mexico, the Special Tax on Production and Services (IEPS) on sugar-sweetened beverages introduced in 2014 was intended to reduce the consumption of soft drinks and other sugar-sweetened beverages to improve the health and quality of life of the Mexican population. It was expected that, on the one hand, the government would increase revenue and, on the other, discourage the excessive consumption of these products, thus addressing the problem of overweight and obesity.

PRONAC 2021–2024 recognizes that 'in Mexico, the trend observed in per capita sugar consumption is downward, which means that, even though

NOM-051
→ Training and communication programme aimed at the end consumer → Physical activity campaign for the population and nutritional orientation to integrate healthier diets → Establish a differentiation in the type of sweetener used in foods and beverages
Tax on beverages containing sugar
→ Development of a proposal for a differentiated tax based on the percentage of sugar contained in the products → Product reformulation programme to reduce sugar content
Impact monitoring
→ Rigorous studies on the effectiveness of regulations, training and communication programmes, as well as physical activation campaigns
Active participation in the development of regulations
→ Information to legislators and authorities → Evaluation of international experiences to reform regulations → Co-ordination with Conadesuca and State Health Secretariats

Figure 4 Key elements concerning regulatory issues. Source: Elaboration.

the population continues to grow, total sugar consumption in the domestic market will not grow and is even likely to decrease' and that 'campaigns against sugar consumption are having an effect on consumers'. Due to the reduction in consumption, PRONAC 2021–2024 only suggests 'the use of surplus sugar (cane) to obtain ethanol to be used as an oxygenate in gasoline throughout the national territory. Sugarcane can also be used to produce various products other than sugar that could have a market' (DOF, 29 December 2021). Given declines in sales, it is essential to take up the recommendations from the regulatory workshops in their entirety (Fig. 4). The sugar industry also needs to develop communication strategies and channels to reach relevant audiences, including involving nutrition specialists to create technical information in a language that is appropriate and understandable to the general public.

7 Labour issues relating to sugarcane agri-business development

Various international organizations have historically singled out Mexico for failing to promote genuine freedom of association, for low wages and poor treatment of labour. This has led to inclusion of labour issues in various treaties involving include Mexico. The North American Free-Trade Agreement (NAFTA) agreement (reinforced by the Trans-Pacific Partnership Agreement), e.g. includes the requirement to ratify the core requirements of the International Labour Organization (ILO) (MLGG Abogados, n.d.):

- That each country be obliged to comply with and enforce its legal framework;
- That this legal framework should uphold the 11 core ILO labour principles in the three countries; and

- A mechanism for dispute resolution and arbitration panel.

The Treaty between Mexico, the USA and Canada (UMSCA) strengthens the protection of labour rights and expressly incorporates the 1998 ILO Declaration on Fundamental Principles and Rights at Work (Articles 23.3 (Labor Rights) and 23.6 (Forced or Compulsory Labour)). According to Roel (2020), the commitments of the parties to the UMSCA include the elimination of all forms of child labour and forced labour. These two issues are of particular importance for the sugarcane sector since sugarcane is included in the list of goods that have been produced using child or forced labour prepared by the Bureau of International Labor Affairs, which means that it is under observation.

An additional element to consider is that, for the first time, a profile of Mexico was included in 2020 in the Worst Forms of Forced Labor Report prepared by the U.S. Department of Labor. A common problem in the sector is the number of young minors who have families to support. Since they have no alternative sources of income, they feel the need to work in the sugarcane sector despite their age. A problem identified by a workshop on this topic is labour shortages in sugarcane fields which can make it harder to avoid using under-age labour.

The World Vision organization has worked on the development of guidelines to address labour problems in sugarcane fields. The key issues to take into account are (Castillo, 2020):

- COVID-19 approach strategy in the sugarcane sector;
- Definitions of child labour, trafficking, labour exploitation and forced labour. Actions considered include not hiring children and adolescents, encouraging their enrolment into school and providing childcare services to children of workers;
- Guidelines for responsible recruitment to prevent child labour and forced labour; and
- Occupational Safety and Health (OSH) guidelines to improve the working conditions of agricultural day labourers.

PRONAC 2021-2024 addresses this issue by agreeing (4.1.5); 'To co-operate with non-governmental organizations in certification processes for the eradication of child labour in sugarcane fields', but without generating a clear proposal for action. It is advisable to use the guidelines that have already been developed, and that sugarcane groups and the World Vision organization work together to address the labour problem in sugarcane fields comprehensively, as well as considering the recommendations illustrated in Fig. 5.

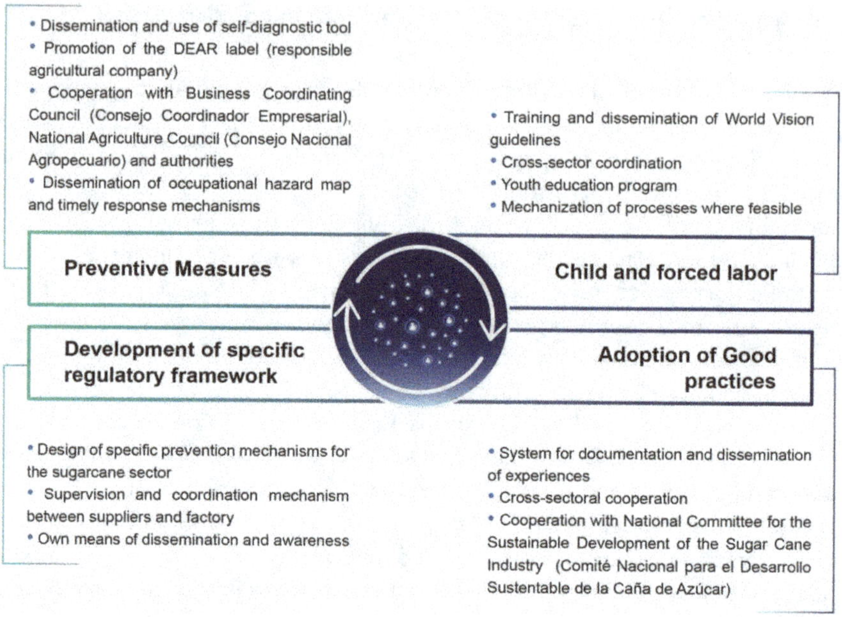

Figure 5 Synthesis of labour regulation elements for the Agenda. Source: Elaboration.

8 Pricing issues in the sugarcane industry

High levels of sugar consumption have been linked globally to health problems, including diabetes, overweight and obesity. In response to this situation, countries with high sugar consumption are taking steps to reduce intake. Before the coronavirus outbreak, sugar inventories were being reduced, but with current trade volatility, the current status of international inventories is uncertain (OECD-FAO, 2020). According to Fernández (2020), the key players in the international market are:

- Brazil is the world's leading producer and main exporter, representing 30% of world trade. World sugar prices are therefore highly dependent on what happens in this country;
- India is the world's second largest producer with a 16% share and the third largest global exporter (at 10% of global trade). India has a very distinctive sugarcane sector: there are many sugar mills and sugarcane production is highly subsidized; and
- Thailand is the third largest global producer but the second largest exporter, so significantly affects the global market.

The global picture of sugar prices over the last two decades shows that sugar consumption is highly correlated with production. However, during years when production exceeds consumption, inventories/stockpiles are generated and world sugar prices fall.

For the 2021 season in Brazil, it was expected that the fall in fuel prices would impact the sugar market since this country allocates a large amount of its sugarcane production to ethanol. However, if prices for that market are not attractive, sugarcane is used for sugar production, which could have a small downward effect on prices. Overall, it is estimated that international market prices will remain relatively stable in the next few years.

Two-thirds of the Mexican sugar market is destined for domestic consumption. The domestic price is calculated based on an index determined by 23 supply locations, plus a 6.4% surplus price. The remaining third of sugarcane produced is surplus and must be sold elsewhere to maintain domestic prices. There are three main markets for the production surplus:

- 50% is sold to the USA;
- 18% goes to the manufacturing, maquiladora (duty free) and export services (IMMEX) sector; and
- the remaining 32% is sent to the world market.

The price of sugar on the world market is much lower than that of the domestic and USA markets. The world market price on the 10 September 2020 was US$262.57 per ton, while the USA market reported a price of US$593.04 and the Mexican market reported a price of US$778.80 per ton. These prices are protected, within the framework of the Treaty between Mexico, the USA and Canada (USMCA), through a tariff of US$360 per ton which guarantees a price of US$507 per ton. Selling surplus sugar on the world market is a last resort since may not be profitable.

Workshops looking at maintaining income and profitability in the sector emphasized factors such as optimizing agro-ecological conditions in crop fields, improving production efficiency, product diversification of final products and better coordination between producers, mills and other stakeholders in the sugar value chain. It is generally recognized that, rather than local improvements, an integrated solution across the supply chain is needed to see significant and sustainable improvement. PRONAC 2021–2024 proposes Priority Strategy 1.3 'Contribute to the generation of mechanisms that provide certainty to the sugarcane agro-industry with respect to the IMMEX and Sugar Reexport Program'. However, its recommendations are limited to verification, follow-up and control of exports and imports, without measures that lead to systemic, long-term improvement.

9 The role of diversification in sugarcane agri-business development

As already mentioned, although sugar continues to be the main product from the industry, many derivative products can also be obtained from sugarcane. However, diversification must be based on scientific knowledge of the composition and functional properties of co-products, by-products and derivatives. It also need to take account of the volume and type of waste materials produced by sugar mills, how they might be further processed (e.g. through dedicated biorefineries for processing waste materials into a range of products) and the markets for the resulting sugarcane derivative products. This requires an integrated, holistic approach that takes account of past experiences (Aguilar et al., 2017; Enriquez, 2020). The workshops recommended promoting diversification, formation of productive clusters with a long-term vision, and the promotion of an industrial policy that establishes incentives for private investment and specific instruments to promote innovative projects for a wide range of products that can be manufactured from sugarcane and its derivatives.

According to PRONAC 2021-2024, 'the strategy to strengthen the competitiveness of the sugarcane agro-industry consists of lowering production costs per ton in the crop fields and in the factories, as well as making sugar marketing more efficient to reduce the intermediation margin between wholesalers and the final consumer'. All these improvements require investment which is lacking in this sector. It is necessary to look for new business which requires a comprehensive vision that is not reflected in PRONAC 2021-2024.

10 Financing sugarcane agri-business development

PRONAC 2021-2024 has, as its central axis, the proposal to promote research and innovation. However, it recognizes that 'with respect to sources of financing for research and technological development work, these are scarce and the different Funds and Trusts in which the Mexican government has invested resources in research and development are ephemeral or temporary. The agro-industrial sugarcane sector has not been a priority in most of these programmes and strategies'. It is, therefore, essential to consider the role of financing. Currently, the supply of economic support is limited to guarantees and interest rate reductions.

The projects recommended in the Innovation Agenda and PRONAC 2021-2024 require a different vision and business models on the part of financial institutions. PRONAC 2021-2024 establishes Priority Strategy 2.2, 'Promote access to financing for actors involved in the sugarcane agro-industry to improve infrastructure and increase productive efficiency', but does not define how this strategy is to be fulfilled. Mexico's current circumstances suggest that it is the private sector that will be responsible for promoting the changes required by this industry since public resources are increasingly limited, both for development and technology transfer projects and for investment in infrastructure and productive improvement.

11 Conclusion

To improve the sector's competitiveness, technological developments must be incorporated into all links of the production chain. Although the lack of technology in crop cultivation is more evident, sugar mills are not oblivious to the need to incorporate technological developments that will make them more efficient.

To drive innovation projects, the collaborative work of producers and companies with higher education and research entities is necessary to help disseminate, adopt and adapt technologies. This holistic thinking is fundamental for the industry to progress. The incorporation of technologies in the different links of the production chain can lead to a sustainable model for the benefit of all actors in the system, with significant effects not only on yields and other production variables but also on the timely response to the challenges of a changing market.

Sugarcane and its main derivative are traditional products. However, they are directly affected by new consumer habits, the demands and requirements of new markets and new technologies. In order to face these challenges, actors in the chain should consider the following aspects:

- Diversify products by taking advantage of the potential offered by sugarcane[2];
- Look for market niches where higher profit margins can be obtained[3]; and
- Rapid incorporation of technologies throughout, suggesting open innovation may be the best alternative to shorten the gaps as rapidly as possible.

Investment is required to face these challenges but also full awareness that such investment is profitable and will contribute to the long-term sustainability of this industry. Unfortunately, PRONAC 2021–2024 appeared late (3 years into the current government administration) and with a 3-year perspective. The strategies presented in this programme are attractive but not viable in a context characterized by the reduction of resources for investment.

Financing must incorporate technologies and promote product diversification and new business models. However, current financial institution

2 The sugarcane value chain includes at least five elements: sugar (in its different forms), electric power, alcohol, fodder and cellulosic materials; however, until now, almost all the focus has been on obtaining sugar.

3 Brown sugar, e.g., is considered a 'healthier' product and its price in the market is higher; advertising campaigns that encourage consumers to choose this type of product could be a good strategy.

schemes are insufficient, so different instruments are needed to trigger fundamental changes. Therefore, an effective strategy of public-private arrangements must be developed to finance technological and organizational innovation in this industry. All this depends on a specific innovation policy for this agro-industry. Under the current conditions of withdrawal of state support for scientific and technological activities, such a policy must be based on the co-ordinated efforts of the various private actors in this productive ecosystem.

Finally, in general, some of the elements presented in this chapter for the sugarcane agro-industry can be applied to other traditional agro-industrial value chains, such as coffee or cocoa, because they share certain common characteristics, such as: lack of co-ordination among stakeholders; technological backwardness; failure to take advantage of market opportunities for differentiated products; scarce financing and erratic public policies. To achieve specificity in the chains, the proposed methodology can be replicated with minor adjustments to create specific agendas for innovation.

12 Where to look for further information

Information System on Sugar Cane Agroindustry Research, https://www.siiba.conadesuca.gob.mx/siiaca/Consulta/.

Sugar and Sweeteners Market Outlook, https://www.ers.usda.gov/topics/crops/sugar-and-sweeteners/market-outlook/.

13 References

Aguilar, N. (2020). Retos que se deben afrontar en el corto plazo para mejorar la productividad, el uso sustentable del agua y el manejo integral del cultivo de la caña [presentación realizada en la mesa de trabajo sobre sustentabilidad, productividad, desarrollo tecnológico e innovación, de las mesas de trabajo colaborativas para la elaboración de la *Agenda de innovación intersectorial para el desarrollo de la agroindustria de la caña de azúcar para Oaxaca, San Luis Potosí y* Veracruz]. México.

Aguilar, N., Debernardi, T. and Herrera, H. (2017). Subproductos, coproductos y derivados de la agroindustria azucarera. *Agroproductividad* 10 (11), 13–20. Recuperado de http://siiba.F.gob.mx/siiaca/Consulta/verDoc.aspx?num=994

Aguilar, N., Galindo, G. and Fortanelli, J. (2011). Factores de competitividad de la agroindustria de la caña de azúcar en México. *Región y sociedad*, 23(52). doi:10.22198/rys.2011.52.a188.

Alcoholera Zapopan (2020). Ingenio etanolero. Recuperado de https://alcoholera-zapopan.com/fabricacion/.

Cámara de Diputados del H. Congreso de la Unión (2005). *Ley de Desarrollo Sustentable de la Caña de Azúcar. Diario Oficial de la Federación, DOF*, 22 de agosto de 2005. Recuperado de http://www.ordenjuridico.gob.mx/Documentos/Federal/html/wo12073.html.

Castillo, O. (2020). Lineamientos para el cumplimiento de los derechos laborales y prevención de COVID-19 en el sector azucarero, [presentación realizada en la mesa de trabajo sobre aspectos laborales realizada dentro de las mesas de trabajo colaborativas para la elaboración de la *Agenda de innovación intersectorial para el desarrollo de la agroindustria de la caña de azúcar para Oaxaca, San Luis Potosí y Veracruz*]. México.

Comité Nacional para el Desarrollo Sustentable de la Caña de Azúcar [Conadesuca] (2020). Mapa interactivo COVID 19. Recuperado de https://conadesuca.maps.arcgis.com/apps/webappviewer/index.html?id=4768bcb6904f4c18a239e623dfdbac5d, Noviembre 15 de 2020.

Conadesuca (2018). Sinfocaña, Reportes-Campo, Dato Acumulado. Recuperado de https://www.siiba.conadesuca.gob.mx/infocana/Consulta/ReportesP.aspx?f=2&acu=1

DOF (2021). Programa Nacional de la Agroindustria de la Caña de Azúcar 2021-2024, Diario Oficial de la Federación, México, 29 de diciembre de 2021.

Enríquez, M. (2020). Precios, diversificación y articulación de la cadena de valor de la agroindustria, [presentación realizada en la mesa de trabajo sobre sustentabilidad, productividad, desarrollo tecnológico e innovación, de las mesas de trabajo colaborativas para la elaboración de la *Agenda de innovación intersectorial para el desarrollo de la agroindustria de la caña de azúcar para Oaxaca, San Luis Potosí y Veracruz*]. México.

Fernández, C. (2020). Tendencias de precios internacionales de azúcar, [presentación realizada en la mesa de trabajo sobre sustentabilidad, productividad, desarrollo tecnológico e innovación, de las mesas de trabajo colaborativas para la elaboración de la *Agenda de innovación intersectorial para el desarrollo de la agroindustria de la caña de azúcar para Oaxaca, San Luis Potosí y Veracruz*]. México.

Flores, C. (2020). Mejoramiento genético de la caña de azúcar [presentación realizada en la mesa de trabajo sobre sustentabilidad, productividad, desarrollo tecnológico e innovación, de las mesas de trabajo colaborativas para la elaboración de la *Agenda de innovación intersectorial para el desarrollo de la agroindustria de la caña de azúcar para Oaxaca, San Luis Potosí y Veracruz*]. México.

García R. (2020). Panorama general del sector agroindustrial de la caña de azúcar, [presentación realizada en la mesa de trabajo sobre sustentabilidad, productividad, desarrollo tecnológico e innovación, de las mesas de trabajo colaborativas para la elaboración de la *Agenda de innovación intersectorial para el desarrollo de la agroindustria de la caña de azúcar para Oaxaca, San Luis Potosí y Veracruz*]. México.

Mandel, E. (1987). El capitalismo tardío. Ediciones Era. Segunda reimpresión. D. F., México. 575 p.

MGGL Abogados (s.f.). La entrada en vigor del USMCA: Aplicación e impactos en materia laboral. Recuperado de http://www.mggl.com.mx/post/la-entrada-en-vigor-del-t-mec-aplicaci%-C3%B3n-e-impactos-en-materia-laboral.

Müller, G. (1992). Transformaciones productivas y competitividad: aspectos conceptuales y metodológicos, [ponencia presentada en el seminario-taller Política Tecnológica y Competitividad Agrícola en América Latina (págs. 14–15)]. Montevideo: Instituto Interamericano de Cooperación para la Agricultura.

Organización para la Cooperación y el Desarrollo Económico [OECD, por sus siglas en inglés] and Organización de las Naciones Unidas de la Alimentación y la Agricultura

[FAO por sus siglas en inglés] (2020). Sugar. En OECD y FAO, Agricultural Outlook 2020–2029. Recuperado de http://www.agri-outlook.org/commodities/oecd-fao-agricultural-outlook-sugar.pdf.

Osoyo, A., Aquino, P. and Fernández, J. (2017). *Agenda nacional de investigación, innovación y transferencia de tecnología de caña de azúcar.* México: Comité Nacional para el Desarrollo Sustentable de la Caña de Azúcar.

Roel, L. (2020). Mapa Riesgo laboral USMCA Sector agropecuario, [presentación realizada en la mesa de trabajo sobre aspectos laborales realizada dentro de las mesas de trabajo colaborativas para la elaboración de la *Agenda de innovación intersectorial para el desarrollo de la agroindustria de la caña de azúcar para Oaxaca, San Luis Potosí y Veracruz*]. México.

SAGARPA. (2012). Programa Sectorial de Desarrollo Agropecuario, Pesquero y Alimentario 2013-2018. D. F., Diario Oficial de la Federación (Official Gazette), December 13, 2012. Mexico, 50–112 pp.

Santoyo, H. (2020). Factores críticos para mejorar la productividad en la agroindustria de la caña, [presentación realizada en la mesa de trabajo sobre sustentabilidad, productividad, desarrollo tecnológico e innovación, de las mesas de trabajo colaborativas para la elaboración de la *Agenda de innovación intersectorial para el desarrollo de la agroindustria de la caña de azúcar para Oaxaca, San Luis Potosí y Veracruz*]. México.

Secretaría de Agricultura y Desarrollo Rural [Sader] (2020). Caña de azúcar, un cultivo que proviene de pequeños productores. Recuperado de https://www.gob.mx/agricultura/articulos/cana-de-azucar-un-cultivo-que-proviene-de-pequenos-productores?idiom=es.

Secretaría de Agricultura y Desarrollo Rural [Sagarpa] y Comité Nacional para el Desarrollo Sustentable de la Caña de Azúcar [Conadesuca] (s.f.). Programa Nacional de la Agroindustria de la Caña de Azúcar, Pronac, 2014–2018. Recuperado de http://www.cndsca.gob.mx/DocumentosEficProductiva/8.%20PRONAC/PRONAC%202014-2018.pdf.

Secretaría de Agricultura y Desarrollo Rural [Sagarpa] y Conadesuca (2015). Ficha técnica del cultivo de la caña de azúcar. Recuperado de http://conadesuca.gob.mx/DocumentosEficProductiva/1.%20Campo/Ficha%20T%C3%A9cnica%20Ca%C3%B1a%20de%20Az%-C3%BAcar.pdf.

Servicio de Información Agroalimentaria y Pesquera [SIAP] (2018). La producción de caña de azúcar supera los 55 millones de toneladas en 2018. Recuperado de https://www.gob.mx/siap/articulos/la-produccion-de-cana-de-azucar-supera-las-55-millonesde-toneladas-en-2018.Roel.

Sistema de Información Agroalimentaria de Consulta. Siacon. https://www.gob.mx/siap/documentos/siacon-ng-161430

Tratado entre México, Estados Unidos y Canadá [USMCA] (s.f.). Capítulo 23. Gobierno de México. Recuperado de https://www.gob.mx/cms/uploads/attachment/file/465805/23Laboral.pdf.

Vázquez, V. (2020). Modificación a la NOM-051. Etiquetado para alimentos y bebidas no alcohólicas preenvasados, [presentación realizada en las mesas de trabajo colaborativas para la elaboración de la *Agenda de innovación intersectorial para el desarrollo de la agroindustria de la caña de azúcar para Oaxaca, San Luis Potosí y Veracruz*]. México.

Zafranet (2013). Congreso de Oaxaca aprueba instaurar el día del cañero. Recuperado de https://www.zafranet.com/2013/02/congreso-de-oaxaca-aprueba-instaurar-el-da-del-caero/.

Chapter 11

Supply chain resilience capabilities in European food supply chains: the impact of COVID-19 in the Netherlands

Renzo Akkerman and Rene Haijema, Wageningen University, The Netherlands; Marvin Kunz, Wageningen Economic Research, The Netherlands; Sander de Leeuw, Wageningen University, The Netherlands; Viet Nguyen, Wageningen University/Amsterdam University of Applied Sciences, The Netherlands; and Harmke Schellekens and Gideon van der Stelt, Wageningen University, The Netherlands

http://dx.doi.org/10.19103/AS.2023.0122.21

1 Introduction

Food supply chains are an essential part of our society. From the primary production of agricultural products to the retail and foodservice settings through which food products are distributed, these extensive supply chains provide a wide range of products to satisfy the daily demands for human nutrition. In addition, these food supply chains also provide a significant contribution to the economy and to employment across the world (Beckman and Countryman, 2021).

Disturbances in these food supply chains can have significant consequences for the economy as well as for local and global food availability. The COVID-19 pandemic has clearly impacted food supply chains in 2020 and 2021. For consumers, this was mostly visible in their experiences with empty shelves in retail due to hoarding behaviour and the closure of cafes and restaurants due to regional and national lockdowns. Also, during the pandemic, the popular media often reported news on upstream supply chain impacts such as the closure of meat-processing facilities due to COVID-19 outbreaks among staff (e.g. Guardian, 2020a) or the occurrence of agricultural surpluses due to mismatches in supply and demand (e.g. Los Angeles Times, 2020; Washington Post, 2020). Due to the often-limited shelf life of food products, these surpluses cannot just be stored unlimitedly to balance supply and demand over time, but these surpluses lead to significant food waste (FAO, 2020), even though in some cases these products could be repurposed, e.g. for distribution through food banks (FEBA, 2020).

The consequences of the COVID-19 pandemic are not limited to these examples from the start or end of the food supply chain, even though these examples were the most visible to the public. The impacts were felt throughout the supply chain as for instance labour shortages and unavailability of packaging material impacted many upstream production and distribution activities. The increase in online retail sales grew significantly, but suppliers to the foodservice sector were left with inventories of products whose shelf lives expired. As food supply chains are often global, transportation systems were also affected by changing customs procedures, which in some cases caused food products waiting at borders to perish. Finally, many food-processing industries had to cope with challenges due to changes on both the supply and demand side of their businesses. The expectations are also that food supply chains will structurally change due to the experiences from the COVID-19 pandemic, even though it is still unclear how this will develop over time (e.g. Poppe, 2020) and some researchers have furthermore stressed restraint in relation to potentially damaging protectionist policies (e.g. Aday and Aday, 2020).

The COVID-19 pandemic has also been a large-scale test of supply chain resilience and a reason for increased demand for research on resilience, as has been clearly illustrated in the increase of recent COVID-19-inspired scientific studies on supply chain resilience. Sodhi et al. (2023) discuss research topics that need to be addressed to improve supply chain responsiveness to future pandemics. Also, Craighead et al. (2020) stress that this both requires restoring supply chain processes as well as changing processes to be better prepared for future supply chain disruptions. Specifically, for the food supply chain, there are some general discussions and case studies on the impacts of the COVID-19 pandemic (e.g. Garnett et al., 2020; Hobbs, 2020, 2021; Burgos and Ivanov, 2021). For instance, Bina et al. (2022) demonstrated that relying on larger production sites (in the context of beef processing) could lead to larger disruptions. Stevens and Teal (2023) also show that horizontal diversification strategies helped increase resilience. Furthermore, Capodistrias et al. (2022) show how also food banks showed resilience in the way they acted as a food supply chain actor during the pandemic. It is beyond the scope of this chapter to provide a full literature review here.

The challenge to recover from disruptions such as COVID-19 and in building resilience to cope with future events that are similar lies in the capacity to build capabilities to reduce the impact of large-scale events such as the pandemic. Chowdhury and Quaddus (2016) argue that this requires the managerial and organizational capability to respond to and recover from such events. Such dynamic capabilities can help firms sense opportunities and threats to adapt the way they deploy resources and seize the opportunities such that needs dictated by the environment are met (Teece et al., 1997).

Dynamic capabilities are different from 'ordinary' process and operational capabilities in the sense that they can combine and adjust existing capabilities ('microfoundations') or those that focus on, e.g. expansion and new product development that take place under uncertainty (Teece, 2018). In this chapter, we focus on the dynamic capabilities employed by companies active in the food supply chain in reacting to an extreme situation – the COVID-19 pandemic – through sensing, seizing, and/or reconfiguring (or transforming) (Teece, 2007).

The recent review by Chowdhury et al. (2021) on COVID-19-related supply chain research emphasizes a lack of theoretically grounded empirical work. Our study particularly aims to understand the 'why' and 'how' behind capabilities employed to counteract the COVID-19 pandemic. This is in line with the call of Brusset and Teller (2017) who suggest to supplement the more quantitative studies populating the DCV framework with qualitative approaches to help better understand the mechanisms of why and how certain capabilities help improve supply chain resilience performance. Although there have been empirical studies focusing on how food supply chains coped with the COVID-19 pandemic, also using a DCV lens such as Kähkönen et al. (2023), a food supply chain focused study eyeing COVID-19 effects from farm to fork has not yet been published. With this study we aim to start filling this gap.

The aim of this chapter is therefore to (i) present and discuss actual supply chain management responses used by companies in the food supply chain during the recent COVID-19 pandemic, (ii) categorize these responses in terms of dynamic capabilities, and (iii) contribute to further development of dynamic capability theory and practice in relation to high-impact supply chain disturbances like pandemics. We base this research on an empirical study of stakeholders from across the food supply chain in the Netherlands. The results provide interesting insights in the practical use of resilience capabilities, and also show that despite initial concerns, food supply chains were able to use their capabilities to cope relatively well with the disruptions caused by the pandemic.

The remainder of this chapter is organized as follows. In Section 2, we first briefly discuss the relevant literature on supply chain resilience, providing the theoretical framework for our subsequent empirical study. Section 3 then introduces the research approach we used to identify, analyze, and discuss the cases in our empirical study. Section 4 subsequently discusses the supply chain responses in relation to the different supply chain resilience capabilities. In Section 5, we then present our discussion regarding the theory and practice of supply chain capabilities in food supply chains and relate these to dynamic capabilities, followed by our conclusions in Section 6.

2 Supply chain resilience: identifying resilience capabilities

Supply chain resilience is a topic that has been widely studied in the supply chain literature (e.g. Christopher and Peck, 2004; Chopra and Sodhi, 2004, 2014). As in the seminal work by Christopher and Peck (2004), we refer to resilience as 'the ability of a system to return to its original state or move to a new, more desirable state after being disturbed'. In the last few decades, a lot of research has been done in relation to supply chain resilience. We do not aim to provide a full review here; interested readers are referred to the classic papers mentioned above for a more general discussion of supply chain resilience and to recent reviews on the identification of relevant supply chain capabilities supporting resilience (e.g. Kamalahmadi and Parast, 2016; Ali et al., 2017; Datta, 2017; Kochan and Nowicki, 2018). In the following, we do however provide a brief overview of the basic principles behind the supply chain resilience theory and the supply chain capabilities that have been identified in this context.

Following the definition given above, supply chain resilience is often considered to be a more reactive characteristic of supply chains. However, to be able to react to disturbances, the preparation of organizations for potential disturbances also plays a role. Ponomarov and Holcomb (2009) therefore distinguish three phases related to supply chain resilience: readiness, response, and recovery (as illustrated in Fig. 1). In the readiness phase, the focus is on capabilities that can be developed before disruptions occur. In the response phase, the focus is on capabilities that help mitigate the impact of disturbances. Finally, in the recovery phase, the focus is on capabilities that support a quick transition back to a normal state after a disruption.

In the different phases, different resilience capabilities are required, even though there might also be some overlap. For instance, one of the most discussed resilience capabilities is flexibility, and even though this is mostly seen as a key capability during the response phase, it is clearly also a capability that needs development in the readiness phase and that support a fast recovery.

For the analysis, we perform later in this chapter, we particularly build on the core resilience capabilities identified by Stone and Rahimifard (2018), who performed a recent and comprehensive study of supply chain resilience literature. In Table 1, we summarize the main supply chain resilience capabilities to structure the discussion of our empirical results in this chapter. We chose to exclude the core capability 'security' identified by these authors, as this capability is linked to the prevention of intentional disturbances of supply chains (e.g. deliberate contamination of food as an act of terrorism), and the prevention of such intentional actions is not relevant in relation to the pandemic response activities discussed in this chapter. Table 1 presents our definitions for each of the capability categories, which are mainly based on the work by Stone and Rahimifard (2018). During our research, we also included a category 'other' to identify capabilities we could not fit to these existing categories.

3 Supply chain resilience in practice: the COVID-19 pandemic

The empirical study presented in this chapter is explorative in nature, focusing on the impact of and responses to a contemporary event, the COVID-19 pandemic,

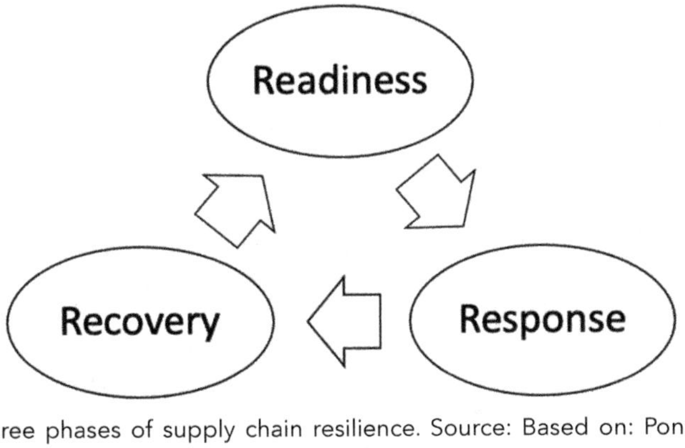

Figure 1 Three phases of supply chain resilience. Source: Based on: Ponomarov and Holcomb (2009).

Table 1 Overview and definitions of key supply chain resilience capabilities

Capabilities	Definition
Early warning	Using foresight to extend preparation time and decrease time spent on reacting to a disruption.
Flexibility	The ability of an organization or supply chain to adapt with minimum time and effort to changing operating environment and customer requests (e.g. switch suppliers, substitute ingredients, outsource processes, share materials, equipment, and staff between sites, the ability of staff to fulfil multiple roles, supply-chain-wide alternative options achieved through partnerships, and the levels of control over market position).
Redundancy	The ability to use surplus and back-up material and capacity (e.g. surplus raw materials and finished inventory, back-up production and storage facilities, surplus pathways between supply chain links, and the extent to which elements are replaceable).
Collaboration	Two or more actors working together to generate advantages that could not be achieved individually with the aim to reduce uncertainties and integrate systems (e.g. shared forecasting, postponement and risk sharing, cooperation, and partnership).
Visibility	The ability to see structures, processes, and products from one end of the supply chain to the other (e.g. channels for the sharing of risk information, IT infrastructure, frameworks guiding how information is delivered to the right people at the right time).
Agility	The ability to respond quickly to unpredictable changes in supply and demand by changing configuration at tactical level (e.g. logistics capabilities and manufacturing flexibility).
Risk-aware culture	The infrastructure a firm has in place to manage risk (e.g. using contingency planning or enterprise risk management programmes).
Adaptability	The ability of a system to adapt incrementally or to completely transform in response to a changing operating environment.

Source: Based on: Stone and Rahimifard (2018).

in the food industry. We interview stakeholders in the food supply chain and domain experts, with the aim to cover as wide a spectrum of stakeholders in the food supply chain as possible. The unit of analysis is the supply, processing, and distribution processes of an individual company (i.e. the company interviewed). We refer to these companies as 'cases'. We furthermore used expert interviews to allow for a cross-check of our findings and to inform interviews. Our investigation aims to answer 'what' and 'how' questions related to what capabilities were employed to counteract COVID-19 and how they were employed, confirming the suitability of an interview-based approach to studying these cases (Voss et al., 2002; Yin, 2018). The approach also allows us to build on the theory presented in the previous section, and potentially refine and elaborate on this theory based on our empirical findings (cf. McCutcheon and Meredith, 1993; Ketokivi and Choi, 2014). Below, we briefly discuss the details of our case selection and analysis.

3.1 Case selection

As we seek to understand the way that different stakeholders in the food supply chain were affected by the pandemic and the different resilience capabilities implemented to deal with it, our case selection aimed to include companies operating in different stages of the supply chain. Building on the general structure of food supply chains (based on Akkerman et al., 2010), we included companies operating in the primary sector, the processing industry, distribution and wholesale, retail, and foodservice. Here, the processing industry includes companies that make consumer products based on products from agriculture and livestock. This includes for instance the dairy industry and meat-processing companies, which are sometimes integrated with the activities of the primary sector. In the distribution stage, we include logistics service providers, even though many distribution activities are performed by other supply chain actors such as the processing industry and retailers, and as such also considered. With retail, we mean all supermarkets and specialty shops (such as bakers and butchers), and with food service, we imply all out-of-home food consumption and preparation, from restaurants and cafes to company cafeterias and event catering.

In total, 26 cases were selected, representing primary producers (8), processing industries (9), distributors (3), retailers (2), and foodservice companies (4). Some of the cases represented here cover more than one stage of their activities. This specifically concerns companies in the primary sector also performing a significant amount of processing (e.g. in the meat sector) and many processing industries managing their own distribution activities. Within the first supply chain stages, organizations with different product groups were selected to allow for potential differences in the impact of the pandemic or the resilience capabilities used for different product types. The list for instance includes meat, dairy, and vegetables in terms of primary products, as well as beverages, sauces, and snacks in terms of the processing industry. We did not specifically select small or large firms, but both ended up being present in our selection. To ensure anonymity, we do not specifically refer to individual companies in this chapter but will often refer to companies in a certain supply chain stage when discussing the results.

3.2 Data collection

During the months of October and November 2020, semi-structured interviews of 1 h long were conducted with the individual case companies described above. These interviews were conducted with staff expected to be familiar with the supply chain impacts of the pandemic. Due to COVID-19 restrictions, the interviews were held online, except for 1 interview where we were allowed to interview in person within the restrictions set. Interviewees were mostly senior managers from supply chain departments but also included general managers or company owners in situations where a supply chain department did not exist (e.g. in some cases in the primary sector).

In these interviews, we discussed the impact of the COVID-19 pandemic on the flow of goods through the company (from sales and distribution back to purchasing), during different stages of the pandemic. We discussed both the impact of the pandemic as well as the strategies that the companies employed in response. We particularly focused on the strategies used to deal with the changing environment to be able to analyze the results from the perspective of supply chain resilience. To assure attention for research ethics procedures, a literature-based case study protocol was used for consistent coverage of categories. The semi-structured approach allowed interviewees to provide extra information as appropriate and enabled freedom of expression.

3.3 Data analysis and validation

After collecting the data, qualitative content analysis was performed (Mayring, 2014). Qualitative content analysis alleges to incorporate two conflicting methodological principles, by applying both theory-guided investigations while maintaining openness. This allows us to deploy categories that emerge out of data (Bryman, 2016), a technique also known as 'open coding' (see also Hendry et al., 2019) or 'deductive category application' (Mayring, 2014).

To validate the findings of the analysis, preliminary results were shared with two organizations with a more general overview of the Dutch food supply chain. Specifically, this concerned a senior sector expert from a Dutch bank with a significant focus on agriculture and food business, as well as a director of a Dutch non-profit federation focused on the food industry. Feedback from these industry experts was used to add to and refine the preliminary outcomes. In addition, a workshop was organized in collaboration with TKI Dinalog, the Dutch Institute for Advanced Logistics that sponsored the research. In this workshop, which was held in January 2021, the preliminary outcomes of this research were presented to and discussed with an audience of representatives from universities and companies. Feedback from this workshop has been incorporated into the findings.

With the previously defined categories and the feedback provided by the two expert organizations, we started to analyze the content of the conducted interviews in a more detailed manner. We compiled a list of all the individual mitigation strategies and structured them so that they could be easily categorized. To this end, we defined a coding scheme based on existing literature on supply chain resilience capabilities, which would allow us to conceptualize the different mitigation strategies in a way that would highlight their similarities (i.e. we used the capabilities shown in Table 1 as categories). Three different researchers, independently of each other, coded the strategies according to the predefined categories and then compared their results. Initially, only a subset of the methods was coded. In doing so, we used the suggestion for the inter-rater reliability process of Hallgren (2012). The differences between the codes given were minor but did lead to an insufficiently low level of agreement. After an in-depth discussion, most disagreements were resolved, allowing for a more comprehensive understanding of the nonagreements, and improving the quality of the coding scheme (as suggested by Burla et al., 2008). As a consequence of this discussion, we agreed to add an additional category to the coding scheme. While the additional category was not initially a part of the existing literature, we decided that it was a valuable addition to the coding scheme. This process took place in line with our deductive category application process and recent contributions to method development that also emphasize that unanticipated mechanisms can be revealed in case study research (Eisenhardt, 2021) and that categorization can and should be used more actively, for instance by using the empirical evidence to elaborate on existing theory (Grodal et al., 2021). To be able to capture the range of strategies employed by companies in response to the pandemic, we then also extended the framework of core supply chain resilience capabilities we use. This framework is presented in the next section.

Once all the mitigation strategies were coded, further discussions of the results and the removal of disagreements resulted in a Fleiss' kappa value of 0.929. While indicating an excellent fit, we do acknowledge that this Kappa value is the result of a few rounds of discussion. Fleiss' Kappa has been suggested as an appropriate alternative to Cohen's Kappa for 3 or more coders (Fleiss et al., 2013).[1]

4 Supply chain resilience capabilities in response to COVID-19: early warning and flexibility

In this section, we discuss how companies responded to the COVID-19 pandemic. Since the interviews were held between September and December 2020, we cover both responses during both the first wave of the pandemic in the Netherlands (March–April 2020) and the second wave (after the summer holidays of 2020). This section is structured around the supply chain resilience capabilities identified in Section 3, Table 1. We discuss how each of these capabilities played a role in the companies' responses to the pandemic.

1 Both Cohen's kappa and Fleiss' kappa are statistical measures to assess the level of agreement between respondents or raters. Cohen's kappa only works for at most two raters, whereas Fleiss' kappa works for any number of raters. A kappa value of 1 for Fleiss' kappa would mean complete agreement, and any value between 0 and 1 would represent an increasing scale of agreement.

4.1 Early warning

The essence of 'early warning' systems is that certain information (i.e. foresight) can be obtained in advance and used to better prepare for a forthcoming supply chain disruption. After COVID-19 was widely spread in China, the pandemic subsequently affected other countries with various time delays, making it possible to learn from what happened in regions that were affected earlier.

By 9 March 2020, Italy had, for instance, registered 7375 cases while in the Netherlands there were fewer than 500 registered cases at that time (Statista, 2020). International companies with activities in China and the south of Europe made use of knowledge and experience acquired from their early affected locations. For example, one multinational food processing company organized online meetings in April 2020 with responsible managers in Italy so that the lessons learned could be communicated to managers of locations in Northern Europe. During these meetings, attention was paid to the management lessons related not only to individual facilities but also to the whole supply chain. However, as stated by a supply chain manager of a multinational company: 'We mostly learned about how to manage problems at the plants, but not enough about the supply chain perspective.' This clearly demonstrates that although at that time useful information was available at the level of individual facilities, the full supply chain impacts were still unclear.

Companies operating in China were able to use lessons learned at an even earlier stage. One of the interviewed companies working in the fruits and vegetables sector indicated that their crisis consultations had taken place already in January 2020 because they expected Europe to face the same situation as China. Based on its experiences in China, this company conducted a risk inventory study and initiated preparations (including discussions on production volume decisions for crops that had to be sown in March).

Although lessons could be learned from experiences in other countries, predictions still proved difficult to make. Many companies felt they were still forced to act in a reactive manner. Various Dutch trade associations however provided information with estimates of market developments (such as expected sale volumes in different markets) to their members. Several interviewees argued that these estimates were relatively accurate, and they were widely appreciated in the sector. In some cases, it was even the only available source of information. Consolidating knowledge thus helps during times of crisis.

4.2 Flexibility

Flexibility is among the most cited capabilities in the supply chain resilience literature. Indeed, many of the responses to the COVID-19 crisis collected from the interviews are also related to flexibility. In the interviews, we encountered flexibility in (i) sourcing, (ii) workforce, (iii) logistics, and (iv) product mix. Below, we elaborate on each of these aspects.

4.2.1 Sourcing flexibility

Multi-sourcing is a well-known strategy to achieve supply chain resilience under disruptions. Some examples of sourcing flexibility came up in the interviews, but in situations in which this flexibility was not already there, it turned out to be difficult to set up on short notice. Clearly, for agricultural products, this is also challenging due to the lead times involved in adjusting production volumes.

In some cases, differences in national policies were beneficial. An interviewed international meat-processing company stated that they did not experience supply shortages during the pandemic because when a lockdown was imposed in Germany, their Dutch supplier managed to deliver the requested volume.

Nevertheless, implementing a multi-sourcing strategy affects not only the sourcing costs factor but also other supply chain processes. Adding suppliers during a pandemic can be challenging. An interviewed company expressed that even though specifications are the same, glass from another supplier often causes issues in production and packaging lines, and the process of streamlining such operations might take too long in relation to the supply chain disruption. Several interviewees mentioned that during the pandemic their companies did however have to start using new suppliers, particularly ones located in Europe.

4.2.2 Workforce flexibility

During the pandemic, supply chain activities mostly had to be carried out with the available existing workforce. In some companies, part of the workforce had more work than others due to the pandemic. As a result, people sometimes carried out different functions during the crisis.

For example, one of the interviewees indicated that there were not enough people to apply floor marking to set physical distancing between workers, so management staff did that themselves. At another company, account managers had less to do for a while and were temporarily added to the customer service team. In another case, management, sales, and technical staff could not visit customers and were therefore deployed on the production floor. In addition to solving capacity issues, this also impacted employee morale, as exemplified by a supply chain manager's remark that 'It has a positive impact if the CEO is standing next to you, pouring the powder in the tank'.

At other companies, office staff were sent home as much as possible. Several companies invested early in equipment for working from home (e.g. laptop computers), and systems were expanded to make this possible. One of the interviewees also acknowledged the role of technological development in the realization that 'if this crisis would have happened two years ago, we would have had a problem'.

Many companies tried to balance their capacity across production lines and even production sites by re-distributing workers across sites. It was however acknowledged that this flexibility might not have been possible without the lack of urgency caused by the pandemic and that this might have not been possible to ask from employees under normal circumstances. In the reallocation of employees, transportation sometimes turned out to be a bottleneck. Additional buses were needed to maintain a safe distance among passengers.

There are also differences between the first and second waves of the pandemic in terms of human resource capacity. Companies indicated that there were fewer staffing problems in the first wave than in the second wave because there were more infections in the second wave: many workers employed on flexible contracts went back to their home countries and full-time workers were more likely to get tested and quarantined in the second wave; thus, staffing became increasingly problematic.

4.2.3 Logistics flexibility

For most of the companies that were interviewed, customer demand increased significantly due to the shifts from food service to retail, which also led to increased demand for logistics capacity. Depending on the flexibility of the specific capacity, this often turned out to be challenging to deal with.

Particularly in the first weeks of the crisis, many companies had to scale up tremendously with workers and transportation resources to compensate for additional demand. It also meant that workers had to work more on weekends and public holidays; several companies therefore gave (financial) bonuses to staff for these situations. Transport capacity that was no longer needed in the foodservice channel could partly be used in the retail channel. Some companies that had their own trailers had to hire extra trailer capacity during peak times. Smaller companies and in particular self-employed drivers were hired for this purpose. Also, supporting infrastructures were sometimes insufficient, as demonstrated in a statement such as, 'We added additional portable toilets due to the increasing number of drivers passing through our distribution centre'.

Several distribution centres saw opportunities to scale up their capacities. Some interviewees indicated that work schedules for workers needed to be significantly changed, for example with increases from three to five shifts. There were however significant differences between types of workers in terms of flexibility: workers for specialized work such as forklift drivers in warehouses or equipment operators in the processing industry proved difficult to scale up. Order pickers were however reasonably flexible in terms of hours and shifts.

The urgency of the situation also helped, as one of the interviewed supply chain managers noted that 'fortunately the staff is aware of the seriousness of the situation, and they are very flexible'.

When the situation returned to somewhat more normal circumstances after the first pandemic wave, just before summer 2020, there was an upward demand trend in the foodservice industry. Finding staff for distribution centres in this industry became a challenging task. As one interviewee stated, '...retailers were doing well and many of our flexible staff had gone there'.

The crisis also made some companies think about planning for a permanent pool of employees and a flexible pool with a temporary employment agency - with the crisis putting more emphasis on the flexible pool.

4.2.4 Product mix flexibility

Many companies were forced to reconsider their product mix during the pandemic, as demand for different products experienced significant shifts. For instance, the industry saw a decrease in demand for products and packaging sizes used in the foodservice sector. This was often complicated by the fact that food supply chains rely on agricultural raw materials that have long lead times, meaning that adjusting the supply to shifts in demand is impossible in the short term.

In industries where demand decreased significantly, which was mostly related to the drop in demand from the foodservice sector due to lockdowns, raw materials were processed into more generic products with longer shelf life if possible. In the dairy industry, for example, milk that would normally have been processed into products for the foodservice sector (e.g. whipped cream, mozzarella) was now processed into milk powder. Also, butter was converted into butter oil. In the meat industry, beef was ground instead of sold as a steak. This type of flexibility was sometimes limited by production line capacities or market opportunities.

Companies working with vegetables as raw materials also used flexibility in product mix and composition as much as possible. These companies were able to build on a significant level of existing flexibility based on having to deal with uncertainties resulting from fluctuating harvest yields. For instance, it is customary that in years of high yields, certain vegetables are included in larger quantities in packages of pre-cut mixed vegetables. This flexibility was used to its full extent during the pandemic. Interviewees also indicated that the product mix was changed or even enlarged. They indicated, among others, that during COVID more tomatoes were being processed into pasta sauce than under normal circumstances and bell peppers were being processed into wine.

In terms of packaging, several interviewees also indicated that they altered the product packaging in response to the crisis. Although an increasing number of vegetables was delivered to retailers and sold to consumers with limited - and often no - packaging material before the pandemic, COVID-19 resulted in a resurgence of plastic packaging in shops for hygiene reasons (regardless of the discussion of whether and how long the virus can survive on the packaging or on the food products themselves). Just as fewer unpackaged vegetables were offered for hygiene reasons, there was a decrease in the practice of in-store slicing of meat products in retail environments to reduce contamination. Interviewed companies indicated that, at least for meat products, the centralization processing of a wide product range is a development that has been going on for some time and was accelerated by the COVID-19 pandemic.

5 Supply chain resilience capabilities in response to COVID-19: redundancy

Redundancy refers to the ability to deploy additional capacity, materials, or end products. At the level of an individual organization, this may relate to having additional inventory or capacity. In terms of a supply chain, redundancy can also relate to the presence of alternative transport routes. Like some of the types of flexibility, redundancy was used when possible but increasing redundancy as a response to the pandemic was often difficult.

Several of the interviewed companies did have some form of strategic inventory. One meat-processing company was able to get by during the first wave of the pandemic because it had sufficient buffer stock. During the summer period (after the first wave of the pandemic), production was scaled up further to bring the buffer inventory back up to the initial level. In fact, it is a common strategy for some processing companies (even in non-crisis situations) to hold considerable buffer inventory to cope with the high demand volatility due to the prevalence of promotions and marketing campaigns in food retailing. As stated by one of the interviewees, 'the fact that promotions are so common in our sector has resulted in an overcapacity that has helped us in our response to this crisis.'

Several companies also attempted to order extra products from suppliers yet indicated that those suppliers also had their own problems during the crisis. In some cases, these were secondary products (additives such as spices and sauces), for which new suppliers sometimes demanded large minimum order quantities. These additional purchases in turn led to the need to rent additional storage space, which was not always possible.

In some cases, companies were also forced by the market to store extra materials and products. Due to delivery problems, the lead times at suppliers of packaging materials increased, for one of the interviewees from 1 week to 8-10 weeks, whereupon they decided to immediately store packaging materials and labels for 6 months. Many companies were also proactive to coordinate this with their customers. For example, one retailer indicated that a strategic partner immediately stepped up by stocking, where possible, products and packaging materials, as well as the protective equipment needed for personnel to handle the product. If possible, retailers temporarily increased their stock levels, partly to help regular suppliers that suffered from decreased demand from the foodservice sector. Similarly, in the processing industry, stock levels were increased where possible.

However, in many cases, it also proved impossible to build up inventory: high demand usually meant that all production output was sold immediately. For some product categories such as fresh meat, there was additional demand during the initial lockdown due to an increase in barbecue activity caused by a combination of people staying at home and the nice Spring weather during that period. Shifting demand patterns also led to prolonged shortages for some other product categories, e.g. bakery raw materials such as yeast because many people had taken up home baking during lockdowns.

In some situations, customers were forced to order ahead or build up stock to ensure their long-term survival. In the wine supply chain, for example, products really must move downstream in the supply chain at some point. Also, barley that is still in storage because of decreased demand for beer must eventually make way for the new harvest. Smaller companies in such situations were especially affected; they were more quickly forced to get rid of stock to make room for new stock.

Finally, for many food products, interviewees also acknowledged that even if it would be worth the efficiency losses or higher costs of redundancy, buffering is limited by typical shelf-life constraints found in the food industry.

6 Supply chain resilience capabilities in response to COVID-19: visibility and collaboration

6.1 Visibility

Visibility refers to the extent to which structures, processes, and products are visible in the supply chain. In our study, interviewees discussed a need for increased visibility in the form of intraorganizational (i.e. between departments in a company) or interorganizational (i.e. with suppliers and customers) consultation and coordination.

In most of interviewed companies, there was regular (often daily) coordination between production and sales departments as well as with customers or suppliers to be able to better forecast and meet demand. Sales and operations planning (S&OP) cycles were accelerated, typically from 4-weekly cycles to weekly cycles. A few interviewed multinational companies even set up international 'control towers' in addition to frequent meetings and calls between purchasing, production/packaging, and sales. In retail, demand was even tracked on an hourly basis to adjust ordering/replenishment decisions based on the latest information.

Not all companies were able to move fast enough with such changes to their supply chain planning and monitoring. One interviewed foodservice company indicated that there was a late switch to daily inventory monitoring, even though their decreased sales volume made inventory management more critical, especially for perishable products.

Due to the time-intensive nature of the shorter planning cycles and more frequent consultations, it quickly became relevant to also consider when to stop these activities, as they required a lot of time from the involved parties. Another consequence of these frequent consultations was that sometimes high-level checks were built into working flows. One company shared that even the country director had to frequently approve air shipments for products that normally came by truck. About 10 weeks after the start of the crisis, several interviewed companies stopped these frequent meetings and calls. Other companies did not stop until after the 2020 summer, depending on how strongly the company was related to changes in market demand.

6.2 Collaboration

Collaboration here refers to two or more parties carrying out joint activities to achieve goals that could not be achieved on an individual basis. In the previous sections, good existing supply chain relationships were sometimes the reason a certain kind of flexibility could be achieved, and as such it is also a capability that strengthens other capabilities (which our interviewees might not always have connected to each other). Our interviews specifically revealed an increase in the contact between buyers and suppliers.

During the interviews, it was indicated multiple times that increased efforts were made on supply chain planning together with suppliers and/or customers. More importantly, the frequency of this planning process increased dramatically: forecasts were discussed more regularly, and priority lists were determined. As stated by one of the interviewees, 'due to the pandemic, we are now in contact with our suppliers on a daily basis, sometimes even two or three times per day.'

The interviews also revealed that there are various ways of organizing supply chain collaboration. A few interviewed companies deliberately avoided hard contractual agreements so that, for example during this pandemic, they had fewer purchase obligations in case of decreased demand and consequently less surpluses and food waste within the company. This strategy did however lead to losses at their suppliers, which were partly (but voluntarily) compensated by the companies.

7 Supply chain resilience capabilities in response to COVID-19: agility

Agility-related responses during the COVID-19 pandemic concern the quick changes of configuration at the tactical level to be able to cope with changes in supply and demand. Several adjustments to, especially, logistical processes and sales channels were mentioned in the interviews.

Many of the interviewed companies indicated that adjustments were made to their logistics processes. For instance, retailers adjusted frequencies and time windows for their deliveries to stores to increase flexibility for stores. This prevented backroom storage at the store level from becoming overfull. In addition, more responsive ordering became possible in this way. In several interviews, it was indicated that such operational changes in logistics were possible without much discussion or explanation because everyone understood such needs. For example, requests from logistics service providers for extra steps in order preparation at shippers (more than the standard preparation under normal circumstances) that could speed up the process were also honoured by shippers. However, at some point during the pandemic, the expectation was that the logistics situation would return to normal, and as one interviewee expressed 'COVID will soon not be a sufficient reason for deviations'.

Adjustments were sometimes also needed in terms of international distribution networks. For instance, when South Africa went into lockdown, alcohol sales were also banned, as this was expected to lower the non-COVID-related demand for health care (Guardian, 2020b). This meant that shipments of alcoholic beverages that were on their way to South Africa had to be rerouted and sold to customers in other countries.

Furthermore, it was also important to be able to switch quickly between types of sales channels: if demand from one country or sector falls away, it may be necessary to try to supply additional products to other countries or sectors because of products otherwise perishing or limited storage capacity. For instance, in the case of the meat processing industry, there was insufficient capacity to freeze the surplus resulting from the demand fallout in the foodservice channel, so there was constant attention to being able to sell surpluses. Some processing companies were successful in negotiating with retailers to take over surplus volumes from the foodservice channel because retailers had a high increase in demand. For other products, higher sales were generated in already existing channels. For example, a larger volume of eggs from the Netherlands was sold to the pharmaceutical industry for use in vaccine development. Retailers also witnessed an increased demand for eggs as their sales increased due to the popularity of home baking. Shifting product flows from food service to retail was possible for products that can be sold through both channels. This was, however, not possible for all products, such as brand-specific sauces used in restaurant chains, which created problems with product expiration and waste.

One solution specific to the Netherlands was that wholesalers were given the opportunity to sell their products directly to consumers; this absorbed part of the loss in demand caused by the diminishing of sales to foodservice customers. This was appreciated by consumers because it gave them access to products that were normally more difficult to obtain. However, not all wholesalers were equipped for such direct sales to consumers. In addition, many packages are bulk packaging that is not suitable for consumer use.

Bringing surplus food to food banks was also considered by some companies. However, in the case of refrigerated products, this proved to not always be desirable due to the possibility of food safety issues resulting from interruptions in the cold chain.

8 Supply chain resilience capabilities in response to COVID-19: risk-aware culture and adaptability

8.1 Risk-aware culture

Our interviews showed that supply chain resilience has been always a point of attention by companies in food supply chains due to the high demand and supply volatility in this sector. However, disruptions caused by the COVID-19 pandemic surpass every past disruption in terms of complexity.

As stated by an interviewee from a multinational company, 'very unlikely scenarios do not normally get much attention'. Several interviewees however mentioned that the crisis has initiated discussions about resilience strategies. The most discussed strategy was about establishing local and regional supply, exemplified in statements such as 'if a product is slightly more expensive but comes from a location nearby, it might be worth considering it after all.'

Risks are however not limited to the supply side, and a few interviewed companies also considered risk on the demand side, inspired by the dramatic demand shifts from the foodservice sector to the retail sector. A logistics service provider indicated that it is necessary to think about how the customer portfolio should be set up: 'if most of a production site is dependent on a single customer, which in turn is largely dependent on one market, the question is whether such a situation is desirable'.

Another notable example was the duck meat supply chain, in which the foodservice sector is by far the largest sales channel. In this specific case, a drastic choice was made to stop the entire production pipeline, even no new eggs were hatched to prepare for future sales, due to a high volume of existing stocks in the pipeline that could cover demand for a long time.

8.2 Adaptability

The degree to which a company can structurally adapt supply chain processes to changing market conditions, i.e. adaptability, was also important in relation to digitalization of workplaces, process automation, and the move to online sales channels.

Many interviewed companies invested in hardware and software to support working from home. It remains to be seen in the future whether this will become a more normal part of working environments.

In the packaging industry, a lot of attention has been paid to innovation in the use of materials for the benefit of existing but also new sectors. For example, this has led to the use of cardboard for separation screens or new boxes for e-commerce, but also to the development of personal protection equipment. Investments in automated processing of products have taken off to decrease the dependence on manual labour in processing (e.g. peeling machines for shrimp).

Many companies, especially in the foodservice industry, have been looking for an increased presence in online food sales. For example, one of the foodservice companies we interviewed launched an aggregator website for its restaurant customers, supporting local restaurants with an online portal. McDonalds partnered with the food delivery service Uber Eats to increase the number of locations from which meals can be ordered online and delivered. Not everyone relied on large platforms; many smaller restaurant chains and independent restaurants actively pursued online ordering and delivery by themselves. For some companies and product categories, the move to online ordering or other markets was difficult. For example, in a case of wine supply chains, an interviewee stated that the wine trade is so set in stone and used to face-to-face contact that the volume of online sales was limited. However, the need to make this switch was acknowledged by these companies and the pandemic was a catalyst for the start of a discussion about such a change.

Adapting to situations with different supply-demand dynamics was not always easy. According to an interviewee, it is for instance important to design product (re)allocation or rationing rules in such a way that there is not one country or customer that pulls everything away. In retail environments, we also saw that some companies were forced to ration products with limited supply (in terms of a maximum number of units or packages to be ordered per store).

9 Supply chain resilience capabilities in response to COVID-19: rationalization

The previous sections discuss different resilience capabilities that have been defined in previous literature. In the analysis of our empirical results, an additional capability appeared: the ability to rationalize supply chain operations.

Rationalization refers to the process of organizing a (part of a) business according to a scientific management principle to increase efficiency and effectiveness. Often, rationalization leads to a reduction of variety, for example of products or services offered. Standardization relates to rationalization and aims to reduce variety as well, albeit from a different perspective; standardization is aimed at increasing repeatability and thus doing things the same way as much as possible. Simply speaking, rationalization is about making the same things, standardization about making things the same. When speaking about assortments, for example, rationalization implies the reduction of the variety of packing sizes. Standardization refers to ensuring that a product that is similar in terms of characteristics such as pack type or pack size has the same appearance everywhere (same colour of caps, for example), or that such a product is made in a standardized manner.

Many of the interviewees utilized rationalization to simplify operations, improve capacity utilization, and increase the extent and effectiveness of abovementioned sources of flexibility. In supply chains with increased demand during the pandemic, the emphasis appeared to be mainly on increasing output and improving capacity utilization. Several interviewed companies, especially in the processing industry, indicated that they were limited by their production capacity, and that attention was focused on making more efficient use of this capacity. This was illustrated by statements such as "you want to run large batches in this situation" and "[high-volume] product A is just more important than [specialty] product B now". Labour-intensive and other inefficient production lines were sometimes even stopped in favour of increasing capacity elsewhere. Many interviewed companies offered a smaller assortment and focused on increasing the overall production output.

Less choice in packaging was also offered. One of the interviewed companies indicated that the number of packaging variants was reduced to one size and a total of three SKUs instead of the 20 SKUs in use before the pandemic. This sometimes also led to offering larger packages: with consumers more often having meals at home, there was an opportunity to sell larger packages of products that would normally be sold in relatively small packs.

Rationalization not only happened in the processing industry, but also at in the retail and foodservice sectors. One retailer indicated that the focus was on critical products (e.g. flower sales stopped for a short time) and in case of insufficient transportation capacity, product groups with longer shelf lives were deprioritized (e.g. beer versus fresh products). For restaurant chains, the rationalization extended to the complexity of menus: more menu choices lead to more international transportation or specific ingredients, and thus the more difficult it becomes in a situation like the COVID-19 pandemic.

Organizationally, marketing departments usually determine which products must be produced, often resulting in complex product portfolios.

Several interviewed companies saw a change: flexibility can be achieved by simplifying portfolios, making it easier to scale up in volume. As one of the interviewees put it, "COVID has been an accelerator in the rationalization of our product assortment". In some cases, this type of rationalization also affected supply chain relationships: customers who buy smaller volumes or smaller SKUs no longer had much influence.

Many companies indicated that the focus was on being able to deliver and therefore there was no time for projects such as new product introductions. This reduction in new product introductions was important for the processing industry to be able to use their capacity as much as possible on production to cover demand. However, this was not only an initiative by the processing industry, as retailers also limited or cancelled product introductions and promotions where possible and encouraged suppliers to at least supply a core product range. After all, new products require a great deal of time and attention for both processing industry and retailers.

Some companies also indicated that rationalization was not only focused on increasing production efficiency but also on an increased focus on products that generate the highest margins or offer the prospect of winning market share.

COVID led to a standardization of product packaging. In some cases, retailers adjusted products with neutral packaging as well as packaging that showed multiple (or even other than regular) languages. Under normal circumstances, retailers would require customized packages, but supply shortages resulted in a more lenient approach. Several interviewed companies in the processing industry indicated that they were now able to standardize packaging on more points, for example, by giving all bottles the same colour cap, saving on changeover times in production. In the past, this was a more difficult discussion point with customers as well as between production departments and marketing departments. However, according to one supply chain manager the pandemic resulted in the fact that "...it is now ok for marketing to have small changes, such as using the same cap for multiple SKUs, which would normally not be the case. A cap is a cap."

Not only packaging was considered for standardization. One of the companies interviewed indicated that the crisis was an important push for the use of the same recipe for products across countries. This type of standardization offers the possibility to cope with current and future capacity problems in production by making the shifting of production volumes between locations more feasible.

10 Discussion: dealing with COVID-19

The analysis above shows that the impact of the COVID-19 pandemic was omnipresent in the food supply chain and that companies used a broad array of strategies to cope with the effects of the pandemic. Interestingly, most strategies were reactive. Not many organizations had specific plans readily available for coping with such lasting supply chain disturbances. Possibly, most companies did not pay attention to scenarios as unlikely as the COVID-19 pandemic (as indicated by one of our interviewees), and thus no contingency plans for such unlikely but impactful events were in place. Nevertheless, even well-prepared companies are expected to feel some level of unpreparedness as the COVID-19 pandemic is novel; an effective way to deal with such levels of complexity is very often not known (Sodhi and Tang, 2021; Ali et al., 2021).

Many companies we interviewed used a variety of capabilities to cope with the crisis, though it is interesting to observe the dominance of flexibility and agility capabilities (which underlines the reactive nature of the capabilities used). In fact, we observed companies tapping into sources of flexibility that would not be possible or at least be very difficult to leverage under normal circumstances. Tapping into these capabilities (or further developing these) was now possible due to the urgency experienced during the pandemic. An example is the reallocation of personnel, which would normally have led to employee resistance, but was now seen as a collective effort to deal with the exceptional circumstances. Also, changes to product packaging to ensure supply were possible, which would normally not have been accepted by the marketing department.

The need to increase flexibility and agility to improve preparedness is clear from our analysis. Our research shows that most company responses were highly reactive, and initiatives to decrease or spread risks might require more attention. This could for instance involve an analysis of dependencies on specific suppliers or supply regions, as well as customers or sales channels. For smaller companies, this might be challenging, since building such supply chain resilience capabilities requires substantial time and effort for which human resources might not be available (Ali et al., 2021). For larger companies, this might be easier to accomplish. Larger companies might also play a role in helping smaller suppliers or customers deal with supply chain disturbances, as there is a clear overall supply chain benefit.

Even though the resilience literature emphasizes the use of early warning signals (e.g. Christopher and Peck, 2004; Pettit et al., 2010), especially signals enabled by big data systems (Spieske and Birkel, 2021; Modgil et al., 2022), this was only mentioned sparsely by our interviewees (and particular in relation to learning from operations in countries where the COVID outbreak occurred earlier than in Western Europe). A limited use of early warning signals has also been reported by van Hoek (van Hoek, 2020). Possibly no historical data was relevant at the time because no past disruptions can be comparable to COVID-19 crisis, although the financial crisis of 2008-2009 also had disruptive effects on the supply chain (see e.g. de Leeuw and Wiers, 2015).

Also, for many companies, a more explicit strategy for prioritization and rationing would be beneficial in preparation for situations in which it is not possible to fulfil all demands. Rationing strategies have a long history in the supply chain management literature (e.g. Cachon and Lariviere, 1999), but the sudden need to ration products because of the pandemic turned out to be challenging for many companies, as trade-offs between e.g. profitability and fairness were not straightforward. Including an analysis of prioritization and rationing strategies in supply chain risk assessments would therefore be beneficial.

In contrast, for situations in which supply exceeds demand, the perishable nature of the products in food supply chains provides an additional complexity: you can't just stop material flows in a supply chain and store intermediate product temporarily like in several other industries. Chicklets grow into full grown chickens, just like corn seed becomes corn after a specific time. This leads to situations in which the lead time for potential supply adjustments is quite long, and resilience strategies might require more attention to the identification of alternative markets for products. Essentially, this would provide more flexibility and agility on the demand side of supply chains (e.g. the ability to quickly reallocate products between the foodservice and retail channels as also discussed by Chenarides et al., 2021).

The financial impact of changes in supply chains due to the pandemic was not always equally distributed in supply chains, and the impacts were highly influenced by the type of agreements and contracts between buyers and suppliers in the supply chain. For instance, more flexible supply contracts meant that some companies could easily deal with reduced foodservice demand, but that their suppliers had to deal with surplus products. From a societal perspective, the design of contracts that share the impact of supply chain disturbances in a way that limits the impact on the overall performance of the food system would be beneficial (Duong and Chong, 2020).

Interestingly, companies indicated that governmental measures and guidelines were often not detailed enough for direct implementation, such that a significant amount of time was spent discussing how to implement the measures in their specific company context. Also, in many cases, interaction with the authorities was required to be able to decide more precisely what was and what was not allowed. This calls for more clear guidelines for companies that provide more detail on the measures that have been installed by the government. The ideal time to develop such guidelines would be now since knowledge of the supply chain disturbances caused by the pandemic is still fresh in people's minds.

Finally, it is interesting to observe that the operations and supply chain function was at the forefront in most companies, while, e.g. product development and marketing were getting less attention. This was clearly visible in the rationalization and standardization of product assortments by many of the interviewed companies. Sometimes, such assortment rationalizations were being discussed before the pandemic, but the implementation became essential due to the pandemic. Whether or not these changes to product assortments will remain in place after the pandemic is not clear at the time of writing. Neither is it clear to what extent the classic discussion about conflicting objectives from operations and marketing will change permanently due to the lessons learned during the pandemic.

All in all, it seems that learning from a crisis and capturing that knowledge in actionable plans is difficult. The financial crisis of 2008–2009 showed a considerable drop in turnover across the board, not as severe as some sectors experienced during the COVID crisis (notably the food services) but still considerable. Empirical research on learnings from that financial crisis showed that in such a disruptive situation preparation through flexible planning strategies may be more productive than relying on (reactive) operational flexibility to counteract disruptions after the fact (de Leeuw and Wiers, 2015).

11 Conclusion and future trends

This study provides empirical evidence of how food companies responded to supply chain disruptions during the first (March–April 2020) and second (after summer holidays 2020) waves of the COVID-19 pandemic. Using semi-structured interviews with companies based in the Netherlands, covering all stages of the food supply chain, we collected and mapped the responses to different supply chain resilience capabilities. The findings show a dominance of flexibility and agility-related responses. It also demonstrates the inherent challenges and opportunities to employ specific resilience capabilities due to the long lead times of primary food production and the perishability of food products. Furthermore, our results indicated the increasing importance of rationalization as an effective way to increase supply chain resilience. In several of the cases we analyzed, rationalization was a key strategy in the response phase, used to mitigate the impact of the pandemic on supply chain operations. In our deductive category application process, we therefore also added rationalization as a key resilience capability in the context of the pandemic responses in food supply chains, complementing the capabilities we identified in the literature.

Our research reveals several interesting avenues of future research. First, future research could investigate the development of supply chain contracts that consider the sharing of risks resulting from disruption, especially during long-term supply chain disruptions like the COVID-19 pandemic. Whereas many existing studies discuss revenue-sharing contracts, a risk-sharing contract during disruptions can enhance trust among firms and financially facilitate a fair supply chain collaboration (Duong and Chong, 2020).

Second, we suggest to further investigate two specific capabilities that were identified in our empirical results but do not receive much attention in the supply chain resilience literature: (1) prioritization and rationing strategies, and (2) rationalization and standardization strategies. Both of these strategies are on the operations-marketing interface within companies and would thus require intra-organizational goal alignment. Also, in terms of timing, it would be beneficial to already consider such strategies in an early stage as part of initiatives to increase the readiness to supply chain disturbances. In the pandemic response studied in this chapter, many of the decisions related to these capabilities had to be made on extremely short notice.

Third, we observe limited employment of data and information systems in food companies for disruption management, even for large multinational companies. Real-time information on supply chain activities at all stages is crucial to overcome challenges during a disruption (Belhadi et al., 2021). Furthermore, given the long lead-time of food production and the impossibility to increase primary sector production volumes in the short term, research attention should also be given to digitalization and the use of data in supporting farm business decisions considering extreme disruptions.

Fourth and finally, even though supply chain collaboration only came up explicitly in a few interviews, it did seem to also have an important role in facilitating other capabilities (such as an increased sourcing flexibility in a case where a good relationship with the supplier existed). This mediating role of supply chain collaboration and good supplier relationships in relation to other supply chain resilience capabilities has been identified before in the literature (see e.g. Scholten and Schilder, 2015), but it is an aspect worth mentioning here and worth further investigation.

Our study collected empirical data from interviewees located in the Netherlands. Because different parts of the world experienced the pandemic at different times and food supply chain structures can be quite between markets (e.g. developed countries vs. developing countries), we realize that this is a limitation and that our findings might not generalize to any kind of food supply chain. However, several of the interviewed companies were multinational companies with operations in different continents, and our results therefore still partly reflect global responses. Another possible limitation is that we only report on the situation in 2020, meaning that we are not able to capture any long-term impacts. A future longitudinal study could therefore provide additional insight on food supply chain resilience in relation to long-term disturbances like the COVID-19 pandemic.

12 Acknowledgements

The authors are grateful for the financial support provided by Dinalog - the Dutch Institute for Advanced Logistics as well as for the many stakeholders from the Dutch food industry that took time out of their schedule during a difficult time.

13 Where to look for further information

Since the COVID-19 pandemic, there has been quite some attention to supply chain resilience in light of major disturbances like the one caused by the pandemic. The references used in the introduction of this chapter provide a good starting point for readers interested in this stream of research. Since the finalization of this chapter, many more interesting studies have been published. To complement this chapter, we for instance suggest Ali et al. (2023) and Zhao et al. (2024) for perspectives based on empirical studies in other countries.

Ali, I., Sadiddin, A. and Cattaneo, A. (2023). Risk and resilience in agri-food supply chain SMEs in the pandemic era: a cross-country study. *International Journal of Logistics Research and Applications* 26(11), 1602–1620.

Zhao, G., Vazquez-Noguerol, M., Liu, S. and Prado-Prado, J. C. (2024). Agri-food supply chain resilience strategies for preparing, responding, recovering, and adapting in relation to unexpected crisis: a cross-country comparative analysis from the COVID-19 pandemic. *Journal of Business Logistics* 45(1), e12361.

14 References

Aday, S. and Aday, M. S. (2020). Impact of COVID-19 on the food supply chain. *Food Quality and Safety* 4(4), 167–180.

Akkerman, R., Farahani, P. and Grunow, M. (2010). Quality, safety and sustainability in food distribution: a review of quantitative operations management approaches and challenges. *OR Spectrum* 32(4), 863–904.

Ali, A., Mahfouz, A. and Arisha, A. (2017). Analysing supply chain resilience: integrating the constructs in a concept mapping framework via a systematic literature review. *Supply Chain Management: An International Journal* 22(1), 16–39.

Ali, M. H., Suleiman, N., Khalid, N., Tan, K. H., Tseng, M. L. and Kumar, M. (2021). Supply chain resilience reactive strategies for food SMEs in coping to COVID-19 crisis. *Trends in Food Science and Technology* 109, 94–102.

Beckman, J. and Countryman, A. M. (2021). The importance of agriculture in the economy: impacts from COVID-19. *American Journal of Agricultural Economics* 103(5), 1595–1611.

Belhadi, A., Kamble, S., Jabbour, C. J. C., Gunasekaran, A., Ndubisi, N. O. and Venkatesh, M. (2021). Manufacturing and service supply chain resilience to the COVID-19 outbreak: lessons learned from the automobile and airline industries. *Technological Forecasting and Social Change* 163, 120447.

Bina, J. D., Tonsor, G. T., Schulz, L. L. and Hahn, W. F. (2022). Regional and plant-size impacts of COVID-19 on beef processing. *Food Policy* 108, 102247.

Brusset, X. and Teller, C. (2017). Supply chain capabilities, risks, and resilience. *International Journal of Production Economics* 184, 59–68.

Bryman, A. (2016). *Social Research Methods* (5th edn.). Oxford University Press, Oxford.

Burgos, D. and Ivanov, D. (2021). Food retail supply chain resilience and the COVID-19 pandemic: a digital twin-based impact analysis and improvement directions. *Transportation Research Part E: Logistics and Transportation Review* 152, 102412.

Burla, L., Knierim, B., Barth, J., Liewald, K., Duetz, M. and Abel, T. (2008). From text to codings: intercoder reliability assessment in qualitative content analysis. *Nursing Research* 57(2), 113–117.

Cachon, G. P. and Lariviere, M. A. (1999). Capacity choice and allocation: strategic behavior and supply chain performance. *Management Science* 45(8), 1091–1108.

Capodistrias, P., Szulecka, J., Corciolani, M. and Strøm-Andersen, N. (2022). European food banks and COVID-19: resilience and innovation in times of crisis. *Socio-Economic Planning Sciences* 82(A), 101187.

Chenarides, L., Manfredo, M. and Richards, T. J. (2021). Covid-19 and food supply chains. *Applied Economic Perspectives and Policy* 43(1), 270–279.

Chopra, S. and Sodhi, M. M. S. (2004). Managing risk to avoid: supply-chain breakdown. *MIT Sloan Management Review* 46(1), 53–61.

Chopra, S. and Sodhi, M. S. (2014). Reducing the risk of supply chain disruptions. *MIT Sloan Management Review* 55(3), 73–80.

Chowdhury, M. M. H. and Quaddus, M. (2016). Supply chain readiness, response and recovery for resilience. *Supply Chain Management: An International Journal* 21(6), 709–731.

Chowdhury, P., Paul, S. K., Kaisar, S. and Moktadir, M. A. (2021). COVID-19 pandemic related supply chain studies: a systematic review. *Transportation Research Part E: Logistics and Transportation Review* 148, 102271.

Christopher, M. and Peck, H. (2004). Building the resilient supply chain. *The International Journal of Logistics Management* 15(2), 1–14.

Craighead, C. W., Ketchen, D. J. and Darby, J. L. (2020). Pandemics and supply chain management research: toward a theoretical toolbox. *Decision Sciences* 51(4), 838–866.

Datta, P. (2017). Supply network resilience: a systematic literature review and future research. *International Journal of Logistics Management* 28(4), 1387–1424.

de Leeuw, S. and Wiers, V. C. S. (2015). Warehouse manpower planning strategies in times of financial crisis: evidence from logistics service providers and retailers in the Netherlands. *Production Planning and Control* 26(4), 328–337.

Duong, L. N. K. and Chong, J. (2020). Supply chain collaboration in the presence of disruptions: a literature review. *International Journal of Production Research* 58(11), 3488-3507.

Eisenhardt, K. M. (2021). What is the Eisenhardt method, really? *Strategic Organization* 19(1), 147–160.

FAO (2020). Mitigating risks to food systems during COVID-19: reducing food loss and waste. *Food and Agriculture Organization of the United Nations*, May 11, 2020. Available at: https://doi.org/10.4060/ca9056en.

FEBA (2020). Challenges and urgent needs of European Food Banks due to COVID-19. *European Food Banks Federation Briefing*, April 15, 2020.

Fleiss, J. L., Levin, B. and Paik, M. C. (2013). *Statistical Methods for Rates and Proportions*. John Wiley & Sons, Hoboken, NJ, USA.

Garnett, P., Doherty, B. and Heron, T. (2020). Vulnerability of the United Kingdom's food supply chains exposed by COVID-19. *Nature Food* 1(6), 315–318.

Grodal, S., Anteby, M. and Holm, A. L. (2021). Achieving rigor in qualitative analysis: the role of active categorization in theory building. *Academy of Management Review* 46(3), 591–612.

Guardian (2020a). 'Chaotic and crazy': meat plants around the world struggle with virus outbreaks. *The Guardian*, May 11, 2020.

Guardian (2020b). South Africans cheer lifting of ban on selling alcohol and cigarettes. *The Guardian*, August 18, 2020.

Hallgren, K. A. (2012). Computing inter-rater reliability for observational data: an overview and tutorial. *Tutorials in Quantitative Methods for Psychology* 8(1), 23–34.

Hendry, L. C., Stevenson, M., MacBryde, J., Ball, P., Sayed, M. and Liu, L. (2019). Local food supply chain resilience to constitutional change: the Brexit effect. *International Journal of Operations and Production Management* 39(3), 429–453.

Hobbs, J. E. (2020). Food supply chains during the COVID-19 pandemic. *Canadian Journal of Agricultural Economics* 68(2), 171–176.

Hobbs, J. E. (2021). Food supply chain resilience and the COVID-19 pandemic: what have we learned? *Canadian Journal of Agricultural Economics* 69(2), 189–196.

Kähkönen, A.-K., Evangelista, P., Hallikas, J., Immonen, M. and Lintukangas, K. (2023). COVID-19 as a trigger for dynamic capability development and supply chain resilience improvement. *International Journal of Production Research* 61(8), 2696–2715.

Kamalahmadi, M. and Parast, M. M. (2016). A review of the literature on the principles of enterprise and supply chain resilience: major findings and directions for future research. *International Journal of Production Economics* 171, 116–133.

Ketokivi, M. and Choi, T. (2014). Renaissance of case research as a scientific method. *Journal of Operations Management* 32(5), 232–240.

Kochan, C. G. and Nowicki, D. R. (2018). Supply chain resilience: a systematic literature review and typological framework. *International Journal of Physical Distribution and Logistics Management* 48(8), 842–865.

Los Angeles Times (2020). Rotting food. Hungry masses. Chaotic supply chains. Coronavirus upends the U.S. food system. *Los Angeles Times*, May 5, 2020.

Mayring, P. (2014). *Qualitative Content Analysis: Theoretical Foundation, Basic Procedures and Software Solution*. SSOAR, Klagenfurt.

McCutcheon, D. M. and Meredith, J. R. (1993). Conducting case study research in operations management. *Journal of Operations Management* 11(3), 239–256.

Modgil, S., Gupta, S., Stekelorum, R. and Laguir, I. (2022). AI technologies and their impact on supply chain resilience during COVID-19. *International Journal of Physical Distribution and Logistics Management* 52(2), 130–149.

Pettit, T. J., Fiksel, J. and Croxton, K. L. (2010). Ensuring supply chain resilience: development of a conceptual framework. *Journal of Business Logistics* 31(1), 1–21.

Ponomarov, S. Y. and Holcomb, M. C. (2009). Understanding the concept of supply chain resilience. *The International Journal of Logistics Management* 20(1), 124–143.

Poppe, K. (2020). Covid-19 will change the agri-food system - but how? *EuroChoices* 19(3), 20–25.

Scholten, K. and Schilder, S. (2015). The role of collaboration in supply chain resilience. *Supply Chain Management: An International Journal* 20(4), 471–484.

Sodhi, M. S. and Tang, C. S. (2021). Supply chain management for extreme conditions: research opportunities. *Journal of Supply Chain Management* 57(1), 7–16.

Sodhi, M. S., Tang, C. S. and Willenson, E. T. (2023). Research opportunities in preparing supply chains of essential goods for future pandemics. *International Journal of Production Research* 61(8), 2416–2431.

Spieske, A. and Birkel, H. (2021). Improving supply chain resilience through industry 4.0: a systematic literature review under the impressions of the COVID-19 pandemic. *Computers and Industrial Engineering* 158, 107452.

Statista (2020). The coronavirus in Europe. Available at: www.statista.com/chart/20964/covid-19-cases-europe-map/, accessed March 9, 2020.

Stevens, A. W. and Teal, J. (2023). Diversification and resilience of firms in the agrifood supply chain. *American Journal of Agricultural Economics* 106(2), 739-778.

Stone, J. and Rahimifard, S. (2018). Resilience in agri-food supply chains: a critical analysis of the literature and synthesis of a novel framework. *Supply Chain Management: An International Journal* 23(3), 207–238.

Teece, D. J. (2007). Explicating dynamic capabilities: the nature and microfoundations of (sustainable) enterprise performance. *Strategic Management Journal* 28(13), 1319–1350.

Teece, D. J. (2018). Business models and dynamic capabilities. *Long Range Planning* 51(1), 40–49.

Teece, D. J., Pisano, G. and Shuen, A. (1997). Dynamic capabilities and strategic management. *Strategic Management Journal* 18(7), 509–533.

Van Hoek, R. (2020). Research opportunities for a more resilient post-COVID-19 supply chain - closing the gap between research findings and industry practice. *International Journal of Operations and Production Management* 40(4), 341-355.

Voss, C., Tsikriktsis, N. and Frohlich, M. (2002). Case research in operations management. *International Journal of Operations and Production Management* 22(2), 195–219.

Washington Post (2020). Extra food is rotting on farms while Americans go hungry. *The Washington Post*, June 10, 2020.

Yin, R. K. (2018). *Case Study Research and Applications: Design and Methods* (6th edn.). Sage Publications, Thousand Oaks, CA, USA.

Chapter 12

Improving agri-food supply chains in Africa

Danie Jordaan and Thabile Tsabedze, University of Pretoria, South Africa

1 Introduction

Agri-food value chains across the African continent require integration into global value chains (GVCs) for sustainable recovery and enhanced food security. According to the Organization for Economic Co-operation and Development (OECD). Africa's current participation in GVCs is only around 2.7% (measured in terms of added value trade), underscoring the imperative need for greater integration, in terms of the following:

- Forward integration: where a country provides inputs for another country's export production; or
- Backward integration: where a country sources inputs from abroad for its own export production.

Backward integration provides the greatest opportunity to add value and therefore improve a country's GDP (OECD, 2017). Integrating these value chains into GVCs is vital for providing improved market access, fostering vertical integration and generating increased income, thereby contributing to rural development and overall economic growth. Analysing the existing

http://dx.doi.org/10.19103/AS.2023.0122.17

state of value chains in Africa reveals various constraints and issues that have hindered their development. Notable challenges include restricted land access for women continent-wide, limited access to finance, inadequate infrastructure, intracontinental trade barriers, and the impacts of climate change on agricultural production in particular (Dessalegn and Eziakonwa, 2023).

Developing sustainable agricultural value chains requires a comprehensive approach tailored to the specific needs of different continents and countries. Success in value chains hinges on various factors. Regions and countries must prioritize good governance to ensure political and economic stability, creating a conducive environment for agribusiness development. Maintaining macroeconomic and fiscal stability through government economic policy is essential to providing a stable environment for agricultural and other businesses. Also critical to success is infrastructure development, encompassing transportation, storage, communication, and energy systems, as well as appropriate standards and infrastructure for export of agricultural raw materials and food products, which play a pivotal role in supporting food industries and agribusiness.

Supporting institutions are another vital factor, emphasizing the establishment of robust institutions and capacity building for value chain actors and advocacy groups. This fosters an environment conducive to the growth of agribusinesses. A supportive legal framework, incorporating land tenure systems, investment regulation and trade policies, is equally crucial for the development of thriving value chains. There also needs to be a sufficient emphasis on promoting food safety and quality, coupled with the implementation of supportive systems and policies, particularly to facilitate integration into GVCs which have minimum standards in areas such as traceability, safety and quality (e.g. GLOBALGAP standards for fresh produce). These factors collectively create an enabling environment for the development of value chains and agro-industries, thereby contributing significantly to the growth and prosperity of the food industry (Koning et al., 2013).

2 The development of African agri-food value chains

De Brauw and Bulte (2021) note that agricultural value chains in Africa have transitioned from informal, localized forms of exchange (e.g. with smallholders selling direct to consumers in local markets) to more formal exchange mechanisms (e.g. via formal contracts to supply produce to regional, national and international markets), demonstrating a non-linear progression over time. The colonial period, marked by extractive practices, established institutions focused on extracting valuable goods from Africa, supplying cheap food commodities such as tropical fruits to Europe and elsewhere. Policy changes between colonial and post-colonial eras initially led some smallholder farmers towards more formally organized markets. However, fluctuations in trade and resulting price volatility prompted some producers back into selling surplus produce through informal, local channels or in spot markets, usually on poor terms for smallholder farmers.

Many post-colonial African governments initiated interventionist policies for their agricultural sectors, e.g. by imposing tariffs on exports as a major source of income and/or by subsidizing producers and imposing price controls on food produce to ensure a supply of cheap food to rapidly growing urban populations. This changed with a structural revolution in the 1990s when many governments started to abandon this interventionist approach in favour of liberalizing agricultural production and trade, including deregulation of producer and consumer prices, reducing taxes and subsidies affecting agricultural production, abolition of state monopolies and privatization of government agencies, and removing barriers to private sector investment.

These reforms have been accompanied by broader attempts to liberalize world trade (e.g. to improve access to the lucrative EU market) and to promote regional trade between African countries. The potential for productive transformation in Africa lies in initiatives such as the African Continental Free Trade Area (AfCFTA), which aims to facilitate intra-African trade, reduce barriers to investment, and address infrastructure challenges. The AfCFTA, currently implemented in 17 African countries, seeks to develop regional value chains (African Union Commission, 2022a). Nurturing regional value chains is vital for promoting regional growth and diversification, increasing resilience to future shocks.

This process of liberalization has had mixed results. Whilst it has stimulated investment in and growth in exports of some agricultural produce, it has not led to the hoped-for broader increases in agricultural production and productivity in individual African countries which can both improve rural livelihoods and ensure a sufficient supply of affordable food to local populations (Dorward et al., 2004; Odjo et al., 2023). This has been attributed to a range of factors, including only partial liberalization by some governments, continuing weaknesses in an appropriate supporting environment and infrastructure to allow investment and productivity improvement, a resulting reluctance by private-sector actors to invest and the particular challenges of enabling smallholders (who still dominate agricultural production in Africa) to escape poverty, improve production, incomes and security for their families (Klauser and Robinson, 2021).

This has meant that agricultural value chains in Africa have continued to historically underperform in GVC trade, with African GVC exports averaging 8% of national GDP, compared to 11% for Asia and 14% for other high-income countries during the 2000–2015 period (Black et al., 2021). In addition to the factors mentioned previously, economic shocks such as the 2008 financial crisis, the COVID-19 pandemic and the war in Ukraine, as well as uncertainties in trade agreements, have raised costs, damaged the global economy and affected Africa's GVC exports. This has limited the potential in exploiting Africa's rich mineral and agricultural resources in value chain development (De Melo and Twum, 2021; African Union Commission, 2022b).

3 Challenges facing African agri-food value chains

African value chains continue to face various challenges that impede their development and undermine overall productivity. These challenges, as identified by Balakrishnan and Zaas (2023), encompass inadequate transport and storage facilities, weak market linkages, non-compliance with food safety standards and sanitation practices, financial constraints, a challenging regulatory environment, climate change, and insufficient infrastructure (e.g. a good transport network). Insufficient and substandard storage facilities contribute to substantial post-harvest losses and significant wastage, ultimately leading to diminished agricultural productivity (Stathers and Mvumi, 2020). Weak market linkages restrict farmers' access to markets, potentially resulting in the sale of produce at less competitive prices. Ensuring food safety and proper sanitation remains a significant challenge in numerous African countries, negatively impacting the quality of produce, demand for agricultural products and potential for export.

Complicated regulatory environments within African countries, particularly concerning certification, regulation and compliance processes, further hinder the development of agri-food value chains. Limited financial resources pose barriers to accessing essential production resources such as credit, seeds, machinery (such as tractors), fertilizers, pesticides and irrigation equipment. These financial constraints disproportionately affect women across the continent, limiting their access to crucial production resources, including land and markets, and thereby restricting their participation in the agri-food sector (Balakrishnan and Zaas, 2023).

Infrastructure challenges, such as substandard roads and transportation networks, coupled with a lack of cold storage facilities, pose obstacles to the storage and distribution of agricultural produce, resulting in significant losses and reduced productivity. Additionally, the pervasive issue of climate change and variability negatively impacts the stability and sustainability of the entire agricultural value chain. Effectively addressing these challenges necessitates collaborative efforts from all stakeholders involved in the agri-food system (AUC/OECD, 2022).

4 Improving African agri-food value chains: enablers

As noted earlier, value chains require an enabling environment to function effectively and efficiently. An enabling environment for value chains is especially important for attracting investment, de-risking chains and enabling returns. An enabling environment for value chains characterized by essential, important and useful enablers (Fig. 1) (Christy et al., 2009).

Essential enablers involve the surety of tenure and property rights, infrastructure, and trade policies. Important enablers involve norms, standards, regulations, and services associated with production, value addition and consumption, and financial services that serve the unique needs of value chains. Useful enablers include the ease of conducting business, business development services, and business connections for value chains. The creation of an enabling environment is essential to support the attractiveness of value chains as a source of investment and to optimize growth and efficiency (Christy et al., 2009).

Beyond creating and enabling environment, enhancing value chains in Africa requires a comprehensive strategy focused on scaling up food production, improving food marketing and overhauling farm management systems and infrastructure (Sasson, 2012). The improvement of food value chains requires a multifaceted approach, incorporating advanced farm management practices, modern technology, storage facilities and effective marketing to reduce dependence on food imports and aid in developing inclusive and resilient food systems (Dessalegn and Eziakonwa, 2023).

Secondly, stimulating import diversification and trade liberalization is crucial for stabilizing regional food prices and supply. Empowering producer organizations to reach climate-vulnerable agricultural communities and disseminate information about new technologies can facilitate this process (Fayad et al., 2022). Thirdly, enhancing the productivity of small-scale farmers and improving value chain efficiencies can address downstream challenges. Reinforcing cold chain solutions is essential for reducing value chain losses, a critical component of food systems in Africa (Odusola, 2021).

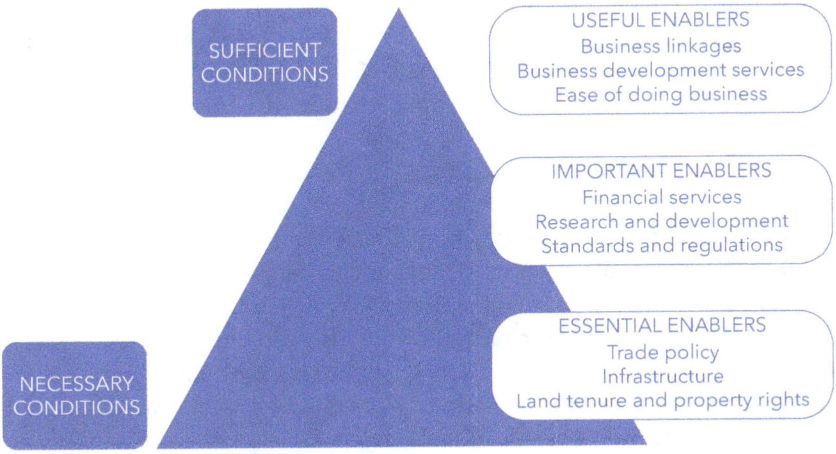

Figure 1 Hierarchy of enablers. Source: Adapted from: Christy et al. (2009).

Intensifying African value chains is pivotal for achieving economic growth. Creating an environment conducive to technology-intensive industries can provide better access to finance solutions, boosting value chain finance for businesses. This, in turn, has the potential to increase wages within the continent, addressing the current disparity, with Africa's minimum wage at $220 per month compared to the US average of $668 (Bhorat et al., 2017). Small- and medium-sized enterprises (SMEs) require increased value chain finance to bridge payment gaps, alleviate financial strain and enhance access to working capital. Despite a reported 40% rise in the value of the value chain finance market in Africa in 2021 and 2022, additional investment is crucial for Africa to become a primary exporter of high-value goods, thereby fostering economic growth, job creation, increased productivity and wages (UNCTAD, 2023). Lastly, initiatives such as the AfCFTA present opportunities for the continent to attract investment and meet rising regional demand. However, limited digital platforms and technology-enabled services for value chains in Africa result from challenges such as the lack of institutional framework and regulation, infrastructure deficiencies, insufficient investment in innovation and information asymmetry (UNCTAD, 2022).

5 Improving African agri-food value chains: interventions

To foster agricultural value chain development in Africa, a strategic focus on market access, value chain efficiency, sustainability and product quality is imperative. Interventions should target the reduction of post-harvest losses, enhancement of market access, infrastructure development, technology adoption and collaboration among stakeholders.

To strengthen GVC participation, African countries must prioritize enhancing value chain resilience, building adaptive capacity and promoting regional value chains (AUC/OECD, 2022). Governments also need to develop trade policies, regulations and economic recovery reforms to provide market access for local producers, including small and medium enterprises (SMEs) across the region (Conde et al., 2015). Governments and stakeholders alike need to invest in research and development, create an innovative environment and promote entrepreneurship. Training programmes addressing sustainable food production, quality control, good agricultural practices and post-harvest handling are crucial for all stakeholders (AGENPO, 2023).

Digitizing agro-value chain activities is also essential for effective interactions among financiers, input suppliers, producers, logistics and transport service providers, storage facilities providers and other value chain partners (Ba, 2016). The significance of digital transformation in Africa cannot be underestimated, particularly in the aftermath of global challenges such as the COVID-19 pandemic, which restricted movement within the continent (AUC, 2022). Partnerships between African private sector companies and international entities should actively support the development of inclusive and local value chains (Omta et al., 2014).

African nations must address key bottlenecks in industrialization and trade to fortify sustainable and resilient participation in GVCs. Implementing interventions such as strengthening regional value chains, digital service delivery and green economy initiatives can be facilitated through ongoing trade policies and economic recovery reforms (Koning et al., 2013; AUDA-NEPAD, 2022b). Designing favourable investment policies to attract more green infrastructure investments is paramount. Strengthening value chains is equally essential for driving rural development, alleviating poverty and enhancing living standards in rural communities by increasing market access, generating higher income opportunities and improving overall well-being (Ajibo, 2023; AUC, 2022). Business-friendly rules, transparent and fair policies, streamlined cross-border investment promotion, cross-border trade-in-services and enterprise regionalization can stimulate the development of regional value chains (Levin and Makgetla, 2021).

Stimulating the development of value chains in Africa necessitates significant infrastructure investments. Improving transportation channels, storage (esp. cold chain infrastructure), communication networks, energy systems and promoting technology-intensive industries are critical (Peters and Sayin, 2021). Attracting investment opportunities through public–private partnerships, implementing fair local regulations and boosting infrastructure development are vital steps (Balakrishnan and Zaas, 2023).

Furthermore, African countries should capitalize on potential opportunities and address challenges to contribute to the continent's economic growth. Interventions such as investing in the development and training of talent in value chains to enhance the knowledge and skills of the labour force and entrepreneurs are essential (AGENPO, 2023). Building strong relationships, including strategic alliances with local partners to navigate cultural complexities and the regulatory environment, is crucial (Goedde et al., 2019). Strengthening regional integration, exemplified by initiatives such as the AfCFTA, offers significant synergies for integrating value chains and facilitating the flow of goods and services across borders (Gayi and Banini, 2019).

6 The food and consumer revolution in the African continent

The food and consumer revolutions in Africa represent significant developments encompassing production, consumption and retail aspects. The food revolution highlights both the promotion of local indigenous foods, the increased consumption of processed foods and the rise of local supermarket chains. Emphasizing expertise in traditional food sectors, particularly in rural African villages, the food revolution has also prioritized sustainable production and strengthened traditional food systems through developments in science and technology (Barrett et al., 2019).

Currently around 70% of Africans still purchase staple foods from traditional local vendors. However, the continent has seen a consumer revolution characterized by the growth of local supermarket chains, a burgeoning consumer middle class, online retail and widespread use of mobile phones (Barrett et al., 2019). Online retail significantly contributes to African value chains by enhancing efficiency, providing extensive market reach for businesses, reducing value chain challenges, accelerating consumer access to diverse products and enabling remote connections for small businesses into GVCs, thereby potentially increasing employment and income levels (Kuteyi and Winkler, 2022; Igue et al., 2021).

The consumer revolution in Africa has transformative effects on the food industry, resulting in changes in consumption patterns and a notable increase in chains of local supermarkets. The shift in consumption patterns is driven by rising incomes, encouraging the increased consumption of highly processed and fast foods while, in some cases, reducing demand for traditional healthy foods and preparation methods (Mitchell, 2021). The growth of local supermarket stores is reshaping the retail landscape in Africa, enhancing convenience and accessibility, particularly in townships (Crush et al., 2018). The growth of urban households has had a particular impact on the food industry as spending in townships has risen twice as fast as in rural areas, driving increased demand for food and the growth of local supermarkets. The consumer revolution also diversifies food demand, fostering opportunities for innovation in the food industry to cater to evolving preferences (Crush et al., 2018). Additionally, there is an increased focus on product quality as consumers expectations of quality have increased, prompting improvements in the food industry and addressing challenges in agricultural value chains such as low-quality storage facilities and poor market linkages (Balakrishnan and Zaas, 2023).

The processed food revolution in Africa over the last 50 years is propelled by factors such as increased income, urbanization, import liberalization and shifts in employment patterns. The rising number of women working outside the home and men commuting to off-farm jobs has shifted consumption towards highly processed foods, including packaged items such as bread and cookies which are manufactured in large-scale processing plants rather than prepared at home or by local businesses such as bakeries (Reardon et al., 2021).

7 Case studies of successful African agri-food value chains

Several African countries have successfully developed agricultural value chains, capable of stimulating economic growth and enhancing food security. Initiatives such as the African Union's New Partnership for Africa's Development (NEPAD) have played a crucial role in providing support and leadership for the development of regional value chains across the continent. Cooperatives and collective action initiatives have facilitated agribusiness, allowing farmers to share resources, knowledge and expertise, leading to increased market access and productivity. Encouraging strategic partnerships between customers and agro-business partners has also improved the competitiveness and efficiency of the agricultural sector in many African countries (Ba, 2016). Effective strategies for improving agricultural value chains include investment in rural services, streamlining certification processes and providing access to context-specific training (Dazé and Dekens, 2016). Lastly, certification schemes and strengthening the enabling environment are identified as secondary factors that contribute to sustainable value chain transformation (Hidayati et al., 2023).

7.1 The wool and mohair value chain in South Africa and Lesotho

The wool value chain in South Africa stands out as highly successful, with the country ranking as the second-largest exporter of apparel wool globally, following Australia. Additionally, South Africa has been the leading producer of mohair since 2016. Contributing approximately 2% to the world's total wool production, South Africa predominantly supplies high-quality fibre to niche markets. The country exports over 75% of the 45 000 tons of raw wool it produces, making wool the highest-earning export in the agricultural sector, particularly in the Eastern Cape province, responsible for 34–38.6% of the national wool production since 2005 (CWSA, 2020).

The South African Wool and Mohair Buyers Association (SAWAMBA) supervises most overseas wool sales, and the industry is efficiently organized under Cape Wools SA, ensuring sustainable practices, animal welfare and fair labour standards. Educational programmes, offered through institutions such as Grootfontein College of Agriculture and the Sheep and Wool Academy for Southern Africa, further contribute to the industry's success (Hertantyo, 2022).

In Africa, South Africa and Lesotho jointly account for more than 70% of the world's mohair output. Industry leaders are exploring advanced technologies from EU companies for further processing of wool into clothing. Despite the potential industrial applications, there is a need to expand mohair volumes, stimulate local breeding programmes and address the current shortfall in raw material availability. This lack of raw material hinders local production of wool products which, at present, operates at less than 50% of total capacity. Short-term efforts should focus on exploring high-value opportunities, stimulating breeding programmes and increasing mohair fibre volumes. The wool and mohair sector in South Africa has benefited from empowerment programmes targeting rural communities, enhancing herds and improving market access. These initiatives are pivotal for increasing local wool output to justify capital investments in local processing plants. The industry has also adopted responsible production standards, showcasing a positive development in sustainable farming practices (Letebele, 2020).

In Lesotho, the wool and mohair industries significantly impact the economy, contributing to income generation and export revenue. In 2018, wool and mohair constituted the largest agricultural export and the fifth contributor to Lesotho's GDP, representing approximately 3.37% of overall exports. South Africa and Lesotho's wool and mohair exports were valued at $271 007 and $24 883, respectively, in 2018, positioning South Africa as the second-largest wool supplier globally after Australia. These exports translate into increased revenue for farmers, elevating household income, improving livelihoods and fostering economic growth in both countries (Public Eye, 2019).

The South African wool industry has strategically positioned itself to capitalize on various market opportunities, supported by programmes designed to enhance the sector's key attributes. To ensure effective organization, the industry aligns itself with Cape Wools SA, a robust body overseeing the interests of various value chain participants, including breeders, small-scale farmers, wool producers, shearing services and exports (Fig. 2). This affiliation fosters collaboration among stakeholders, minimizing the risk of conflicting interests and promoting unified efforts in development projects. Market positioning and recognition are integral aspects, with South Africa earning a reputation for producing high-quality merino wool, primarily marketed through monthly auctions in Port Elizabeth (Cape Wools SA, 2019).

Differentiation is a cornerstone in the industry's pursuit of excellence. Cape Wools SA guarantees adherence to international standards, emphasizing sustainable wool production and animal welfare. The product complies with fair and ethical trade principles, including traceability, crucial requirements for the EU markets. Traceability is maintained from production through the auction floor, ensuring verification of compliance with sustainable production standards. In pursuit of differentiation, the wool undergoes high-quality processing locally before reaching international markets. Notably, the industry focuses on supplying niche markets, recognizing the diverse applications of wool qualities, from carpets to fine apparel (Fig. 3). Conducting focused market studies allows the industry to align product grades with specific customer needs, targeting high-grade audiences and capitalizing on niche markets to achieve higher profit margins (Cape Wools SA, 2019).

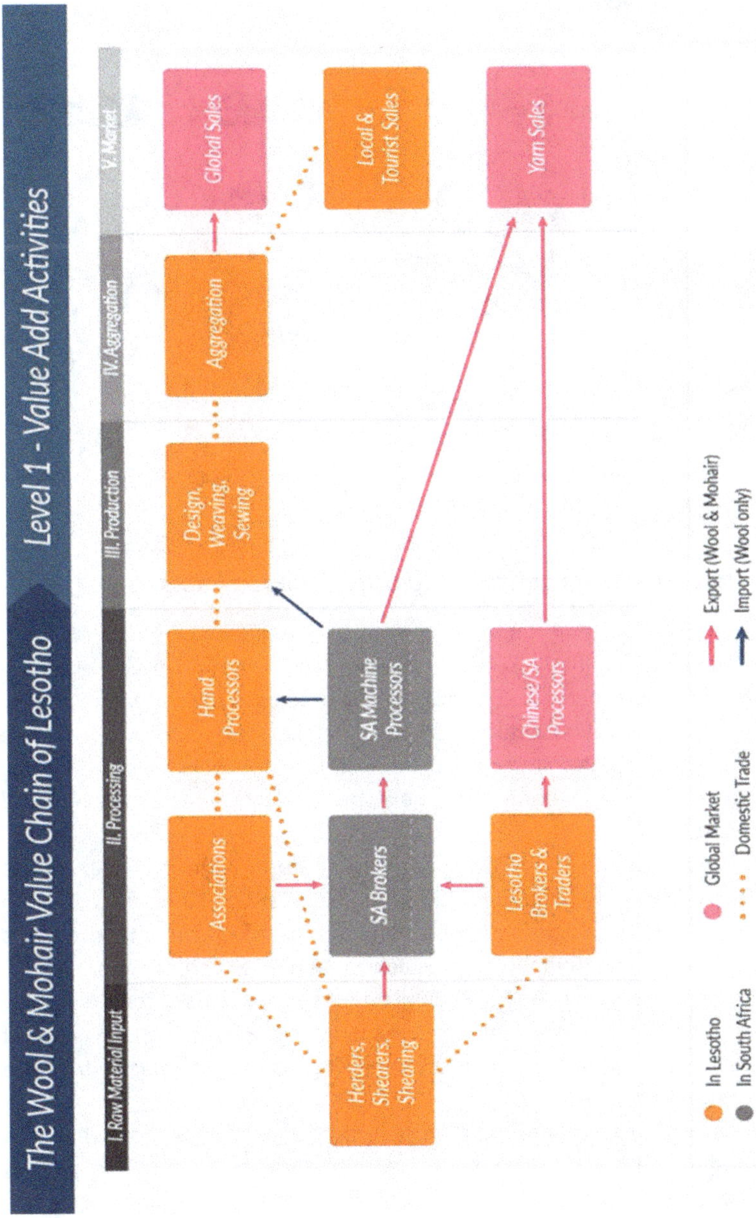

Figure 2 Lesotho and South Africa wool and mohair value chain stages. Source: Adapted from: Millennium Challenge Corporation (2022).

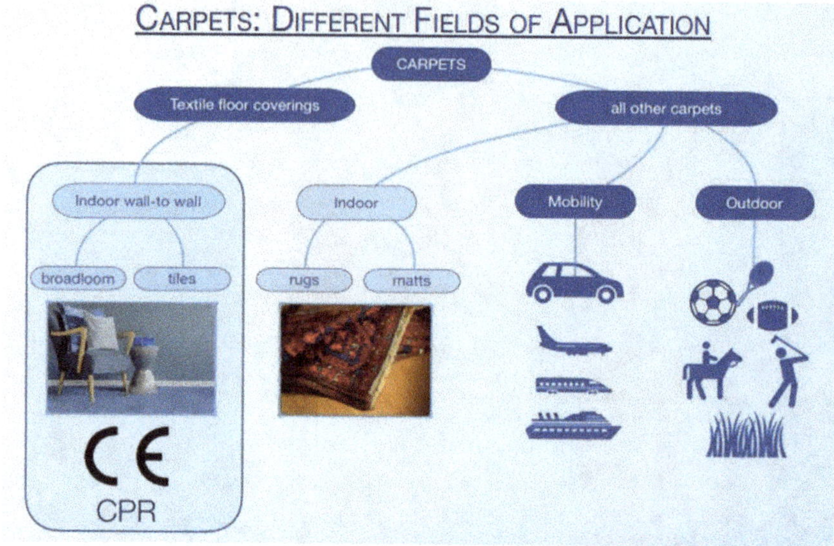

Figure 3 Processing of wool into different types of carpets in the EU. Source: Adapted from: ECRA (2021).

7.2 Finance and insurance for agricultural value chains in Nigeria

Numerous successful initiatives have been introduced and implemented in Nigeria's finance and insurance sector to support agricultural value chains. Historically, Nigeria's agricultural sector faced challenges such as inadequate infrastructure, poor market linkages, and limited use of technology, resulting in low productivity, lack of competitiveness, and heavy dependence on food imports (FAO, 2021).

One notable initiative is the 'Feed the Future Nigeria Agribusiness Investment Activity', spearheaded by the Cultivating New Frontiers in Agriculture (CNFA) organization (https://www.cnfa.org/). This initiative aims to enhance agricultural finance access for smallholder farmers and micro, small, and medium enterprises (MSMEs). The focus is on creating an enabling environment for agricultural industry growth by improving investment opportunities, infrastructure, and addressing logistical and land ownership constraints (Shaibu and Tasie, 2022).

The Markets II Project in Nigeria, implemented in 18 states, concentrates on facilitating market access for various agricultural products. This project accelerates access to credit for farmers and SMEs, enhancing agricultural value chains nationwide (Ajani and Igbokwe, 2014).

The Nigerian Incentive-based Risk Sharing System for Agricultural Lending (NIRSAL) plays a pivotal role in institutional lending for the country's agricultural sector. With a focus on policy reformation, NIRSAL channels financial resources through government and Central Bank-sponsored initiatives. This $500 million public liability organization, fully owned by the Central Bank of Nigeria, aims to provide farmers with financial products and reduce loan risks for local banks. The NIRSAL financing initiatives, including Risk Sharing Facility (RSF), Technical Assistance Facility (TAF), Insurance Component (IC), Agricultural Bank Rating System (ABRS), and Bank Incentive Mechanisms (BIM), absorb up to 75% of the risk, ensuring sustainable financing for agriculture. NIRSAL's success includes providing credit guarantees for over 454 agricultural projects, paying out interest rebates, and guaranteeing a substantial portion of Nigeria's Ministry of Agriculture's Growth Enhancement Scheme (GES) projects (Polycarp, 2018).

Lastly, successful initiatives for Nigerian agricultural value chains involve crop insurance and agricultural digitalization. The Pay-at-Harvest (PAH) insurance scheme, facilitated by the Shell Foundation, aims to increase insurance access for farmers. Noteworthy digitalization efforts are led by AgroMall, an agritech company affiliated with The Global Systems Mobile Telecommunications Association (GSMA) Innovation Fund for Digitization of Agricultural Value Chains. AgroMall leverages digital technologies to provide advisory and agri Digital Financial Services (agri DFS) such as loans, insurance, and digital agronomic services, ultimately supporting farmers in improving profit margins (Zechner, 2021).

7.3 Kenya's flower value chain

The Kenyan flower value chain stands as a notable success story in the evolution of value chains in Africa. This particular value chain involves smallholder growers engaged in the cultivation of flowers destined for the global market. Kenya has emerged as a significant global exporter of cut flowers, with smallholder farmers playing a pivotal role in the country's overall flower production (Kenya Flower Council, 2023). According to IIED (2010), these smallholder growers contribute substantially to the production of various summer flowers, which are subsequently sold either by auction in The Netherlands (the leading player in the European floriculture market) or direct to retail customers in Europe or the United States, e.g. via companies such as Wilmar which supplies over 5000 Kenyan smallholder farmers with subsidized seed and agronomic advice as well as finding and negotiating with overseas customers (https://www.wilmar.co.ke/).

While smallholder flower farmers in Kenya have achieved success, they grapple with challenges in the global market, including inadequate infrastructure, limited access to credit, and a deficiency of technical knowledge. However, Bolo (2010) highlights that Kenyan smallholder growers have enhanced their global market position through the establishment of robust partnerships with buyers and other stakeholders within the value chain. Consequently, there has been an increase in production and revenue, with an average income from exports of Kshs 68000 per year, constituting approximately 60% of total household income for smallholder flower farmers for the year 2008/2009 (Fig. 4) (IIED, 2010).

Collaborative ventures between farmers and exporters play a pivotal role in elevating the capacities of smallholder farmers, acting as a key catalyst for achieving sustainability within agricultural value chains. These partnerships are instrumental in addressing various challenges encountered by farmers, such as securing market access and fostering innovation among smallholders. In the context of Kenyan smallholder flower producers, these partnerships have notably facilitated their integration into the high-value floriculture export industry.

Implementation of these partnerships has led to a notable enhancement of grower production capabilities. Farmers have acquired knowledge of new technologies, gained access to improved crop varieties, refined production techniques and acquired compliance with environmental standards. Additionally, the partnerships have contributed to an augmentation of value addition capabilities. Through these collaborations, farmers have acquired proficiency in new bouquet arrangement techniques, improved packaging methods, reduced spoilage of flowers and learned strategies to prolong the shelf life of their produce. Lastly, the partnerships have been instrumental in bolstering growers' marketing capabilities by lowering production costs, imparting advanced marketing strategies and ensuring compliance with evolving market standards (Bolo, 2010).

7.4 Rwanda coffee global value chain

The coffee sector in Rwanda has played a pivotal role in significantly reducing poverty in the country in recent years. The government's National Coffee Strategy, coupled with market liberalization, has led to a notable decline in the

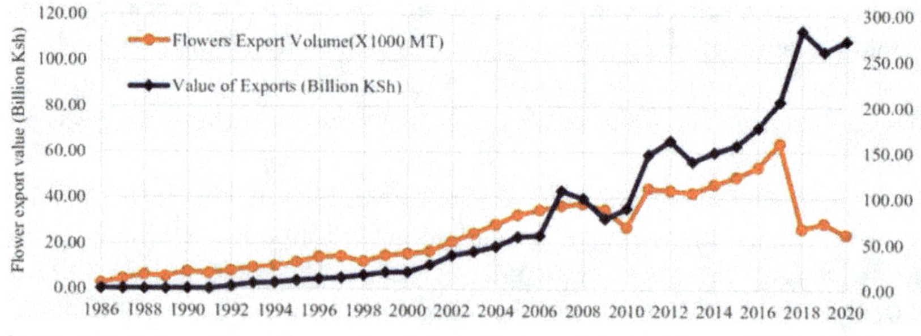

Figure 4 Statistics of flower export volumes. Source: Adapted from: KNBS (2020).

proportion of the population living in poverty, dropping from 69.1% in 2005 to 56.5% in 2016 (Macrotrends, 2023). This positive impact provides communities with enhanced opportunities to augment income and savings, thereby improving living standards and overall quality of life. The proliferation of coffee cooperatives in the country has not only contributed to economic growth but has also played a role in mitigating ethnic tensions.

The coffee industry stands as a key catalyst for economic and social recovery in Rwanda, transforming it from a sector characterized by poor-quality production cycles to a highly competitive export industry (Fig. 5) (USIDFC, 2023). As a result, the Rwandan coffee value chain has emerged as a potent tool in the fight against poverty, generating increased incomes for coffee farmers and creating new employment opportunities (Boudreaux, 2007).

In recent years, Rwanda's coffee value chain has experienced considerable success, contributing significantly to the country's agricultural export value. In 2017 coffee represented approximately 7% of total exports and 20% of the overall value of agricultural exports in Rwanda (LOGIC, 2018). In 2020 this increased to more than 24% of total agricultural exports, comprising around 19% of the country's overall agricultural export value (Daniels, 2021).The country has demonstrated substantial potential for enhancing coffee quality and improving processing efficiency, thereby increasing its competitiveness in global markets (Clay, 2018).

Successful initiatives from both private enterprises and governmental efforts, particularly the prioritization of the coffee value chain and the establishment of the Farmer Support Centre, have helped develop Rwanda's coffee industry (Oirere, 2022). A pivotal factor in this success has been the transformation of the coffee sector from a low-quality, low-production cycle industry to a highly competitive one. Notably, the active involvement of extraordinary women, exemplified by the Hingakawa women's group, has played a remarkable role in the success of the coffee value chain (Hameed and Harpst, 2014). The Rwandan coffee industry serves as a noteworthy example of successful value chains in Africa, playing a vital role in the country's economy by significantly contributing to rural and foreign exchange income. This includes access to global markets through partnership with major international companies, including Starbucks, the world's biggest chain of coffee houses (https://athome.starbucks.com/learn/how-we-approach-ethical-sourcing).

The Rwandan coffee industry has achieved this position by improving quality, volume and reliability of supply, particularly the production of high-quality specialty blends which has enabled the country to command higher market prices in the ready-to-drink coffee export sector. Rwanda's specialty coffee has successfully competed on the global stage, with Starbucks as a major buyer. The Rwandan government has actively implemented quality control programmes to enhance competitiveness in international markets.

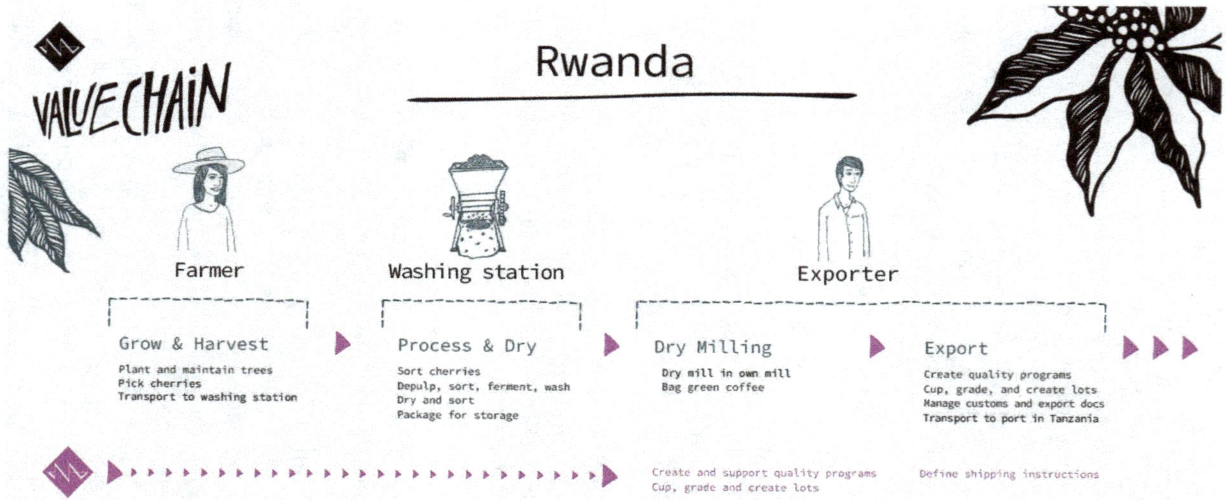

Figure 5 The value chain process for coffee in Rwanda. Source: Adapted from: Nordic (2023).

The partnership between Starbucks and Rwanda's coffee industry stands as a testament to the prosperity of Rwanda's coffee GVC (USIDFC, 2023).

7.5 The Rural Cattle and Dairy Opportunities (MERCADO) Project in Mozambique

The rural cattle and dairy value chains in Mozambique, supported by the Mozambique Expansion of Rural Cattle and Dairy Opportunities (MERCADO) project, have demonstrated success in enhancing dairy production and improving the livelihoods of farmers. Funded by the United States Department of Agriculture (USDA) and administered by the Mozambican organization Land O'Lakes Venture 37, the project aims to boost productivity in Mozambique's dairy sector and expand trade opportunities. Successful outcomes include strengthened collaboration among stakeholders, the provision of dairy cows to farmers, training initiatives and the establishment of centralized milk collection, processing and distribution centres. Additionally, the project has played a pivotal role in capacity building and training for farmers, contributing to the enhancement of policy and regulatory frameworks in the cattle and dairy sector (O'Donovan, 2017). The project has also created regional and international networking opportunities, fostering links and business opportunities within the dairy sector. The project has actively participated in regional and local events, such as the Mozambique Agro-Livestock Labour Day Parades, business conferences, the Commercial and Industrial Fair (Feira Agro-Pecuária) and the Comercial e Industrial de Moçambique (FACIM), raising the profile of the sector and creating new business opportunities (O'Donovan, 2021).

The success of the project can be seen in a significant increase in net income for many dairy farmers with remarkable growth in milk production (Daniels, 2021). Some farmers have tripled their net income while achieving increased milk output (Fig. 6) (AUDA-NEPAD, 2022). The project has significantly enhanced the productivity of the dairy sector, benefitting approximately 10 000 value chain actors, including smallholder farmers, commercial breeders, feed manufacturers, mini-agribusinesses, cooperatives, processors and retailers. This surge in productivity is attributed to effective training and capacity-building initiatives. The MERCADO scheme has provided dairy farmers with enhanced farm management skills, improved dairy practices and facilitated access to inputs. Additionally, the project has strengthened access to finance for a wide range of value chain actors, including input suppliers, producers, aggregators, processors and retailers (Venture 37, 2023).

8 Conclusion

Successful agricultural value chains in Africa highlight in particular positive outcomes of collaborative efforts and strategic interventions in enhancing

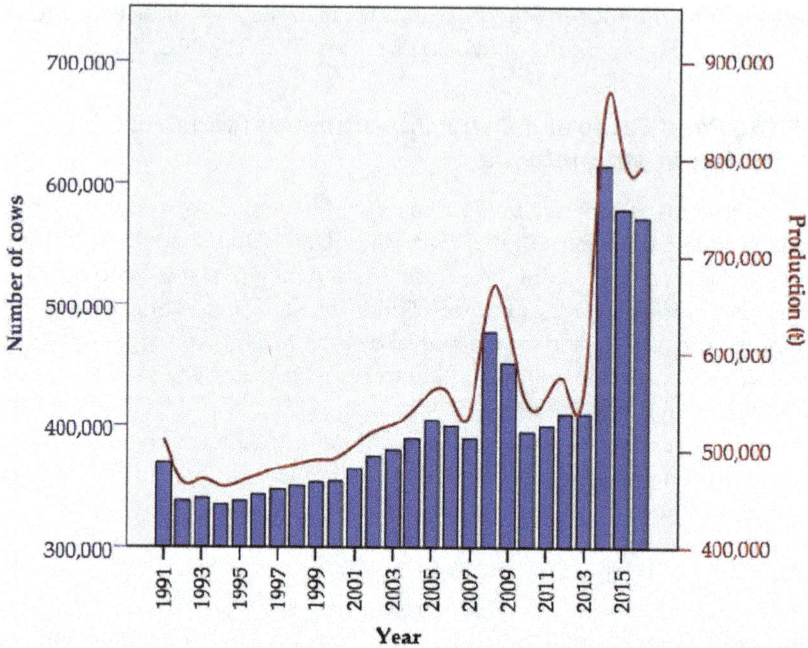

Figure 6 Number of cattle and milk production in Mozambique from 1991 to 2016. The red line represents the variation in milk quantity and the bars represent the number of cows. Source: Adapted from: Cambaza (2018).

the sustainability, efficiency and resilience of agricultural value chains in the continent. The objectives and impact of these value chains are shown in Table 1.

The transformation of the agricultural sector and its integration into GVCs is imperative for sustainability, enhanced food security, increased income and accelerated economic growth in Africa. To achieve this, a substantial proportion of micro- and small-scale farmers must shift towards more export market-oriented practices, producing high-quality products meeting global standards. However, multiple challenges impede the development of value chains in the region, including issues such as inadequate infrastructure, an unfavourable regulatory environment, financial constraints, climate change, limited access to inputs, poor adherence to food safety standards and sanitation issues, among others.

Addressing these challenges requires a strategic focus on public policy reform to bolster the development of Africa's agri-food value chains. Key strategies include fostering good governance, establishing robust partnerships between policymakers and the private sector, improving infrastructure, promoting digital transformation, investing in skills development and implementing fair investment policies. Good governance can be achieved through policy interventions addressing constraints, enhancing market

Table 1 Comparison of case studies

Case study	Objective/focus of the case study	Impact of the case study
Wool and mohair industry of South Africa and Lesotho	The industry ensures sustainable production, on-farm practice standards, animal welfare audits and fair labour practices. The industry also offers different educational programmes including through various institutions	Improved quality, access to markets, responsible farming standards, market compliance, increased supplies
Finance and insurance for agricultural value chains in Nigeria	To develop an enabling environment for the agricultural industry growth through enhancing investment opportunities, improving infrastructure, addressing logistical and land ownership constraints	Accelerated access to reasonable credit for farmers and small- and medium-sized enterprises (SMEs), loan risk reduction, increased access to markets for agricultural products, crop insurance, agricultural digitalization
Kenya flower value chain	Formation of strong partnerships with buyers and other value chain stakeholders to facilitate coffee export	Increased production and revenue, improved farmer–exporter partnerships, new technologies, improved varieties, improved production techniques, compliance to environmental standards, access to world market
Rwanda coffee industry	Transformation of the coffee sector from poor quality with low production to a highly competitive export industry	Monetization of the rural economy and foreign exchange income, access to world markets such as Starbucks, increased women involvement, improved quality, increased market prices
Expansion of Rural Cattle and Dairy Opportunities (MERCADO) in Mozambique	The project aims to increase the productivity of dairy sector in Mozambique and expand trade opportunities	Increased productivity and trade, training and capacity building for farmers, policy and regulatory reforms, strengthened stakeholder collaboration, distribution of dairy cows to farmers, construction of milk collection and distribution centres

linkages, promoting inclusive business models, improving institutional quality, supporting agricultural research and development, and leveraging digital technologies (UNDP, 2023; AUDA-NEPAD, 2022; Amanor, 2019).

Effective government policies will play a pivotal role in establishing an enabling environment for the sustainable development of agricultural value chains in Africa. This, in turn, will enable African countries to cultivate an environment conducive to the success of agribusiness value chains, fostering enhanced value addition to agricultural raw materials and facilitating active and profitable participation in GVCs.

9 References

African Union Commission (AUC). 2022a. *Africa's Development Dynamics 2022 Regional Value Chains for a Sustainable Recovery: Regional Value Chains for a Sustainable Recovery*. Paris: OECD Publishing.

African Union Commission (AUC). 2022b. *Strengthening Regional Value Chains in the African Continental Free Trade Area*. Paris: OECD Publishing.

AGENPO. 2023. Developing sustainable supply chain networks for African agribusiness. https://www.linkedin.com/pulse/developing-sustainable-supply-chain-networks-african-agribusiness/. Downloaded: 20 December 2023.

Ajani, E. N. and Igbokwe, E. M. 2014. A review of agricultural transformation agenda in Nigeria: The case of public and private sector participation. *Research Journal of Agriculture and Environmental Management* 3(5), 238–245.

Ajibo, C. C. 2023. AfCFTA and regional value chain development: Confronting the barriers and changing the orthodoxy. *Journal of World Trade* 57(4), 577–592.

Amanor, K. S. 2019. Global value chains and agribusiness in Africa: Upgrading or capturing smallholder production? *Agrarian South: Journal of Political Economy* 8(1-2), 30–63.

AUC/OECD. 2022. Integrating value chains in West Africa and the agri-food industry. In *Africa's Development Dynamics 2022: Regional Value Chains for a Sustainable Recovery*. Paris: OECD Publishing. https://doi.org/10.1787/e68c02a0-en. Downloaded: 08 January 2024.

AUDA-NEPAD African Union Development Agency. 2022a. *Mozambique Government, Partners Join Forces to Improve Dairy Farming*https://www.nepad.org/news/mozambique-government-partners-join-forces-improve-dairy-farming.

AUDA-NEPAD African Union Development Agency. 2022b. *Strengthening Competitiveness in Africa's Agricultural Value Chain Using Smart Technologies*. AUDA-NEPAD, Midrand, South Africa.

Ba, M. N. 2016. Strategic agricultural commodity value chains in Africa for increased food: The regional approach for food security. *Agricultural Sciences* 7(9), 549–585.

Balakrishnan, U. and Zaas, D. 2023. *Challenges and Opportunities in Agricultural Value Chains*. American Institutes for Research, Arlington.

Barrett, C., Reardon, T., Swinnen, J. and Zilberman, D. 2019. *Structural Transformation and Economic Development: Insights from the agri-Food Value Chain Revolution*. Cornell University. http://barrett.dyson.cornell.edu/files/papers/BRSZ%2013%20Aug%202019.pdf.

Bhorat, H., Kanbur, R. and Stanwix, B. 2017. Minimum wages in sub-Saharan Africa: A primer. *The World Bank Research Observer* 32(1), 21–74.

Black, A., Edwards, L., Ismail, F., Makundi, B. and Morris, M. 2021. The role of regional value chains in fostering regional integration in Southern Africa. *Development Southern Africa* 38(1), 39–56.

Bolo, M. O. 2010. Learning to export: Building farmers' capabilities through partnerships in Kenya's flower industry. In: *ISDA* (p. 11). Montpellier: Cirad-Inra-SupAgro. https://hal.archives-ouvertes.fr/hal-00526145.

Boudreaux, K. 2007. State power, entrepreneurship, and coffee: The Rwandan experience. Mercatus Policy Series. *SSRN Electronic Journal* 15. https://doi.org/10.2139/ssrn.1336628.

Cambaza, E. 2018. A glance at the Mozambican dairy research. *African Journal of Agricultural Research* 13(53), 2945–2956.

Cape Wools SA. 2019. National Wool Market Research Report Season 2019-2020. https://www.capewools.co.za/.

Chemonics. 2023. Enabling environment for agriculture. https://chemonics.com/technical-areas/enabling-environment-agriculture/. Downloaded: 29 December 2023.

Clay, D. C., Bro, A. S., Church, R. A., Ortega, D. L. and Bizoza, A. R. 2018. Farmer incentives and value chain governance: Critical elements to sustainable growth in Rwanda's coffee sector. *Journal of Rural Studies* 63, 200–213.

Conde, C., Heinrigs, P. and O'Sullivan, A. 2015. Tapping the potential of global value chains for Africa. *Europe* 57, 50–59.

Crush, J., Nickarnor, N., Kazembe, L. N. and Wagner, J. 2018. Revisiting Africa's supermarket revolution: The case of Windhoek, Namibia. *Development Southern Africa* 38(2) 230–247.

Daniels, S. 2021. Rwanda's Coffee Industry: Colonialism and the Impact of Fair-Trade Coffee. *Oswald Research and Creativity Competition* 40, 1-12.

Dazé, A. and Dekens, J. 2016. *Enabling Climate Risk Management along Agricultural Value Chains: Insights from the Rice Value Chain in Uganda*. International Institute for Sustainable Development (IISD), Winnipeg.

de Brauw, A. and Bulte, E. 2021. African agricultural value chains: A brief historical overview. In: *African Farmers, Value Chains and Agricultural Development. Palgrave Studies in Agricultural Economics and Food Policy*. Cham: Palgrave Macmillan. https://doi.org/10.1007/978-3-030-88693-6_2.

de la Cadena, M. V. 1900. *A New Pronouncing Dictionary of the Spanish and English Languages: Spanish-English* (Vol. 1). New York: D. Appleton.

De Melo, J. and Twum, A. 2021. Prospects and challenges for supply chain trade under the Africa Continental Free Trade Area. *Journal of African Trade* 8(2), 49–61.

Department of Agriculture, Forestry and Fisheries (DAFF). 2015. *A Profile of the South African Wool Market Value Chain*. http://www.daff.gov.za/. Downloaded: 2 December 2023.

Dessalegn, H. and Eziakonwa, A. 2023. *Food Security: Strengthening Africa's Food Systems*Brookings Institution, United States of America.

Dorward, A., Kydd, J., Poulton, C., Stockbridge, M., Al Hassan, R. and Zulu, B. 2004. Agricultural liberalisation in sub-Saharan Africa. DFID EC-PREP Project Centre for Development and Poverty Reduction – Imperial College London, London, UK.

European Chamber of Commerce and industry of Southern Africa. 2020. *Opportunities in the South African Cotton & Wool Value Chains*, Johannesburg.

European Carpet and Rug Association (ECRA) 2021. *Leading the Carpet Industry Towards Circular Economy, a 2030 Strategic Approach*. https://circulareconomy.europa.eu/platform/sites/default/files/ecra-2030_action_planhq.pdf. Downloaded: 27 February 2024.

FAO. 2021. *The Federal Republic of Nigeria Resilience Strategy 2021-2023. Increasing the Resilience of Agriculture-Based Livelihood. The Pathway to Humanitarian-Development-Peace Nexus*. https://doi.org/10.4060/cb5856en. Downloaded: 14 January 2024.

Fayad, L. K., Lanci, L., Mitra, P., Muehlschlegel, T., Okou, C., Spray, J., Tuitoek, K. and Unsal, F. 2022. *Climate Change and Chronic Food Insecurity in Sub-Saharan Africa*. International Monetary Fund Publishing Services, Washington, DC.

Gayi, S. K. and Banini, J. K. 2019. Does Africa have what it takes to upgrade in global value chains? *Logistics and Global Value Chains in Africa: The Impact on Trade and Development*, 169–216.

Goedde, L., Ooko-Ombaka, A. and Pais, G. 2019. *Winning in Africa's Agricultural Market*. Global Editorial Services, New York.

Hameed, S. and Harpst, E. 2014. *Supporting Value-Chain Growth Has Promise in Post-Conflict Countries, As Rwanda's Coffee Industry Shows*.

Hansen, J. W., Araba, D., Hellin, J. and Goslinga, R. 2017. A Roadmap for Evidence-Based Insurance Development for Nigeria's Farmers. CCAFS Working Paper no. 218. CGIAR Research Program on Climate Change, Agriculture and Food Security, (CCAFS).

Hertantyo, S. 2022. Regenerative farming is the future of farm-to-fashion wool systems in South Africa. https://twyg.co.za/regenerative-farming-is-the-future-of-farm-to-fashion-wool-systems-in-south-africa/. Downloaded: 22 December 2023.

Hidayati, D. R., Garnevska, E. and Childerhouse, P. 2023. Enabling sustainable agri-food value chain transformation in developing countries. *Journal of Cleaner Production* 395, 136300.

Igue, C., Alinsato, A. and Agadjihouédé, T. 2021. E-commerce in Africa: Issues and challenges in M. *Adapting to the Digital Trade Era: Challenges and Opportunities*, 118–139. https://doi.org/10.30875/dbe546d6-en

International Institute for Environment and Development (IIED). 2010. *Value Chain Analysis of Flowers Produced for Export by Smallholder Growers in Kenya – 1*. Baseline Survey. Fintrac Inc. https://pdf.usaid.gov/pdf_docs/PA00KC7X.pdf. Downloaded: 21 December 2023.

Kenya Flower Council. 2023. Annual report 2023. https://kenyaflowercouncil.org/.

Kenya National Bureau of Statistics (KNBS). 2020. *Economic Survey*. Kenya National Bureau of Statistics, Nairobi.

Klauser, D. and Robinson, M. (Eds) 2021. *The Sustainable Intensification of Smallholder Farming Systems*. Cambridge: Burleigh Dodds Science Publishing.

Koning, G., Da Silva, C. A. and Mhlanga, N. 2013. *Enabling Environments for Agribusiness and Agro-Industries Development*. Rome: Food and Agriculture Organization of the United Nations.

Kuteyi, D. and Winkler, H. 2022. Logistics challenges in sub-Saharan Africa and opportunities for digitalization. *Sustainability* 14(4), 2399.

Letebele, B. 2020. The potential contribution of EU business in the value chains for wool, mohair and cotton in South Africa. *EU Draft Position Paper*.

Levin, S. and Makgetla, N. 2021. Regional value chains and industrialisation: The southern African experience. *Transforming Agriculture: Harnessing Regional Value Chains and Industrial Policy for Development* 38, 1–30.

Logic, A. 2018. *Value Chain Analysis for the Coffee Sector in Rwanda*. https://www.cbi.eu/sites/default/files/cbi_vca_rwanda_coffee.pdf

Macrotrends. 2023. Rwanda poverty rate 2000-2023. https://www.macrotrends.net/countries/RWA/rwanda/poverty-rate. Downloaded: 26 December 2023.

Millenium Challenge Corporation. 2022. *Value Chain Mapping for Key Products in Lesotho's Creative Industries*. Creative Learning Inc., Chevy Chase.

Mitchell, C., Karl-Waithaka, Z., Unnikrishnan, S. and Oyekan, T. 2021. Transforming Africa's food systems from the demand side. *Boston Consult Group* 30, 9–17.

Nordic. 2023. *Transparency in Rwanda*. https://nordicapproach.no/origins/Rwanda. Downloaded: 13 January 2024.

O'Donovan, F. 2017. *Shaping the Future of Mozambique's Dairy Sector*. https://agrilinks.org/post/shaping-future-mozambiques-dairy-sector#:~:text=Through%20engagement%20with%20the%20government,dairy%20policy%20and%20regulatory%20frameworks.

Odjo, S. P., Traore, F. and Zaki, C. (eds). 2023. *Africa Agriculture Trade Monitor 2023.*. AKADEMIYA2063 and International Food Policy Research Institute (IFPRI), Kigali, Rwanda and Washington DC.

Odusola, A. 2021. *Africa's Agricultural Renaissance*. Cham: Springer International Publishing.

OECD. 2017. *African Economic Outlook 2017*. Paris: OECD Publishing.

Oirere, S. 2022. *Success in Coffee Erasing Rwanda's Sad History*. October Inter Co. Ltd., Bangkok.

Omta, S. W. F., van Elzakker, B. and Schoenmakers, W. W. M. E. 2014. *IOB Review-Riding the Wave of Sustainable Commodity Sourcing. Review of the Sustainable Trade Initiative IDH, 2008-2013* (No. 397). The Hauge: Ministry of Foreign Affairs.

Peters, T. and Sayin, L. 2021. Sustainable cold chain development. *Cold Chain Management for the Fresh Produce Industry in the Developing World*, 55–68. https://doi.org/10.2147/JMDH.S385466.

Polycarp, I. M. 2018. Enhancing institutional lending to the agricultural sector by derisking the value chain: The case of Nigerian Incentive Based Risk Sharing System for Agricultural Lending (NIRSAL). *Bullion* 42(1), 3.

Public Eye. 2019. *Lesotho's Wool and Mohair Saga: What Does Statistical Data Say?* Public Eye news, Maseru.

Reardon, T., Tschirley, D., Liverpool-Tasie, L. S. O., Awokuse, T., Fanzo, J., Minten, B., Vos, R., Dolislager, M., Sauer, C., Dhar, R., Vargas, C., Lartey, A., Raza, A. and Popkin, B. M. 2021. The processed food revolution in African food systems and the double burden of malnutrition. *Global Food Security* 28, 100466.

Sasson, A. 2012. Food security for Africa: An urgent global challenge. *Agriculture and Food Security* 1(1), 1–16.

Scope Insight. 2023. *The Top Five Challenges You Need to Know, and Which Frustrate Agricultural Input Retailers in Emerging Markets*. https://scopeinsight.com/the-top-five-challenges-you-need-to-know-and-which-frustrate-agricultural-input-retailers-in-emerging-markets/. Downloaded: 27 December 2023.

Shaibu, M. and Tasie, O. 2022. *Feed the Future Nigeria Agricultural Policy Activity*. Department of Agricultural, Food and Resource Economics, Michigan State University, Michigan.

Stathers, T. and Mvumi, B. 2020. Challenges and initiatives in reducing postharvest food losses and waste in sub-Saharan Africa. In *Engineering in Medicine: Preventing food losses and waste to achieve food*. pp. 729–786. Burleigh Dodds Science Publishing, Cambridge.

United Nations Conference and Trade Development (UNCTAD). 2022. Annual report. https://unctad.org/publication/unctad-annual-report-2022/. Downloaded: 19 December 2023.

United Nations Conference on Trade and Development (UNCTAD). 2023. The economic development in Africa report 2023. https://unctad.org/system/files/official-document/aldcafrica2023_en.pdf.

United Nations Conference and Trade Development (UNCTAD). 2023. UNCTAD report: Africa's rise as a global supply chain force. https://unctad.org/press-material/unctad-report-africas-rise-global-supply-chain-force. Downloaded: 21 December 2023.

United National Development Programme (UNDP). 2023. *Policy Brief: Strengthening Agricultural Value Chains & AfCFTA Linkages in South Sudan: Opportunities and Challenges*. United Nations Development Programme, Juba, South Sudan.

United States International Development Finance Corporation (USIDFC). 2023. *Modernizing Rwanda's Coffee Industry*. Development Finance Corporation, Washington, DC.

Venture. 2023. *Mozambique Expansion of Rural Cattle and Dairy Opportunities*. IMPAQ International LLC, Washington, DC.

Zechner, K. 2021. *Empowering Smallholder Farmers in Nigeria: An Introduction to AgroMall*. GSMA Association, London.

Chapter 13

Improving agri-food supply chains in Asia

Yijun Li, China Agricultural University, China and Wageningen University, The Netherlands; and Ting Meng, Jingjing Wang, Lu Liu, Mo Li, Xin Zhang and Wen-Feng Cong[1], China Agricultural University, China

1 Introduction: nutrition security and sustainability in Asia

Food and nutrition security faces unprecedented risks and challenges in Asia, and globally, including climate change, COVID-19, environmental degradation, trade friction and regional conflicts; consequently, food security is becoming increasingly important.

Food insecurity exists in many developing countries, with Asia being home to almost 65% of the world's undernourished population (Zhou and Wan,

1 Corresponding author. E-mail: wenfeng.cong@cau.edu.cn.

http://dx.doi.org/10.19103/AS.2023.0122.18

2017). Moreover, the COVID-19 pandemic has exacerbated food insecurity and malnutrition. Data and reports from the United Nations Food and Agriculture Organization show that 720–811 million people worldwide faced hunger in 2020, while more than half of those were in Asia (FAO et al., 2021). Furthermore, the growth of income, urbanization and population has led to a substantial rise in food demand. Income growth and urbanization are continuing in China, India and Southeast Asia, while population growth is expected to continue, especially in South Asia (FAO, 2018). Therefore, reviewing the challenges and threats to the agri-food supply chain in Asia and pathways to solutions is more relevant than ever.

1.1 Human health and nutrition

The food supply system in Asia is under unprecedented strain as a result of the global pandemic and regional conflicts. Although economic and food system revolutions and changes in dietary patterns have been substantial, undernutrition, overweight/obesity, hidden hunger and malnutrition rates have remained high in all regions, making it difficult to achieve the sustainable development goal (SDG) 2 aiming at eliminating hunger and all forms of malnutrition. Asia has recorded the highest rate of increase in moderate or severe food insecurity, from 21.3% in 2019 to 24.6% in 2021, with the highest rate in the southern subregion (UNICEF, 2021).

In 2021, Asia experienced the second largest occurrence of undernourishment, after Africa, with more than half (425 million) of the world's hungry residing in Asia (UNICEF, 2021). Child malnutrition continues to be a problem, with more than half of the world's stunted children (81.7 million or 54.8%) living in Asia (Micha et al., 2020), and the increase in childhood obesity between 2000 and 2020 was considerable, particularly in Eastern and South-eastern Asia, causing developmental problems and lasting ill health (FAO et al., 2021; Micha et al., 2020). In addition, over 30% of Asian women suffer from anaemia (Micha et al., 2020). The double burden of malnutrition is particularly concentrated in South Asia and East Asia, with the prevalence of overweight and obesity among the poor in the majority of East Asian nations (headed by China and Indonesia) rising substantially (Popkin et al., 2020; Damsgaard et al., 2016). Between 2012 and 2016, adult obesity increased in every subregion, with none on track to meet the 2025 world health assembly (WHA) objective (Micha et al., 2020). These many types of malnutrition result in new challenges in various locations and have led to an increase in diet-related non-communicable diseases (NCDs) such as diabetes, heart disease and stroke. Furthermore, South Asia's low-income nations are disproportionately afflicted by hypertension.

1.2 Resources and environment

Asian agriculture is responsible for approximately two-thirds of the global agricultural gross domestic product (GDP) and also plays an important role worldwide (Mendelsohn, 2014). Eleven countries in Asia contribute approximately 87% to the total global rice production, with the export of eight of these countries comprising around 35% of the global rice export. Global food security is highly impacted by rice production in Asia, in general, and particularly in India and China (Bandumula, 2017). Compared with other emerging regions of the world, in recent years Asia has experienced a slower decline in agricultural employment compared with its output, rapid growth in labour and land productivity, and a shift of the agricultural output from traditional to high-value products (Briones and Felipe, 2016).

Although global crop productivity has more than doubled over the past few decades, negative impacts on the environment, biodiversity, soil quality and air quality are common (Tilman et al., 2011; Godfray and Garnett, 2014). The food production system is the largest cause of global environmental change, with agriculture occupying approximately 40% of global land (Foley et al., 2005) and food production being responsible for up to 30% of global greenhouse gas (GHG) emissions (Vermeulen et al., 2012) and 70% of freshwater use (Molden, 2013). The Green Revolution resulted in Asian farmers accounting for two-thirds of global agricultural productivity (Mendelsohn, 2014). However, the overuse and misuse of nitrogen and phosphorus fertilizers can cause eutrophication and dead zones in lakes and coastal zones (Diaz and Rosenberg, 2008; Jat et al., 2020). The present situation is anticipated to worsen with climate change, with rising temperatures and changing monsoon rainfall patterns projected to cost India 2.8% of its gross domestic product (Mani et al., 2018). Asian agriculture has become a global 'hotspot' for contemporary and future climate vulnerability (Jat et al., 2020). We are facing the challenge of feeding 10 billion people a healthy and sustainable diet by 2050, while a rising number of environmental systems and processes are being pushed beyond safe boundaries by food production requirements (Willett et al., 2019).

2 Challenges and solutions across the supply chain: crop production

Improving agri-food supply chains is crucial for food and nutrition security, resource conservation and environmental sustainability. Figure 1 illustrates the pathways and corresponding solutions to improve agri-food supply chains in Asia via food production, food storage, distribution, processing and consumption stages. In each supply-chain stage, we explore the challenges and threats driving the transformation of the agri-food supply chain. We also reveal the key points for each stage in the process of transition to a more nutritious,

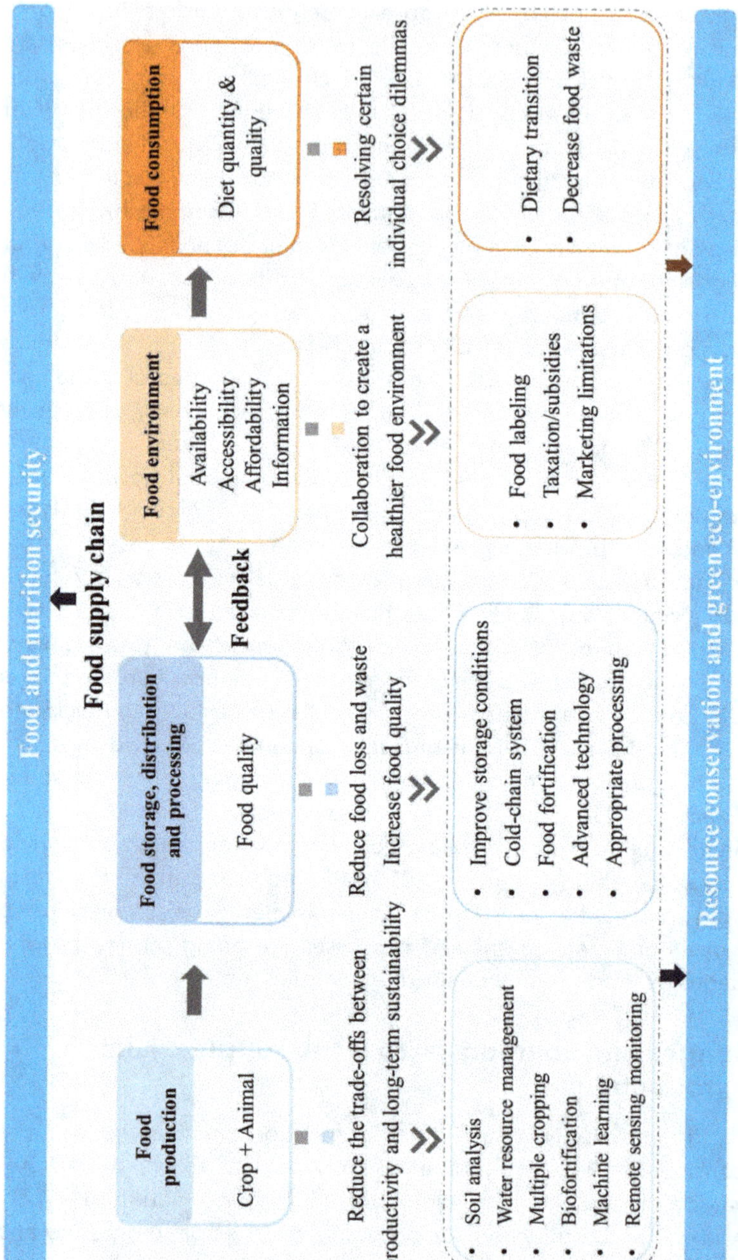

Figure 1 Pathways to solutions for improving agri-food supply chains in Asia.

environmentally friendly and economically sustainable path, with specific interventions discussed around every key point. Subsequently, the case studies later in the chapter will further demonstrate the potential of the interventions and their effects on successfully improving the sustainability of the agri-food supply chain and supporting its transformation. This section starts by looking at crop production.

The Green Revolution began in Asia in the 1960s, with the primary aim of introducing high-yielding varieties of cereals, e.g. rice, wheat and maize, to alleviate poverty and malnutrition with a high investment in infrastructure, crop research, market development and appropriate policy support (John and Babu, 2021; Pingali, 2012). Between 1970 and 1995, although the population increased by 60%, food production rose faster. Consequently, cereal and calorie availability per person increased by nearly 30%, and at the same time wheat and rice became cheaper (Hazell, 2009). By increasing the supply of food and reducing the prices of staple foods in Asia, the Green Revolution improved the nutrition of poor people and the absolute number of poor people declined by 28% from 1975 to 1995 (Brainerd and Menon, 2014). In conclusion, the Green Revolution lifted many populations out of poverty, jump-started economic growth and saved large areas of other fragile lands and forest wetlands from conversion to cropland (Hazell, 2009).

Despite great success in increasing food production, Asia still faces enormous food security challenges (John and Babu, 2021). At present, agriculture faces the unprecedented challenge of securing food supplies for a rapidly growing human population while seeking to minimize the adverse impacts of agriculture production on the environment and reduce the use of non-renewable energy and resources (Garnett and Godfray, 2012). A shift towards more sustainable agricultural production entails the adoption of system-oriented strategies, which include farm-derived inputs and productivity based on ecological processes and functions (Garnett and Godfray, 2012). Sustainable agricultural systems also involve traditional knowledge and entrepreneurial skills of farmers (Beintema et al., 2009). The greatest challenge for agriculture is to reduce the trade-offs between productivity and long-term sustainability (Niggli, 2015). Improving the sustainability of food systems means not only continuing to increase food production but also enhancing food nutrition, adapting to global climate change and minimizing GHG emissions and environmental impact (Davis et al., 2019; Godfray et al., 2010). However, sustainable agriculture offers entirely new opportunities by emphasizing the productive values of natural, social and human capital, all assets that Asian countries either have in abundance or that can be regenerated at a relatively low financial cost (Pretty and Hine, 2000).

Soil plays a critical role in securing food production and consequently, it is necessary to monitor the soil at the field scale to obtain information regarding soil health to make precise decisions at different crop developmental stages. The main objective of soil analysis is to measure the content of nutrients present in the soil to increase fertilizer use efficiency and decrease pollution to air and water (Martínez-Fernández et al., 2016). Due to the increase in irrigation requirements, freshwater will only be available to those countries that have adopted the best water resource management practice. To overcome the water shortage problem, new irrigation techniques, including sprinkler and drip irrigation methods, are used to reduce losses and water shortage (Motoshita et al., 2018).

Plant growth, development and reproduction mainly depend on receiving the necessary nutrients from fertilizers or organic amendments. New technology is used to estimate the spatial-temporal application of nutrients for the fulfilment of fertilization, which is relatively less laborious and achieves maximum efficiency (Ihtisham et al., 2020). In addition, the need for increased crop yields has led to growing interest and research focused on agricultural intensification, which has a myriad of environmental impacts (Tilman, 2020). Intercropping can bring the benefits of intensification within a reasonable footprint (Li et al., 2020). It is expected that the future of agriculture will develop into a progressive sector when integrated systems appreciate artificial intelligence and big data capabilities. Advanced technologies, such as robots in agricultural production processes, cloud computing, artificial intelligence and big data calculation, can usher in a different era of super-convergence in different farming sectors (Khan et al., 2021a).

3 Challenges and solutions across the supply chain: storage, distribution and processing

The impact from each process in the food supply chain are added together to give their cumulative environmental impact and nutritional output. Over 40% of food loss occurs at the post-harvest handling and processing stages in developing countries, which is particularly true for China (Lu et al., 2022).

During the process of food storage and transportation, reducing food loss is an effective way to simultaneously achieve food and nutrition security as well as reduce the environmental externalities of the food supply chain (HLPE, 2014; Luo et al., 2020; Gibson et al., 2020). Many researchers have endeavoured to identify the possible factors that influence food loss and waste (FLW) and suggest possible interventions. In the following section, we first focus on revealing the development of interventions aimed at food loss reduction during storage and transportation phases for non-perishable and perishable foods. Then, from the operations management perspective, we describe the current topics, themes and methods applied to the FLW of the supply chain.

The major contributor to food loss occurring at the storage stage for non-perishable food, e.g. cereals and dry beans, is inadequate storage conditions (Luo et al., 2020; Kumar and Kalita, 2017). In China, warehouses and bags are widely used storage facilities for rice and wheat. Nevertheless, local storage facilities like storage baskets, tanks, and jars are generally used for corn storage (Luo et al., 2020). The lack of appropriate storage conditions leads to the greatest food loss for corn in China, compared with rice and wheat.

Temperature, humidity and the biophysical properties of grains determine how much energy will be applied to grain storage. Energy input is regarded as a solution that can prevent grain from being affected by pests, mould and respiration (Puri, 2016). In the context of the crop production stage in India, energy allocated to provide suitable storage conditions for grain only accounts for 8–16% of the entire energy input (Gibson et al., 2020); however, the potential for loss reduction is huge. According to the estimation, if grain losses can be saved and targeted to the nutrition deficiency population in India, protein, calorie, zinc and iron deficiency can be reduced on average by 46%, 27%, 26% and 11%, respectively (Gibson et al., 2020). Similarly in China, if the loss rate of household storage falls to the advanced rate (1.00%), 379 000 hectares of land, 2.66 billion m^3 of water resource, 510 600 tons of GHG emissions could be reduced and 5.87 million people could be fed in 1 year (Luo et al., 2020).

Perishable foods, e.g. fruits, vegetables and animal-sourced foods, comprise the largest portion of total food loss (Xue et al., 2021; Parfitt et al., 2010). The application of cold-chain logistics is identified as an efficient investment in preventing perishable food losses (Han et al., 2021). An efficient and complete cold-chain system is necessary to maintain perishable foods at an appropriate temperature and humidity range. Any interruption during the cold-chain stages will greatly reduce the effectiveness of the whole system.

Han et al. (2021) explored current research topics related to cold-chain logistics for perishable foods. Efficiency optimization, integrity and accurate control as well as green and sustainable cold chains are major topics. During the storage stage, the refrigerated warehouse (RW) plays a critical role in reducing food loss and balancing supply and demand. Three research streams were recently investigated: one stream focused on balancing the electric power demand of cold storage with the conservation of product quality; the second stream investigated the influence of storage atmospheres on food quality; the third stream explored the optimization of refrigerating systems. For the transport stage, refrigerated transport (RT) is used to maintain stable temperature and humidity conditions during the delivery period. Advanced innovations such as cloud computing (for data processing and mining) and artificial intelligence (data interpretation) will be helpful in the upgrade of RT.

From the operations management perspective, Luo et al. (2022) analysed the topics, research themes and methods of FLW along the food supply chain by reviewing 346 articles (see Fig. 2). For research topics, the causes, measures, barriers and drivers of FLW have been widely investigated, especially by studies focused on one or several supply-chain stages. Regarding research themes, sustainability and environmental impact, food supply chain management and activity analysis are the main explored themes. Concerning methodology applied to FLW, around 30% of studies quantitatively analysed the FLW problems; among these, statistical analysis and life-cycle assessment (LCA) are mainly used for empirical research.

Food processing refers to activities that transform a raw product into edible food, including milling, heating, canning, drying or mixing (Nordhagen, 2020). Insufficient investment in infrastructure and auxiliary equipment was one of the reasons behind the loss of processing in China. Over the past few decades, the use of advanced technology and equipment has greatly reduced loss and waste during the processing of staple foods, but the problems still exist in most small-scale staple food processing industries.

There is another primary reason that the lack of relevant technology leads to a low utilization rate of food processing by-products. During the process of harvesting, selling on the market and then serving to consumers, the inedible parts of fruits and vegetables are thrown away, which leads to kitchen waste that is difficult to deal with, and the whole process also leads to a waste of resources. The large-scale processing and non-damaging treatment technology of fruits and vegetables should be improved and applied to increase the edible rate and resource utilization rate, which is conducive to the centralized treatment of fruit and vegetable wastes, such as technology involved in precooling, peeling, preservation and drying.

In addition, food processing activities such as excessive peeling and cutting are known to not only lead to a quantitative loss but also compromise the micronutrient quality. For example, minerals and vitamin B are easily lost

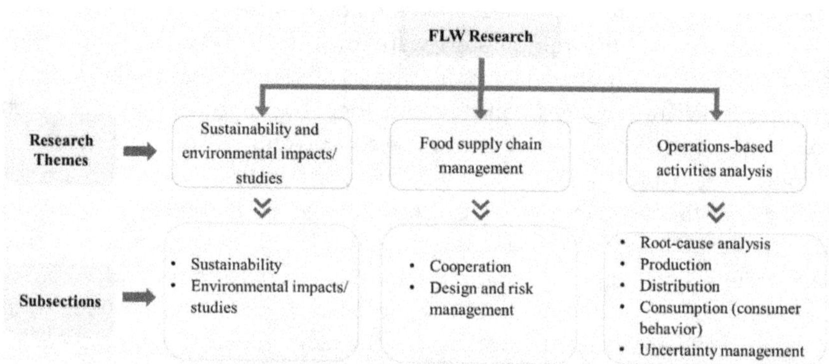

Figure 2 The structure of the research stream related to FLW (food loss and waste). Source: Adapted from: Luo et al. (2022).

in refined grain flour after milling compared with whole-grain flour. However, in China, the definition of whole-grain food is unclear, and relevant standards and labelling are insufficient. There is no effective supervision, resulting in the uneven quality of whole-grain products in the market. New processing technology with high efficiency and low cost should be developed to enhance the quality of whole-grain products and improve nutrient retention.

4 Challenges and solutions across the supply chain: consumption

Food environments, dietary patterns and food choices are influenced by globalization, urbanization and wealth growth. Along with economic and social expansion, advancements in technology and the food supply chain have made agricultural goods in the agri-food system more affordable and diverse for consumers. Throughout the consumption stage, consumers make decisions regarding purchasing, preparing and consuming food based on their local food environment. The retail and food environments, notably the rapid expansion of supermarkets and fast-food chains, altered consumer food preferences and eating patterns. Household consumption of most food is heavily commercialized in metropolitan areas, and the retail and food environments, particularly the rapid proliferation of supermarkets and fast-food chains, have influenced consumer food choices and dietary patterns (WHO, 2013; Baker and Friel, 2016).

However, there are still several challenges throughout the stages of food consumption. First, the ongoing shift in diet exerts pressure on agricultural production, individual nutrition, the environment and climate. In recent years, the Asian diet has become more calorically dense, with the increased consumption of salt, fats and oils, and sugar (Fan et al., 2017). In the majority of middle-income nations, packaged and ultra-processed foods, which do not conform to the World Health Organization's (WHO) definition of a healthy diet, comprise an increasing proportion of diets (Micha et al., 2020; Baker and Friel, 2016). The nutrition transition that goes along with it exacerbates the double burden of malnutrition and NCDs. It also affects natural resources, climate change and other costs that are hard to measure (Reardon and Zilberman, 2018). Urbanization, population and wealth growth will continue to raise consumer demand for food that is more nutrient-dense and safer; yet, how to feed Asia's numerous and enormous cities remains an open subject (Fan et al., 2021b). Second, the food environment does not encourage healthy eating choices among customers. Consumers' food choices are often influenced by food availability, accessibility, cost and desirability. Along the food supply chain, the food processing, retailing and food service sectors are key drivers of the nutrition shift in Asia (Baker and Friel, 2016), but the effort they put into providing nutritious and sustainable food remains limited. Third, the cost-effectiveness of healthy diets remains a concern. Asia is home to more than half of the world's population unable to afford a nutritious meal in 2020 due to the high cost of nutritious diets and substantial income disparities (1.89 billion out of 3.07 billion) (UNICEF, 2021).

To deal with the many problems in terms of people and planetary health, different parts of the food supply chain should work together to create a healthier food environment and encourage consumers to change their habits. Information campaigns on modifying habits, such as dietary transition and decreasing food waste, have been shown to be effective in increasing consumer knowledge and demand for nutritious and healthful food, as well as in resolving certain individual choice dilemmas (Fan et al., 2021a). To achieve this, consumer-oriented approaches like food labelling, taxation/subsidies and marketing limitations could be implemented. By investing in technical breakthroughs, food fortification and nutrition-sensitive interventions, the food industry and the corporate sector may contribute to improved nutrition. The labelling and marketing of foods should be monitored to ensure that consumers are able to make informed decisions without being misled. With simple nutrition information, such as front-of-package labelling, consumers can make better-informed food choices (Hawkes et al., 2015; Julia et al., 2015). In order to safeguard children and all other age groups from the negative effects of food marketing, these practices should be re-evaluated and revised. A number of Asian nations have enacted food marketing restrictions based on best practices. For instance, the Philippines has regulated marketing and legal protection for newborns and young children (Reeve et al., 2018), while Turkey has broadcast laws to protect children under 18 years old (Bosi et al., 2018). In addition, greater emphasis should be placed on raising the equity among consumers by reducing the price of food. As long as the price of nutritious and healthy food remains high, low-income and disadvantaged groups may be unable to afford it. In these circumstances, initiatives to promote consumer demand, such as unemployment insurance and other social safety nets, can be crucial for protecting the most vulnerable populations and can also lead to increases in welfare and food and nutrition security (Nordhagen et al., 2021).

5 Challenges and solutions across the supply chain: a system perspective

Several leading institutions have explored solutions to improve the agri-food supply in China. CCICED (2022) (China Council for International Cooperation on Environment and Development) proposes four categories of solutions to transform the agri-food system in China, i.e. producing more (and more nutritious) food, protecting nature, reducing agri-food system inefficiencies and pollution, and restoring degraded lands. AGFEP (2021) (Academy of Global Food Economics and Policy) found that agricultural GHG emissions in China can be mitigated to a large degree through technological innovations, reduction of FLW, and transformation of dietary patterns, even while ensuring long-term food security. Transforming the agri-food supply, China demands a shift towards more sustainable and healthy diets, long-term prevention and treatment of agricultural non-point-source pollution, and proactive participation in the governance of global food and agriculture. PwC (2019) (Pricewaterhouse Coopers) emphasizes the importance of technology innovation during the process. Asia will not solve its food challenges through a continuation of traditional farming practices or via linear modernization. New technologies will need to be deployed to increase yields, reduce the environmental impact of farming, improve food safety, traceability and nutritional value, reduce waste, shorten the supply chain and bring food to consumers in their increasingly urban settings.

Technology will be essential in helping to solve this food challenge; however, this requires investment and collaboration across the food and agricultural industry, including corporates, start-ups, investors, academia and government (PwC, 2019). Therefore, policies, governance and institutional innovations are recommended to facilitate such concerted transformation and improvement of the agri-food supply chain. Over the next 10 years, an estimated cumulative investment requirement of US$800 billion is above existing levels. The majority (around US$550 billion) will help to satisfy the demand for better quality food (including safer, healthier and more sustainable food). The remainder (around US$250 billion) will drive increased quantities of food to feed Asia's growing population. PwC (2019) pointed out that government co-ordinated ecosystems and the development of regional agri-food tech hubs are critical catalysts for investment and innovation. Khan et al. (2021b) suggest that government food authorities, global agencies, non-government organizations and food businesses collaborate to make the food supply chain resilient in low-income countries of Asia.

Policy recommendations proposed by CCICED (2022) are to develop a national agri-food system transformation strategy, repurpose agricultural fiscal incentives and finance, promote healthy diets and consumption, accelerate the agri-food system transformation through private sector leadership and incentivize alternative sources of protein and food technology and green international foodvalue chains. Regarding reforming the agricultural support policies, AGFEP (2022) explores how to reposition China's agricultural support policies in the new development era, and their recommendations include:

- review agricultural support policies to promote a win-win situation of nutrition and health security and green and low-carbon development;

- support policies should be optimized according to the concept of 'Big Food';
- adjust agricultural support policies to promote the development of green and low-carbon technologies in agriculture;
- enhance investment in high-standard farmland construction and green agricultural R&D (research and development) and extension;
- better use of China's fiscal expenditure on agriculture to narrow the income gap between urban and rural populations; and
- reducing the minimum purchase price to the amount that can cover the total cost of agricultural production and combining the minimum price with full-cost insurance.

6 Case studies: production structure optimization in India

Several Asian countries have taken steps at regional and national levels to improve the sustainability and food nutrition of their local food systems. This section provides an overview of case studies from India, China, and Singapore on their efforts to improve production structure, joint distribution, e-commerce and supply-chain innovation, as well as their food policies to enhance the food environment. The discussion starts by looking at India.

From 1947 to 1960, food production in India was insufficient as there was a growing population and food availability was 417 g per day per person (Nelson et al., 2019; Ghosh, 2004). The Green Revolution began in the 1960s aiming to introduce high-yielding varieties of cereals to alleviate poverty and malnutrition with a high investment in research, infrastructure, market development and appropriate policy support (Pingali, 2012; John and Babu, 2021). The total production of cereal crops tripled with only a 30% increase in the cultivated land area, helping India move from a state of importing grains to a state of self-sufficiency (Brainerd and Menon, 2014).

However, India had faced substantial negative consequences involving the environment, resources and the health of its growing population (Davis et al., 2018; Bhattacharya et al., 2015; Anupam and Khajuria, 2016) as a result of unsustainable production measures including the increased usage of pesticides (Neethu et al., 2016). In addition, the majority of Indian crops are cereal, which accounts for almost 50% of the dietary water footprint, causing severe stress on water resources (Kayatz et al., 2019; Bkab et al., 2019). Moreover, there was reduced cultivation of indigenous varieties of rice, millets, lentils, etc., but an increased harvest of hybrid crops (Taylor, 2020). As a result, the modern food production system failed to provide a diverse diet but increased calorie consumption with fewer micronutrients (Kumssa et al., 2015).

Sustainable food systems aim to provide sufficient and nutritious food while maximizing climate resilience and minimizing resource demands as well as negative environmental impacts. However, the Green Revolution prioritized the single objective of maximizing production over other nutritional and environmental dimensions. In India, diversifying crop production to include more coarse cereals, such as millets and sorghum, can make the food supply more nutritious, reduce resource demand and GHG emissions, and enhance climate resilience without reducing calorie production or requiring more land (Davis et al., 2019). The extent of these benefits partly depends on the feasibility of switching harvested areas from rice to coarse cereals. National- and state-level strategies considering multiple objectives in decisions regarding cereal production can move beyond many shortcomings of the Green Revolution while reinforcing the benefits. This ability to realistically incorporate multiple dimensions into intervention planning and implementation is the crux of sustainable food production systems worldwide. Similar multidimensional approaches to food production challenges in other parts of the world can identify win-win scenarios where food systems meet multiple nutritional, environmental and climate resilience goals (Davis et al., 2019).

7 Case studies: joint distribution of perishable food in China

Joint distribution, also known as co-operative distribution, was first proposed by the Japanese government in 1977 to improve resource utilization (Liu et al., 2020). This is a distribution mode jointly organized by several different enterprises, e.g. manufacturers, wholesalers and retailers, and is widely conducted in many developed regions such as Europe and Japan. With increasing high-frequency and low-quantity orders, joint distribution implementation can help meet the growing requirements of customized orders, improve the utilization efficiency of distribution facilities and equipment and make the whole process more efficient and intensive. It can also effectively reduce the total delivery distances, contributing to reducing carbon emissions (Maruyama, 2004; Chen and Hsu, 2015).

The typical modes of joint distribution can be categorized by horizontal and vertical integration. Horizontal integration is conducted by several companies at the same stage of the supply chain, for instance, several manufacturers. In terms of vertical integration, different companies along the whole supply chain will typically create an alliance, deliver the products to a joint distribution centre and then deliver the products to final stores.

Joint distribution can effectively satisfy small shipments and timely delivery requirements. With the growing demand for fresh food, the application of joint distribution under the context of the cold chain has been widely considered.

Kuo and Chen (2010) proposed a multi-temperature joint distribution (MTJD) system to minimize the total cost and investment of storage and transportation stages, improve truck usage flexibility and logistics performance and achieve food quality and safety goals under the cold-chain context. Figure 3 demonstrates the MTJD example for perishable food. Multi-temperature food can be stored together in a shared refrigerated distribution centre and then delivered to various stores and markets using multi-temperature trucks. Note that the reason why joint distribution can increase resource utilization efficiency and improve logistics performance is that it changed the previous situation of independent distribution by multiple enterprises. By maximizing truck utilization and promptly gathering the latest demand information, MTJD makes it possible to quickly meet customers' needs at a lower cost.

In the MTJD system, two modules were put forward to tackle the challenges and fulfil requirements. The service-integrated module was designed to standardize the logistics service process and to balance delivery time and diverse demands with high efficiency and low cost. The other is the resource-integrated module, in which the transportation resource planning system was the core engine. This module aimed to integrate resources to minimize total cost and improve operations efficiency. With the application of the MTJD system, cost reduction, improved satisfaction, food safety and quality can be achieved simultaneously.

Not only lower costs and higher efficiency can be realized in the MTJD system, but carbon emission reduction can also be achieved. Chen and Hsu (2015) compared the traditional multi-vehicle distribution system (TMVD) with the joint distribution system in terms of GHG emissions and total cost in the context of fresh foods. They considered 20 fresh food items with multiple temperature-controlled requirements, e.g. fish, duck, pork, beef, vegetables, milk. Figure 4 demonstrates the difference in unit carbon footprint with shipping volume between the traditional distribution system and the joint distribution system. It is noticeable that for range 3 food (livestock, poultry meat and fish

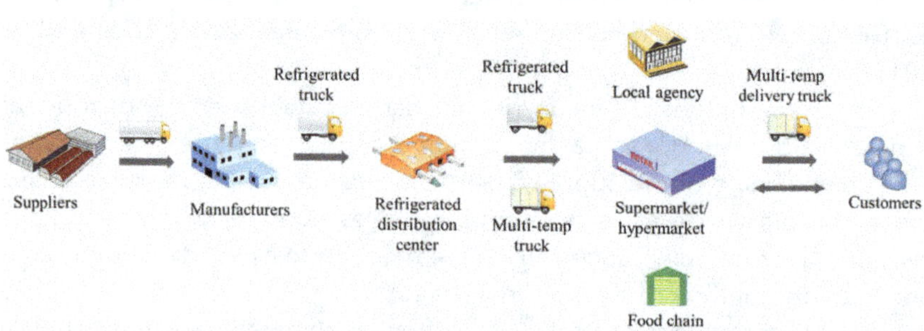

Figure 3 An example of multi-temperature joint distribution for perishable food. Source: Kuo and Chen (2010).

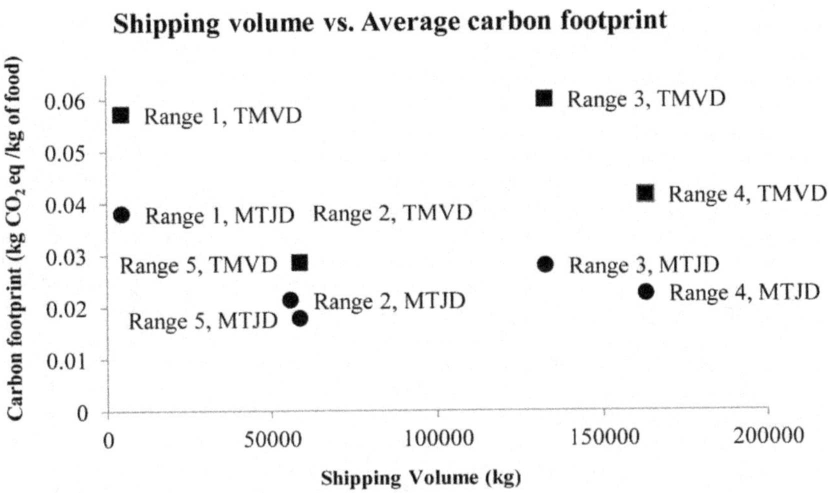

Figure 4 Distributed volume vs. average carbon footprints (source: Adapted from: Chen and Hsu, 2015). TMVD, the traditional multi-vehicle distribution system; MTJD, multi-temperature joint distribution. Ranges 1, 2, 3, 4, 5 represent the temperature ranges <−30°C, −30°C ~ −18°C, −2°C ~ +2°C, 0°C ~ +7°C, > +18°C, respectively.

meat with a temperature-controlled range between -2°C and +2°C), the joint distribution showed great potential to decrease carbon emissions. The higher the distribution volume, the greater the potential reduction of GHG emissions can be achieved by implementing joint distribution.

In conclusion, joint distribution is an effective system to lower total cost, reduce GHG emissions and satisfy dynamic demands.

8 Case studies: e-commerce and supply-chain technology innovation in China

China is a global leader in e-commerce, food delivery and other consumer-facing platforms (AGFEP, 2021). The development of e-commerce has changed the traditional circulation system of food and other agricultural products. Internet technology provides a virtual platform to conduct a direct transaction between individual farmers and final consumers nationwide, leaving mediators and intermediate steps less critical in the food supply chain.

With the support of the internet platform and modern logistics industry, smallholder farmers can access consumers through e-commerce via three modes (Fig. 5), in which e-commerce can boost smallholder farmers' connections with larger markets. First, farmers rely on the e-commerce platform to help them operate independently. In recent years, with the help of the internet, some farmers have set up online stores on third-party e-commerce platforms to have direct contact with consumers so that they can successfully

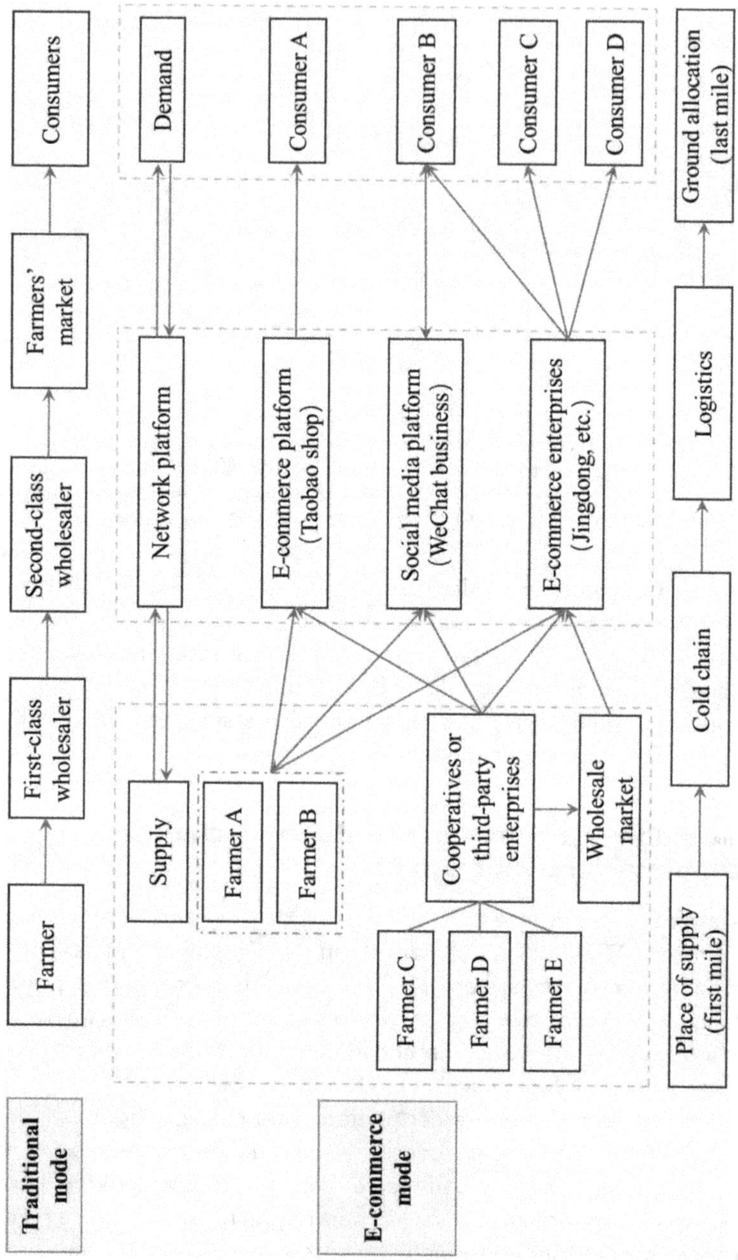

Figure 5 Circulation systems of agricultural products, traditional versus e-commerce modes. Source: Adapted from: AGFEP (2021).

bypass mediators and sell directly to foreign consumers. Second, farmers rely on social media platforms to help them operate independently. Apart from the independent business model of the e-commerce platform, in rural areas of China, some farmers also display their production processes and product information on social media such as WeChat, Weibo and TikTok, thus enhancing their direct connection with consumers. With the help of social media, people as critical actors and social relations as a link, this new business model based on mobile internet is called 'We Business'. Third, farmers work in co-operation with e-commerce enterprises. In rural areas, some farmers work in groups, families or professional co-ops, using the internet to connect with the market indirectly by establishing a co-operative relationship with an e-commerce platform enterprise or an operational service provider and sharing certain digital dividends. Here, we elaborate on how e-commerce enterprises strengthen the connections between smallholder farmers in rural areas and large markets and how management strategies and innovative technologies play an inseparable role in building this connection.

As a leading Chinese e-commerce enterprise, JD (JD Com, Inc.) is also known as a logistics-integrated solutions provider with an extended warehouse network to nearly all counties and districts nationwide until 2022 (JD, 2023). In 2015, JD began investigating the potential of leveraging its e-commerce platform to directly connect farmers' products with consumer markets, aiming to boost farmers' income (Fang and Huang, 2020). JD utilizes its advanced e-commerce platform and logistics network in multiple stages, from agricultural input procurement to the retail and distribution of final products, to explore this potential (Kong and Loubere, 2021). Agricultural product suppliers can directly publish product information on JD's platform, eliminating the need for intermediate stages in the traditional agricultural product supply chain and ensuring that agricultural products can enter the market more quickly and efficiently. As estimated, there was a remarkable 6.4-fold increase in online sales of agricultural and food products from poverty-stricken rural areas, observed from the first quarter of 2016 to the second quarter of 2018 (Fang and Huang, 2020).

The agricultural industry in rural or poverty-stricken areas faces problems mainly due to insufficient processing, decoupling of production and distribution, lagging market sales information, and logistics and transportation barriers, which significantly hinder the linkage between agricultural products and larger consumer markets. To overcome these problems and boost more direct connections between farmers' products and consumer markets, JD contributes to building a whole-process supply-chain system by applying advanced management strategies and innovative technologies. First, brand images of agricultural products are established to differentiate from other similar products and to better integrate into retail markets. JD worked with local enterprises (or groups, families or professional co-ops) that source agricultural products from individual farmers and then assisted these enterprises in branding regional agricultural products (Fang and Huang, 2020; Wu et al., 2023). This is beneficial in shaping the regional brand image of agricultural products and enhancing consumer awareness. Second, tailored marketing strategies are devised for potential consumers through big data platforms and algorithms to increase the possibility of purchase and increase sales. Applying big data platforms and algorithms such as personalized recommendation algorithms, JD conducts precise marketing of agricultural products and identifies potential consumers. To provide farmers with timely information on market demand, JD analyses real-time retail data and shares this information with farmers to help them adjust production and sales strategies. Based on the advanced cold-chain system of JD, the transportation conditions of fresh foods can be improved, breaking through regional restrictions and entering a broader consumer market (Li et al., 2023).

The development of e-commerce has dramatically boosted the connection between smallholder farmers in China and the larger market. It has become an essential way for smallholder farmers to connect to the fast-growing demand of the Chinese urban population, which will help promote the transformation of China's agri-food systems into one characterized by safety, nutrition, sustainability and resilience. In addition, e-commerce enables smallholder farmers to enter the global value chain by reducing their transaction costs for information, thus facilitating their entry into the market.

Some experiences of e-commerce development in China can be shared with other countries in Asia and worldwide. Rural e-commerce needs an ecosystem suitable for the participation of smallholder farmers, composed of network operators, government and service providers. Creating an e-commerce ecosystem suitable for the development of smallholder farmers requires: (1) an expanded and improved infrastructure, including information and communication infrastructure and other facilities; (2) improved ability of smallholder farmers to use the internet effectively; (3) supporting the development of inclusive digital business platforms; and (4) giving full play to e-commerce public service providers. In the process, a benefit distribution mechanism needs to be developed to ensure and enhance the profitability of smallholder farmers participating in e-commerce. Logistics and communication infrastructure, inclusive digital business platforms for smallholder farmers and a new generation of farmers are prerequisites for developing rural e-commerce. Thus, farmers need to be trained with e-commerce insights, knowledge and skills to help them lead other smallholder farmers to join e-commerce. As e-commerce may exclude smallholder farmers who cannot participate in the digital economy or lack the required skills, attention must be paid to the regional disparities and inequalities among farmers caused by the digital divide.

9 Case studies: Singapore's policy innovations for improving whole-grain consumption

Government policies can be critical in shaping the food environment towards health and sustainability. In Singapore, the consumption of whole grain used to be extremely low, with an average of just 0.2 servings per day, and only one in ten Singaporean adults consumed a serving or more of whole-grain foods (Health Promotion Board (HPB), 2004). To increase whole-grain consumption and protect Singaporeans' health, the national HPB has adopted rules to encourage the use of whole grains since 2009. The following strategies were implemented in order to raise the level of whole-grain knowledge among consumers, enhance their ability to recognize whole grains and expand the availability and accessibility of whole grains in the marketplace. The percentage of adults consuming more than one serving of whole grains a day has risen from 8.4% in 2004 to 27.0% in 2010 in Singapore (HPB, 2010).

9.1 Communicating with consumers

After communicating with consumers, HPB revised the dietary guideline and the healthy diet pyramid (increasing the daily whole-grain consumption recommended from 1 serving to 2–3 servings or 50 g) and recognized whole grains to be an essential part of daily meals. HPB also launched an educational campaign titled 'Eat More Whole Grains' to inform the public regarding the health benefits, suggested amounts, where whole-grain items can be found, and how they can be prepared and creatively included in the daily diet. The HPB mandates the serving of whole grains in schools and preschools through the Healthy Meals in School Programme, aiming to help students cultivate habits of eating whole grains and other healthy foods from their youth.

9.2 Regulating food labels and health claims

HPB regulated labels and health claims of whole-grain foods to eliminate identification barriers for consumers. Whole-grain foods which meet the HPB requirements can carry the Healthier Choice Symbol (HCS) with 'higher in whole grains'. In order to use this health claim or label a product as 'higher in whole grains', manufacturers must also declare the whole-grain content in the product's ingredients list. For example, a compliant ingredients list might say, 'Rice (20% brown rice), fish, cheese, salt, flavouring'. The package must also include a statement expressing the amount of whole grain per serving as a percentage of the 50 g/day whole-grain requirement established by the HPB (e.g. '1 serving of [name of product] provides [x]g of whole grain, which meets [x%] of your daily whole-grain requirement').

9.3 Providing market incentives for private sectors to increase the availability and accessibility of whole grains

HPB collaborated with food stores and food outlets to increase the availability of whole grains. Approximately 80% of Singaporeans eat at a hawker centre at least once a week, so the potential nutritional impact of incorporating more whole grains into these food courts is huge. HBP incentivized hawker stalls to adopt the whole-grain options by covering up to 80% of the investment cost (including product and packaging development, marketing, publicity and trade promotion) through the Healthier Ingredient Development Scheme. In addition, restaurants, cafes and quick-service restaurants were also incentivized to offer whole-grain dishes/items to diners. The number of healthy meals sold under the programme climbed from 7.5 million in fiscal year (FY) 2014 to 50 million in FY 2017 (HPB, 2018).

Accessibility to whole-grain foods is also crucial. HPB partners with suppliers to offer discounts on brown rice and drive competition in the market. The collaboration with major supermarket chains such as FairPrice, Cold Storage and Shop N Save outlets features targeted in-store promotions of whole-grain products. In addition, HPB teams up with noodle suppliers and food vendors to develop and market partially whole-grain noodles that would be sold at the same price as their refined grain counterparts (Toups, 2020). By experimenting with different product formulations, a balance was struck among health considerations, taste and manufacturing costs.

All of these strategies to increase whole-grain consumption in Singapore involve government policies and private-public partnerships along the food supply chain, and it creates a win-win business case that benefits all parties involved.

10 Conclusion and future trends in research

We summarized and highlighted effective measures and methods aimed at achieving sustainability, efficiency, resilience and nutritional health across various stages of the supply chain (Table 1). Engaging stakeholders is also essential during an agri-food system transformation.

10.1 Crop production stage

A sustainable food supply chain requires innovating agricultural production technologies as well as promoting the application of these innovative technologies. For example, a new generation of fertilizer, the so-called 'green intelligent fertilizer', is designed to meet the nutrient requirement of crop growth by considering the characteristics of plant growth and the fertilizer, as well as their interactions with climatic and soil conditions. The green

Table 1 Summary of agri-food supply chains in Asia

Supply-chain stages	Successful interventions	Key words	Primary stakeholders	Secondary stakeholders
Production	Multi-objective optimization of crop production structure in India	Sustainable	Farmers	Policymakers
	The Science and Technology Backyard platform in China	Green		Governments
Storage, distribution, processing	Multi-temperature joint distribution	Efficiency	Manufacturers, wholesalers and retailers	Inspection services
	E-commerce and supply-chain technology innovation	Resilience	Food traders	Certification bodies
Consumption	The Health Promotion Board in Singapore	Healthy	Consumers	Scientific community,
		Nutritious		non-governmental organizations and media

intelligent fertilizer can thus enhance crop production, crop quality and reduce environmental pollution (Bruulsema et al., 2020). One example regarding technology adoption is that agricultural scientists living in villages among farmers advance participatory innovation and technology transfer via the Science and Technology Backyard platform or via collaboration with extension agents and agri-business personnel, garnering public and private support (Cui et al., 2018; Zhang et al., 2016).

10.2 Storage, distribution and processing stages

Crucial issues of the storage, distribution and processing stages of the agri-food supply chain are to minimize food loss in post-harvest stages through applying standard facilities and equipment, adopting advanced technology and management measures, to improve the absorption efficiency of nutrients in food as much as possible, and to reduce environmental externalities.

In future, it is important to consider strengthening the resilience of the agri-food supply chain via storage and transportation planning. This will substantially improve our capacity to deal with various challenges such as extreme climate events, warfare, market risks and other unforeseen circumstances.

10.3 Consumption stage

Considerable advances have been made in improving the food environment and ensuring food availability in many Asian countries; however, consumers continue to face challenges in obtaining nutritious food at reasonable prices, and with different food cultures and dietary habits, better food utilization remains a problem. With the changing climate and limited natural resources, it is unclear how our food systems will deliver sufficient and nutritious food to future generations. Food systems must be transformed to improve efficiency, nutrition and sustainability. Given that SDG-12 stresses responsible consumption and production, important participants in the consumption stage, such as consumers, processing, retailing and marketing firms, should be motivated to adopt actions that contribute to healthy and sustainable local and national food systems. To this end, consumers, as influential actors in food systems, can not only improve their own behaviour by adopting sustainable and nutritious diets and reducing food waste, but they can also reward fair and ethical food production by making these changes.

10.4 System perspective

Changes in technologies, policies, institutions and behaviour are vital to transform the agri-food supply chain to achieve the multiple objectives of nutrition, health, sustainability, efficiency, resilience and inclusiveness. Aside from the pathways mentioned previously, the following aspects need to be further addressed when we aim to improve food security and nutrition in Asia through the lens of systems thinking. It includes emphasizing food production and consumption, encouraging public and private sector synergies and promoting policy innovations and public investment.

11 Where to look for further information

Key research in nutrition and food system can be found at the following organizations:

- Food and Agriculture Organization of the United Nations (https://www.fao.org/home/).
- International Food Policy Research Institute (IFPRI) (https://www.ifpri.org/).
- Harvestplus (https://www.harvestplus.org/).
- Geonutrition(http://www.geonutrition.com/).
- Academy of Global Food Economics and Policy (http://agfep.cau.edu.cn/col/col39575/index.html).
- Academy of Global Food Economics and Policy (AGFEP) (http://agfep.cau.edu.cn/).

Good introductions to the food system, food supply chain and nutrition subject:

- Searchinger, T., Hanson, C., Dumas, P., Waite, R., Ranganathan, J. and Matthews, E. (2018). Creating a sustainable food future, WRI: World Resources Institute. United States of America. https://policycommons.net/artifacts/1360128/creating-a-sustainable-food-future/1973409/ on 29 Sep 2022. CID: 20.500.12592/0397jv.
- Willett, W., Rockström, J., Loken, B., Springmann, M., Lang, T., Vermeulen, S. and Murray, C. J. (2019). Food in the anthropocene: The EAT–Lancet Commission on healthy diets from sustainable food systems. *The Lancet* 393(10170), 447–492.
- HLPE. (2017). *Nutrition and Food Systems*. A report by the High Level Panel of Experts on Food Security and Nutrition of the Committee on World Food Security, Rome.
- Uccello, E., Kauffmann, D., Calo, M. and Streissel, M. (2017). *Nutrition-Sensitive Agriculture and Food Systems in Practice: Options for Intervention*. FAO.
- Food Systems Summit Brief.

12 Acknowledgements

This research was financed through the China Scholarship Council (No.201913043), Hainan University, and the Program of Advanced Discipline

Construction in Beijing (Agriculture Green Development) and the National Natural Science Foundation of China (72103188,72061147002). We would like to thank Juanjuan Hu from the China Agricultural University for collecting and organizing the references.

13 References

AGFEP. 2021. *Rethinking Agrifood Systems for the Post-COVID World* [Online]. China: Academy of Global Food Economics and Policy. http://agfep.cau.edu.cn/art/2021/4/30/art_39031_744872.html.

AGFEP. 2022. *Reforming Agricultural Support Policies to Transform Agrifood Systems* [Online]. China: Academy of Global Food Economics and Policy. http://agfep.cau.edu.cn/art/2022/5/24/art_39031_861011.html.

Baker, P. and Friel, S. 2016. Food systems transformations, ultra-processed food markets and the nutrition transition in Asia. *Globalization and Health* 12(1), 80.

Bandumula, N. 2017. Rice production in Asia: Key to global food security. *Proceedings of the National Academy of Sciences India*. 88, 1323-1328.

Beintema, N., Bossio, D. A., Dreyfus, F., Fernandez, M., Gurib-Fakim, A., Hurni, H., Izac, A., Jiggins, J., Kranjac-Berisavljevic, G. and Leakey, R. 2009. *International Assessment of Agricultural Knowledge, Science and Technology for Development (IAASTD): Agriculture at a Crossroads, Global Summary for Decision Makers*. International Water Management Institute.

Bhattacharya, R., Ghosh, B., Mishra, P., Mandal, B., Rao, C., Sarkar, D., Das, K., Anil, K., Lalitha, M., Hati, K. and Franzluebbers, A. 2015. Soil degradation in India: Challenges and potential solutions. *Sustainability* 7(4), 3528–3570.

Bosi, T. L. B., Ergüder, T., Breda, J. and Jewell, J. 2018. *Monitoring Food Marketing to Children in Turkey, 2018*. World Health Organization. Regional Office for Europe.

Brainerd, E. and Menon, N. 2014. Seasonal effects of water quality: The hidden costs of the green revolution to infant and child health in India. *Working Papers* 107, 49–64.

Briones, R. M. and Felipe, J. 2016. Agriculture and structural transformation in developing Asia: Review and outlook. *Working Papers*.

Bruulsema, T., Cakmak, I., Dobermann, A., Gerard, B., Majumdar, K., McLaughlin, M., Reidsma, P., Vanlauwe, B., Wollenberg, E. K. and Zhang, F. 2020. A new paradigm for plant nutrition. *Issue Briefs*.

CCICED. 2022. *Sustainable Agrifood Systems: Meeting China's Food and Climate Security Goals* [Online]. China: China Council for International Cooperation on Environment and Development. https://cciced.eco/wpcontent/uploads/2021/12/SPS-5-Food-and-supply-chains-EN.pdf.

Chen, W.-T. and Hsu, C.-I. 2015. Greenhouse gas emission estimation for temperature-controlled food distribution systems. *Journal of Cleaner Production* 104, 139–147.

Cui, Z., Zhang, H., Chen, X., Zhang, C., Ma, W., Huang, C., Zhang, W., Mi, G., Miao, Y., Li, X., Gao, Q., Yang, J., Wang, Z., Ye, Y., Guo, S., Lu, J., Huang, J., Lv, S., Sun, Y., Liu, Y., Peng, X., Ren, J., Li, S., Deng, X., Shi, X., Zhang, Q., Yang, Z., Tang, L., Wei, C., Jia, L., Zhang, J., He, M., Tong, Y., Tang, Q., Zhong, X., Liu, Z., Cao, N., Kou, C., Ying, H., Yin, Y., Jiao, X., Zhang, Q., Fan, M., Jiang, R., Zhang, F. and Dou, Z. 2018. Pursuing sustainable productivity with millions of smallholder farmers. *Nature* 555(7696), 363–366.

Damsgaard, C. T., Michaelsen, K. F., Molbo, D., Mortensen, E. L. and Sørensen, T. 2016. Trends in adult body-mass index in 200 countries from 1975 to 2014: A pooled analysis of 1698 population-based measurement studies with 19·2 million participants. *Lancet* 387(10026), 1377–1396.

Davis, K. F., Chiarelli, D. D., Rulli, M. C., Chhatre, A., Richter, B., Singh, D. and DeFries, R. 2018. Alternative cereals can improve water use and nutrient supply in India. *Science Advances* 4(7), eaao1108.

Davis, K. F., Chhatre, A., Rao, N. D., Singh, D. and Defries, R. 2019. Assessing the sustainability of post-Green Revolution cereals in India. *Proceedings of the National Academy of Sciences* 116(50), 25034–25041.

Diaz, R. J. and Rosenberg, R. 2008. Spreading dead zones and consequences for marine ecosystems. *Science* 321(5891), 926–929.

Fan, S., Cho, E. E. and Rue, C. 2017. Food security and nutrition in an urbanizing world. *China Agricultural Economic Review* 9(2), 162–168.

Fan, S., Headey, D., Rue, C. and Thomas, T. 2021a. Food systems for human and planetary health: Economic perspectives and challenges. *Annual Review of Resource Economics* 13(1), 131–156.

Fan, S., Teng, P., Chew, P., Smith, G. and Copeland, L. 2021b. Food system resilience and COVID-19 – lessons from the Asian experience. *Global Food Security* 28, 100501.

Fang, L. and Huang, C. C. 2020. Targeted poverty alleviation in China: Evidence from Jingdong e-commerce poverty alleviation. *Poverty and Public Policy* 12(4), 386–396.

FAO, F. 2018. *The Future of Food and Agriculture: Alternative Pathways to 2050*. Rome: Food and Agriculture Organization of the United Nations.

FAO, IFAD, UNICEF & WHO, W. A. 2021. *The State of Food Security and Nutrition in the World 2021: Transforming Food Systems for Food Security, Improved Nutrition and Affordable Healthy Diets for All*. Food & Agriculture Org.

Foley, J. A., Defries, R., Asner, G. P., Barford, C., Bonan, G., Carpenter, S. R., Chapin, F. S., Coe, M. T., Daily, G. C., Gibbs, H. K., Helkowski, J. H., Holloway, T., Howard, E. A., Kucharik, C. J., Monfreda, C., Patz, J. A., Prentice, I. C., Ramankutty, N. and Snyder, P. K. 2005. Global consequences of land use. *Science* 309(5734), 570–574.

Garnett, T. and Godfray, C. 2012. Sustainable intensification in agriculture. Navigating a course through competing food system priorities. *Food Climate Research Network and the Oxford Martin Programme on the Future of Food*. UK: University of Oxford, 51.

Ghosh, J. 2004. *Social Policy in Indian Development*. Social Policy in a Development Context.

Gibson, M. F., Rao, N. D., Slade, R. B., Pereira, J. P. and Rogelj, J. 2020. The role of energy in mitigating grain storage losses in India and the impact for nutrition. *Resources, Conservation and Recycling* 163, 105100.

Godfray, H. C. J., Beddington, J. R., Crute, I. R., Haddad, L., Lawrence, D., Muir, J. F., Pretty, J., Robinson, S., Thomas, S. M. and Toulmin, C. 2010. Food security: The challenge of feeding 9 billion people. *Science* 327(5967), 812–818.

Godfray, H. C. J. and Garnett, T. 2014. Food security and sustainable intensification. *Philosophical Transactions of the Royal Society of London. Series B, Biological Sciences* 369(1639), 20120273.

Han, J.-W., Zuo, M., Zhu, W.-Y., Zuo, J.-H., Lü, E.-L. and Yang, X.-T. 2021. A comprehensive review of cold chain logistics for fresh agricultural products: Current status, challenges, and future trends. *Trends in Food Science and Technology* 109, 536–551.

Hawkes, C., Smith, T. G., Jewell, J., Wardle, J., Hammond, R. A., Friel, S., Thow, A. M. and Kain, J. 2015. Smart food policies for obesity prevention. *Lancet* 385(9985), 2410–2421.

Hazell, P. B. 2009. Transforming agriculture: The green revolution in Asia. In: Spielman, D. and Pandya-Lorch, R. (Eds.), *Millions Fed: Proven Successes in Agricultural Development*, International Food Policy Research Institute, Washington. 25–32.

HLPE. 2014. Food losses and waste in the context of sustainable food systems. *A report by the High Level Panel of Experts on Food Security and Nutrition of the Committee on World Food Security*, Rome

HPB. 2004. *Report of the National Nutrition Survey 2004* [Online]. Singapore: Health Promotion Board. https://www.hpb.gov.sg/docs/default-source/default-document-library/national-nutrition-survey-2004.pdf?sfvrsn=82daeb72_0.

HPB. 2010. *National Nutrition Survey 2010 Singapore* [Online]. Singapore: Health Promotion Board. https://www.hpb.gov.sg/docs/default-source/pdf/nns-2010 report.pdf?sfvrsn=18e3f172_2.

HPB. 2018. *Health Promotion Board Annual Report 2017/ 2018* [Online]. Singapore: Health Promotion Board. https://www.hpb.gov.sg/docs/default-source/annual-reports/hpb-annual-report-2017-2018.pdf?.

Ihtisham, M., Liu, S., Shahid, M. O., Khan, N., Lv, B., Sarraf, M., Ali, S., Chen, L., Liu, Y. and Chen, Q. 2020. The optimized N, P, and K fertilization for Bermudagrass integrated turf performance during the establishment and its importance for the sustainable management of urban green spaces. *Sustainability* 12(24), 10294.

Jat, M. L., Chakraborty, D., Ladha, J. K., Rana, D. S., Gathala, M. K., McDonald, A. and Gerard, B. 2020. Conservation agriculture for sustainable intensification in South Asia. *Nature Sustainability* 3(4), 336–343.

Jd, C. 2023. *2022 Annual Report* [Online]. https://ir.jd.com/static-files/9cc8985d-be5d-476b-81c5-5a0366f09d6c.

John, D. A. and Babu, G. R. 2021. Lessons from the aftermaths of green revolution on food system and health. *Frontiers in Sustainable Food Systems* 5, 644559.

Julia, C., Kesse-Guyot, E., Ducrot, P., Péneau, S., Touvier, M., Méjean, C. and Hercberg, S. 2015. Performance of a five category front-of-pack labelling system to differentiate nutritional quality of foods and to stimulate product reformulation: Application to breakfast cereals. *BMC Public Health* 15, 1–9.

Kayatz, B., Baroni, G., Hillier, J., Luedtke, S., Heathcote, R., Malin, D., Tonder, C. V., Kuster, B., Freese, D., Huettl, R. and Wattenbach, M. 2019a. Cool farm tool water: A global on-line tool to assess water use in crop production. *Journal of Cleaner Production* 207, 1163–1179.

Kayatz, B., Harris, F., Hillier, J., Adhya, T., Dalin, C., Nayak, D., Green, R. F., Smith, P. and Dangour, A. D. 2019b. "More crop per drop": Exploring India's cereal water use since 2005. *Science of the Total Environment* 673, 207–217.

Khajuria, A. 2016. Impact of nitrate consumption: Case study of Punjab, India. *Journal of Water Resource and Protection* 8, 211–216.

Khan, N., Ray, R. L., Sargani, G. R., Ihtisham, M., Khayyam, M. and Ismail, S. 2021a. Current progress and future prospects of agriculture technology: Gateway to sustainable agriculture. *Sustainability* 13(9), 4883.

Khan, S., Razzaq, A., Zhang, Y., Shah, A. and Janjua, L. 2021b. Disruption in food supply chain and undernourishment challenges: An empirical study in the context of Asian countries. *Socio-Economic Planning Sciences*, 82, 101033.

Kong, S. T. and Loubere, N. 2021. Digitally down to the countryside: Fintech and rural development in China. *The Journal of Development Studies* 57(10), 1739–1754.

Kumar, D. and Kalita, P. 2017. Reducing postharvest losses during storage of grain crops to strengthen food security in developing countries. *Foods* 6(1), 8.

Kumssa, D. B., Joy, E. J., Ander, E. L., Watts, M. J., Young, S. D., Walker, S. and Broadley, M. R. 2015. Dietary calcium and zinc deficiency risks are decreasing but remain prevalent. *Scientific Reports* 5, 10974.

Kuo, J.-C. and Chen, M.-C. 2010. Developing an advanced multi-temperature joint distribution system for the food cold chain. *Food control* 21, 559–566.

Li, C., Hoffland, E., Kuyper, T. W., Yu, Y., Zhang, C., Li, H., Zhang, F. and Van Der Werf, W. 2020. Syndromes of production in intercropping impact yield gains. *Nature Plants* 6(6), 653–660.

Li, Y., Tan, C., Ip, W. H. and Wu, C. H. 2023. Dynamic blockchain adoption for freshness-keeping in the fresh agricultural product supply chain. *Expert Systems with Applications* 217, 119494.

Liu, G., Hu, J., Yang, Y., Xia, S. and Lim, M. K. 2020. Vehicle routing problem in cold Chain logistics: A joint distribution model with carbon trading mechanisms. *Resources, Conservation and Recycling* 156, 104715.

Lu, S., Cheng, G., Li, T., Xue, L., Liu, X., Huang, J. and Liu, G. 2022. Quantifying supply chain food loss in China with primary data: A large-scale, field-survey based analysis for staple food, vegetables, and fruits. *Resources, Conservation and Recycling* 177, 106006.

Luo, N., Olsen, T., Liu, Y. and Zhang, A. 2022. Reducing food loss and waste in supply chain operations. *Transportation Research Part E: Logistics and Transportation Review* 162, 102730.

Luo, Y., Huang, D., Li, D. and Wu, L. 2020. On farm storage, storage losses and the effects of loss reduction in China. *Resources, Conservation and Recycling* 162, 105062.

Mani, M., Bandyopadhyay, S., Chonabayashi, S. and Markandya, A. 2018. *South Asia's Hotspots: The Impact of Temperature and Precipitation Changes on Living Standards*. Washington, D.C.: World Bank Publications.

Martínez-Fernández, J., González-Zamora, A., Sánchez, N., Gumuzzio, A. and Herrero-Jiménez, C. 2016. Satellite soil moisture for agricultural drought monitoring: Assessment of the SMOS derived Soil Water Deficit Index. *Remote Sensing of Environment* 177, 277–286.

Maruyama, M. 2004. Japanese distribution channels: Structure and strategy. *Japanese Economy* 32(3), 27–48.

Mendelsohn, R. 2014. The impact of climate change on agriculture in Asia. *Journal of Integrative Agriculture* 13(4), 660–665.

Micha, R., Mannar, V., Afshin, A., Allemandi, L., Baker, P., Battersby, J., Bhutta, Z., Chen, K., Corvalan, C. and Di Cesare, M. 2020. 2020 global nutrition report: Action on equity to end malnutrition. 89023.

Molden, D. 2013. *Water for Food Water for Life: A Comprehensive Assessment of Water Management in Agriculture*. London: Routledge.

Motoshita, M., Ono, Y., Pfister, S., Boulay, A.-M., Berger, M., Nansai, K., Tahara, K., Itsubo, N. and Inaba, A. 2018. Consistent characterisation factors at midpoint and endpoint relevant to agricultural water scarcity arising from freshwater consumption. *The International Journal of Life Cycle Assessment* 23(12), 2276–2287.

Neethu, N., Suman, G., Gajbhiye, V. T. and Manjaiah, K. M. 2016. Novel biopolymer-Nanoorganoclay composites for the decontamination of pesticides from water. *Pesticide Research Journal*. 28, 25-34.

Nelson, A. R. L. E., Ravichandran, K. and Antony, U. 2019. The impact of the Green Revolution on indigenous crops of India. *Journal of Ethnic Foods* 6(1), 8.

Niggli, U. 2015. Sustainability of organic food production: Challenges and innovations. *Proceedings of the Nutrition Society* 74(1), 83–88.

Nordhagen, S. 2020. Food supply chains and child and adolescent diets: A review. *Global Food Security* 27, 100443.

Nordhagen, S., Igbeka, U., Rowlands, H., Shine, R. S., Heneghan, E. and Tench, J. 2021. COVID-19 and small enterprises in the food supply chain: Early impacts and implications for longer-term food system resilience in low- and middle-income countries. *World Development* 141, 105405.

Parfitt, J., Barthel, M. and Macnaughton, S. 2010. Food waste within food supply chains: Quantification and potential for change to 2050. *Philosophical the Royal Society of London. Series B, Biological Sciences* 365(1554), 3065–3081.

Pingali, P. L. 2012. Green Revolution: Impacts, limits, and the path ahead. *Proceedings of the National Academy of Sciences of the United States of America* 109(31), 12302–12308.

Popkin, B. M., Corvalan, C. and Grummer-Strawn, L. M. 2020. Dynamics of the double burden of malnutrition and the changing nutrition reality. ScienceDirect.

Pretty, J. and Hine, R. 2000. The promising spread of sustainable agriculture in Asia. *Natural Resources Forum* 24(2), 107–121.

Puri, M. 2016. *How Access to Energy Can Influence Food Losses. A Brief Overview*. FAO.

PWC, R. A. T. 2019. *Harvesting the Future: The Asia Food Challenge [Online]*. Singapore. https://www.pwc.co.nz/industry-expertise/global-food-supply-and-integrity/afc-report-112019.pdf.

Reardon, T. and Zilberman, D. 2018. *Climate Smart Food Supply Chains in Developing Countries in an Era of Rapid Dual Change in Agrifood Systems and the Climate*. Cham: Springer.

Reeve, E., Thow, A. M., Bell, C., Engelhardt, K., Gamolo-Naliponguit, E. C., Go, J. J. and Sacks, G. 2018. Implementation lessons for school food policies and marketing restrictions in the Philippines: A qualitative policy analysis. *Globalization and Health* 14(1), 8.

Taylor, M. 2020. Hybrid Realities: Making a new Green Revolution for rice in south India. *The Journal of Peasant Studies* 47(3), 483–502.

Tilman, D. 2020. Benefits of intensive agricultural intercropping. *Nature Plants* 6(6), 604–605.

Tilman, D., Balzer, C., Hill, J. and Befort, B. L. 2011. Global food demand and the sustainable intensification of agriculture. *Proceedings of the National Academy of Sciences of the United States of America* 108(50), 20260–20264.

Toups, K. E. 2020. Global approaches to promoting whole grain consumption. *Nutrition Reviews* 78(Suppl. 1), 54–60.

UNICEF. 2021. *The State of Food Security and Nutrition in the World 2021* [Online]. United Nations International Children's Emergency Fund. https://data.unicef.org/resources/sofi-2021/.

Vermeulen, S. J., Campbell, B. M. and Ingram, J. S. I. 2012. Climate change and food systems. *Annual Review of Environment and Resources* 37(1), 195–222.

WHO. 2013. *Global Action Plan for the Prevention and Control of Noncommunicable Diseases 2013–2020* [Online]. World Health Organization. https://www.who.int/publications/i/item/9789241506236.

Willett, W., Rockström, J., Loken, B., Springmann, M., Lang, T., Vermeulen, S., Garnett, T., Tilman, D., Declerck, F., Wood, A., Jonell, M., Clark, M., Gordon, L. J., Fanzo, J., Hawkes, C., Zurayk, R., Rivera, J. A., De Vries, W., Majele Sibanda, L., Afshin, A., Chaudhary, A., Herrero, M., Agustina, R., Branca, F., Lartey, A., Fan, S., Crona, B., Fox, E., Bignet, V., Troell, M., Lindahl, T., Singh, S., Cornell, S. E., Srinath Reddy, K., Narain, S., Nishtar, S. and Murray, C. J. L. 2019. Food in the anthropocene: The EAT–Lancet Commission on healthy diets from sustainable food systems. *The Lancet* 393(10170), 447–492.

Wu, J., Zhang, J. and Zhao, N. 2023. How to boost e-commerce for poverty alleviation? A perspective on competitiveness analysis using online reviews. *Electronic Commerce Research*, 1–32. https://doi.org/10.1007/s10660-023-09708-2..

Xue, L., Liu, X., Lu, S., Cheng, G., Hu, Y., Liu, J., Dou, Z., Cheng, S. and Liu, G. 2021. China's food loss and waste embodies increasing environmental impacts. *Nature Food* 2(7), 519–528.

Zhang, W., Cao, G., Li, X., Zhang, H., Wang, C., Liu, Q., Chen, X., Cui, Z., Shen, J., Jiang, R., Mi, G., Miao, Y., Zhang, F. and Dou, Z. 2016. Closing yield gaps in China by empowering smallholder farmers. *Nature* 537(7622), 671–674.

Zhou, Z.-Y. and Wan, G. 2017. *Food Insecurity in Asia: Why Institutions Matter*. Asian Development Bank Institute.

Part 4

Conclusions

Chapter 14

Understanding and improving global agri-food supply chains in an interconnected world: a systems approach

Rodrigo Romero-Silva, Renzo Akkerman and Sander de Leeuw, Wageningen University, The Netherlands

1 Introduction

Agri-food supply chains (AFSCs) are complex, interconnected networks that span the world, yet they are highly influenced by regional contexts, as the chapters from this book have illustrated. From large-scale farming in countries like the United States to small and medium-sized farmers in Latin America and Africa, each region presents its own set of challenges and opportunities. These do not only result from regional environmental conditions but also political, economic, social, technological, and regulatory aspects. Using the framework introduced in Chapter 1, this closing chapter synthesizes the content of this book by differentiating between the managed system, the managing system, and the information system to provide an understanding of AFSCs across the world.

http://dx.doi.org/10.19103/AS.2023.0122.19

2 The managed system

The managed system is where the transformation of raw materials into finished goods occurs. Farms, (cold) warehouses, transportation resources, as well as retail and food service points all relate to this system, ultimately making up for the network structure of the AFSC. This system has become increasingly intricate due to technological advancements, regulatory changes, climate change, and consumer demands for the availability of a wide range of products throughout the year, while also increasingly emphasizing higher quality, sustainability, and traceability.

Throughout the book, we learned that the configuration of the managed agri-food system differs per region. For instance, the AFSC in the United States is dominated by large players, benefiting from economies of scale and advanced technology, which results in higher productivity. An example of this reality is highlighted in Chapter 2 with the use of automated harvesting equipment in the Midwest grain belt, which allows for rapid and efficient collection of crops. In contrast, regions like Mexico rely heavily on small to medium-sized farmers, particularly in the sugarcane industry, where regulatory constraints also play a significant role due to the past role that the government played in controlling the production of several strategic industries in the country, whereas African AFSCs often rely on smaller-scale farming operations, reflecting a blend of manual and traditional farming practices where limited access to finance and poor infrastructure create operational challenges. In countries like Colombia, coffee is often grown by smallholders who face challenges related to market access and price volatility due to financial constraints.

On the other hand, the agricultural sector in Europe is highly organized and mechanized, focusing on both crop and livestock farming, where the Netherlands is a clear example with its prevalence of greenhouse farming and its dairy industry. Additionally, AFSCs in Asia face significant challenges as the food supply system is under strain due to various factors including regional conflicts such as the war in Ukraine, a shift toward more calorically dense diets, and the global pandemic of COVID-19. In India, for instance, the rice and wheat supply chains are often disrupted by monsoons, while in Singapore, there is a high reliance on imported goods, which has ignited a significant investment in agri-food technologies to increase their self-sufficiency for some critical products.

3 The managing system

The managing system is responsible for ensuring that the output of the managed system meets certain targets. This involves making decisions on production volumes, sourcing strategies, quality control points, and other tactical and operational aspects. Planning and control in agri-food supply chains involve complex decision-making at various stages, including production, harvesting, and processing. Furthermore, the planning and control of agri-food supply chains are complicated by the perishable nature of agri-food products. Recent events, including the need for more sustainable food systems and increased resilience, have intensified the challenges supply chain managers face and have changed the focus of supply chain managers from maximizing profits (or minimizing costs) subject to some quality/safety constraints toward more systems-oriented objectives, such as robustness, sustainability, and food security considerations.

Due to the different levels of technological and planning capabilities, the managing system also differs significantly between regions. For example, in the United States and Europe, the managing system is highly advanced, benefiting from technological innovations that allow for better planning and control. However, uncertainty in production yields (for the United States particularly in corn and soybean farming) often complicates the planning process. Latin America, on the other hand, faces challenges related to regulatory constraints and the perishability of products due to the large percentage of goods going all the way to the United States, the EU, and even Japan (the most lucrative markets). For example, the perishability of tropical fruits like bananas and pineapples creates logistical challenges to make them reach the (northern) markets in optimal conditions. In Spain, olive oil production is often subject to strict quality controls to meet EU standards. However, unpredictable weather patterns can affect olive yields, making it difficult to plan production and react to changes in the plan, as shown by the 2022 droughts in Spain (BBC News 2022). In Asia, the managing system faces the challenge of balancing the need for nutritious food with the high cost of such diets, especially given the income disparities in the region. In China, while the demand for pork products has significantly increased due to the increase in purchasing power, the pork supply chain has been disrupted by African swine fever, requiring stringent management measures. The perishability of seafood in countries like Japan adds another layer of complexity to the managing system due to safety concerns.

The managing system faces the challenge of incorporating various regional particularities into the planning and control of the agri-food supply chain. Factors such as unpredictable weather patterns, regulatory constraints, and economic disparities add layers of complexity to the planning process. For instance, while the United States may rely on advanced data analytics for decision-making, such technologies may not be readily available or applicable in major parts of Latin America or Asia. Similarly, the perishability of products like seafood in Japan or fruits in Latin America requires specialized planning that may not be as critical in other regions. This makes the task of creating a unified, efficient managing system a complex endeavor that requires a nuanced understanding of these regional specificities.

It is worth noting that the managing system might behave differently in steady-state (normal) conditions than under the effect of disturbances, as Chapter 11 demonstrates by showing the approaches that different actors in AFSCs took to overcome the disruptions created by the COVID-19 pandemic. In contrast to normal circumstances where processes and strategies are well-established and predictable, the pandemic required a more reactive and flexible approach. Companies had to leverage capabilities that would normally be difficult to utilize, such as rapid reallocation of personnel and changes in product packaging, which were accepted due to the collective effort to deal with the exceptional circumstances. This shift highlights the need for the managing system to be agile and adaptable, capable of responding to unforeseen challenges while maintaining food safety and quality. However, it also reveals the limitations due to informational constraints, as many companies lacked specific plans for such unprecedented disruptions, highlighting the importance of contingency planning and risk management in the managing system.

4 The information system

The information system collects data generated during the transformation process in the managed system. These data are crucial for the managing system to make informed decisions. The ability to measure and monitor AFSC performance is a pressing issue, particularly considering that, due to climate change, crops might need closer monitoring due to unexpected changes in their maturation process and, consequently, their final quality and safety. Also, due to the large contribution of AFSCs to many key environmental indicators, clear and comprehensive performance measures are needed to be able to decide on and follow up on initiative toward the reduction of these environmental impacts. The complexity of these supply chains, the multitude of inputs and outputs, and the lack of consensus on determining performance make the selection and discussion of appropriate measures (ranging from economic value generation to environmental and social measures) a challenging task. Nevertheless, performance measurement is crucial for informed decision-making and end-to-end supply chain design.

In some value chains in the United States and Europe, advanced information systems utilize technologies for real-time tracking and analytics, aiding in better decision-making. For example, blockchain technology is increasingly used for traceability in the European meat industry. On the other hand, information systems in Africa, Asia, and Latin America are still in the process of incorporating more advanced technologies for better monitoring and traceability. In Thailand, for instance, QR codes are being used to trace the origin of certain fruits, providing consumers with more information about their purchases.

Performance measurement plays a crucial role in ensuring the quality and safety of food in agri-food supply chains. However, the planning system often faces limitations due to informational constraints. In regions with less developed technology infrastructure, the lack of (real-time) data can hinder effective planning and compromise food safety and quality. This is particularly challenging in a global network where regional differences and mismatched infrastructures can create gaps in the managing system, especially when supply networks face significant disruptions. The COVID-19 pandemic exposed the need for real-time data and analytics to make informed decisions quickly. However, the lack of preparedness and the absence of relevant historical data made it challenging for many companies to adapt their performance measurement strategies effectively. This situation underscores the dependency of the information system on technology and its capabilities, as well as the need for robust and adaptable information systems capable of supporting the managing system under both normal and disrupted conditions.

5 Future research trends

As we conclude the book, it is evident that agri-food supply chains are at a critical juncture. The ongoing need to meet the food demands of a growing global population, adapt to shifting consumer preferences, decrease environmental impacts, and respond to disruptive events like pandemics and geopolitical conflicts highlight the need for innovative solutions. Future opportunities for research in agri-food supply chains are then also abundant. Given the complexities and challenges identified in the agri-food supply chains across different regions, several avenues for future research can be suggested.

5.1 Supply chain design for sustainability

Research should focus on the impact of environmentally sustainable practices. There are many initiatives with the aim to be more sustainable, e.g. agroforestry (Lillesø et al. 2011), within the agri-food supply chain, particularly in regions where intensive farming is prevalent. The impact of this type of agricultural production can significantly influence soil contamination and productivity in the future (Gaffney et al. 2019). This research could help us understand the value of sustainable practices (either positive or negative) and determine the optimal configuration of different agri-food supply chain networks based on a holistic analysis of demand and nutrient requirements, geographic supply constraints, and impacts on sustainability. In regions where small farms are prevalent, such as Latin America and Africa, research on sustainability should also consider the social aspect of sustainability, as it is critical to attain economic and environmental sustainability due to the challenges regarding the living standards of small farmers (Giller et al. 2021). Thus, future research of AFSC networks where small farmers play a significant role should consider a triple-bottom-line lens (de Carvalho et al. 2022), which incorporates the social, economic, and environmental aspects of sustainability. This approach can help find the optimal network configuration (using multi-objective mathematical programming, for example) that finds a balance between these three aspects of sustainability (see Zhu et al. 2018 for a review of operational research approaches focusing on these issues).

5.2 Technological innovations

The role of emerging technologies such as blockchain, IoT, and AI in enhancing the efficiency and traceability of agri-food supply chains warrants further investigation, as highlighted in Chapter 9. Furthermore, more studies on the role that digitalization might play in some regions to optimize processes and increase efficiency across the supply chain (Annosi et al. 2020). This efficiency ideally also extends to the environmental dimension, in which an efficient use of resources is arguably even more important. This is especially the case due to the large share of the global food production that is lost or wasted somewhere in the AFSC, clearly leading to many supply chain challenges to address (Akkerman and Cruijssen 2024). These challenges related to the managing system and the information system, however, also extend to the managed system, in which technological developments in, for instance, alternative protein production might significantly alter AFSCs in the future, in turn requiring the managing systems and information systems to adjust.

5.3 Performance measurement analytics across regions and supply chain echelons

The development of standardized performance metrics and indicators to facilitate the traceability, monitoring, and control throughout the supply chain is crucial. This includes considering a minimum technological denominator, which could be supported by the incorporation of new technological standards such as blockchain, as suggested by Chapter 9. In addition, the multi-criteria aspect of performance measurement in AFSCs, due to involvement of many stakeholders and systems in the production and distribution of food, results in a very complex system to measure. Therefore, advanced analytics tools that integrate different measures into a single framework of analysis, such as data envelopment analysis (Priyadarshini and Abhilash 2023), could be very helpful in better understanding the key characteristics of well-performing AFSCs across economic, environmental, and social criteria.

5.4 Supply chain financing in agri-food value chains

While research on supply chain finance is abundant (Xu et al. 2018), its role in the performance of AFSCs has not been sufficiently investigated, particularly from the point of view of the economic viability of small and medium-sized farms (Oostendorp et al. 2019) in the face of large-scale operations as well as increasingly uncertain yields due to climate change, as discussed by Chapter 12 regarding the impact of financing and insurance on African AFSCs. Research on AFSC finance could provide insights into better financing options to reach more equitable, heterogeneous, and ultimately resilient agri-food supply chains, helping to reach a better triple-bottom-line for the entire network and overall better food security for different regions. To this end, simulation has shown to be a powerful tool for modeling the various interactions between actors as well as the system uncertainties present in AFSC (Utomo et al. 2018). New technologies such as blockchain offer many opportunities for further development in this area, as exemplified by recent studies using a game-theoretic approach (Choi 2023; Dong et al. 2023).

5.5 Resilience and risk management for climate change adaptation

Given the increasing impact of climate change on agricultural yields (Jägermeyr et al. 2021), research into adaptive and resilient agri-food supply chains is crucial. Thus, understanding and developing strategies to enhance the resilience of these supply chains in the face of global uncertainties, particularly in regions that are most likely to suffer the consequences of climate change (Giorgi 2006), is necessary to ensure global access to safe food as well as the viability of future agri-food networks. Due to its capabilities to analyze and understand complex, intertwined systems as well as their feedback and delayed effects, future studies could use system dynamics as a modeling approach as it has proved to be a suitable tool for modeling resilient and sustainable aspects of AFSC (Aboah et al. 2021; Amiri et al. 2020; Kotir et al. 2024).

5.6 The role of supply chains in food systems

Due to their differing scope and scale, as characterized in Chapter 1, research that considers the interplay between food systems and food supply chains could provide a better understanding of the design of current and future agri-food supply networks. For instance, future research could focus on investigating how government policies and regulations influence the efficiency and resilience of food supply chains. Furthermore, since AFSC modeling can be focused on various levels of decision-making (e.g. network configuration, and production site) and transactions (e.g. production lots, truck dispatches, or inventory units), more detailed interactions between logistics decisions and the biological systems could be modeled to understand the impact of certain policies on land and water use, biodiversity conservation, nutritional content, and the impact of agricultural practices on ecosystems.

6 Conclusion

In closing this book, we have seen that agri-food supply chains are complex and diverse, shaped by regional differences and global trends. From managing systems under various conditions to the critical role of information systems in ensuring food quality and safety, the chapters of this book have covered a wide range of topics. This book aimed to provide a foundation for understanding these complex systems and inspire further exploration into making agri-food supply chains more efficient, resilient, and sustainable. Looking ahead, research in this field is full of opportunities. The future calls for innovative and collaborative approaches to tackle challenges in sustainability, technology, policy, and climate change, to generate and apply practical improvements and informed decision-making in agri-food supply chains.

7 References

Aboah, J., Wilson, M. M. J., Bicknell, K. and Rich, K. M. 2021. Ex-ante impact of on-farm diversification and forward integration on agricultural value chain resilience: a system dynamics approach. *Agricultural Systems* 189(April): 103043. https://doi.org/10.1016/j.agsy.2020.103043.

Akkerman, R. and Cruijssen, F. 2024. Food loss, food waste, and sustainability in food supply chains. In: *Sustainable Supply Chains* (2nd edn.), Eds. Yann Bouchery, Charles J. Corbett, Jan C. Fransoo and Tarkan Tan. Springer Nature Switzerland AG.

Amiri, A., Mehrjerdi, Y. Z., Jalalimanesh, A. and Sadegheih, A. 2020. Food system sustainability investigation using system dynamics approach. *Journal of Cleaner Production* 277(December): 124040. https://doi.org/10.1016/j.jclepro.2020.124040.

Annosi, M. C., Brunetta, F., Capo, F. and Heideveld, L. 2020. Digitalization in the agri-food industry: the relationship between technology and sustainable development. *Management Decision* 58(8): 1737–57. https://doi.org/10.1108/MD-09-2019-1328.

BBC News 2022. Spain's olive oil producers devastated by worst ever drought, August 29, 2022, sec. Europe. Available at: https://www.bbc.com/news/world-europe-62707435.

Choi, T.-M. 2023. Supply chain financing using blockchain: impacts on supply chains selling fashionable products. *Annals of Operations Research* 331(1): 393–415. https://doi.org/10.1007/s10479-020-03615-7.

de Carvalho, M. I., Relvas, S. and Barbosa-Póvoa, A. P. 2022. A roadmap for sustainability performance assessment in the context of Agri-food supply chain. *Sustainable Production and Consumption* 34: 565–85. https://doi.org/10.1016/j.spc.2022.10.001.

Dong, L., Qiu, Yunzhe and Xu, F. 2023. Blockchain-enabled deep-tier supply chain finance. *Manufacturing and Service Operations Management* 25(6): 2021–37. https://doi.org/10.1287/msom.2022.1123.

Gaffney, J., Bing, J., Byrne, P. F., Cassman, K. G., Ciampitti, I., Delmer, D., Habben, J., Lafitte, H. R., Lidstrom, U. E., Porter, D. O., Sawyer, J. E., Schussler, J., Setter, T., Sharp, R. E., Vyn, T. J. and Warner, D. 2019. Science-based intensive agriculture: sustainability, food security, and the role of technology. *Global Food Security* 23(December): 236–44. https://doi.org/10.1016/j.gfs.2019.08.003.

Giller, K. E., Delaune, T., Silva, J. V., van Wijk, M., Hammond, J., Descheemaeker, K., van de Ven, G., Schut, A. G. T., Taulya, G., Chikowo, R. and Andersson, J. A. 2021. Small farms and development in Sub-Saharan Africa: farming for food, for income or for lack of better options? *Food Security* 13(6): 1431–54. https://doi.org/10.1007/s12571-021-01209-0.

Jägermeyr, J., Müller, C., Ruane, A. C., Elliott, J., Balkovic, J., Castillo, O., Faye, B., Foster, I., Folberth, C., Franke, J. A., Fuchs, K., Guarin, J. R., Heinke, J., Hoogenboom, G., Iizumi, T., Jain, A. K., Kelly, D., Khabarov, N., Lange, S., Lin, T., Liu, W., Mialyk, O., Minoli, S., Moyer, E. J., Okada, M., Phillips, M., Porter, C., Rabin, S. S., Scheer, C., Schneider, J. M., Schyns, J. F., Skalsky, R., Smerald, A., Stella, T., Stephens, H., Webber, H., Zabel, F. and Rosenzweig, C. 2021. Climate impacts on global agriculture emerge earlier in new generation of climate and crop models. *Nature Food* 2(11): 873–85. https://doi.org/10.1038/s43016-021-00400-y.

Kotir, J. H., Jagustovic, R., Papachristos, G., Zougmore, R. B., Kessler, A., Reynolds, M., Ouedraogo, M., Ritsema, Coen J., Aziz, A. A. and Johnstone, R. 2024. Field experiences and lessons learned from applying participatory system dynamics modelling to sustainable water and agri-food systems. *Journal of Cleaner Production* 434(January): 140042. https://doi.org/10.1016/j.jclepro.2023.140042.

Lillesø, J. B. L., Graudal, L., Moestrup, S., Kjær, E. D., Kindt, R., Mbora, A., Dawson, I., Muriuki, J., Ræbild, A. and Jamnadass, R. 2011. Innovation in input supply systems in smallholder agroforestry: seed sources, supply chains and support systems. *Agroforestry Systems* 83(3): 347–59. https://doi.org/10.1007/s10457-011-9412-5.

Oostendorp, R., van Asseldonk, M., Gathiaka, J., Mulwa, R., Radeny, M., Recha, J., Wattel, C. and van Wesenbeeck, L. 2019. Inclusive agribusiness under climate change: a brief review of the role of finance *Current Opinion in Environmental Sustainability*, Sustainability Science: Inclusive Business: A Multi-stakeholder issue 41(December): 18–22. https://doi.org/10.1016/j.cosust.2019.09.014.

Priyadarshini, P. and Abhilash, P. C. 2023. An empirical analysis of resource efficiency and circularity within the agri-food sector of India. *Journal of Cleaner Production* 385: 135660. https://doi.org/10.1016/j.jclepro.2022.135660.

Utomo, D. S., Onggo, B. S. and Eldridge, S. 2018. Applications of agent-based modelling and simulation in the agri-food supply chains. *European Journal of Operational Research* 269(3): 794–805. https://doi.org/10.1016/j.ejor.2017.10.041.

Xu, X., Chen, X., Jia, F., Brown, S., Gong, Y. and Xu, Y. 2018. Supply chain finance: a systematic literature review and bibliometric analysis. *International Journal of Production Economics* 204: 160–73. https://doi.org/10.1016/j.ijpe.2018.08.003.

Zhu, Z., Chu, F., Dolgui, A., Chu, C., Zhou, W. and Piramuthu, S. 2018. Recent advances and opportunities in sustainable food supply chain: a model-oriented review. *International Journal of Production Research* 56(17): 5700–22. https://doi.org/10.1080/00207543.2018.1425014.

Index

www.ingramcontent.com/pod-product-compliance
Lightning Source LLC
LaVergne TN
LVHW010620300926
849162LV00001B/13
* 9 7 8 1 8 0 1 4 6 2 7 1 6 *